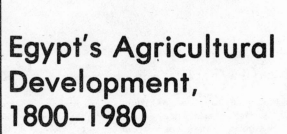

# Egypt's Agricultural Development, 1800–1980

# Westview Replica Editions

The concept of Westview Replica Editions is a response to the continuing crisis in academic and informational publishing. Library budgets for books have been severely curtailed. Ever larger portions of general library budgets are being diverted from the purchase of books and used for data banks, computers, micromedia, and other methods of information retrieval. Interlibrary loan structures further reduce the edition sizes required to satisfy the needs of the scholarly community. Economic pressures on the university presses and the few private scholarly publishing companies have severely limited the capacity of the industry to properly serve the academic and research communities. As a result, many manuscripts dealing with important subjects, often representing the highest level of scholarship, are no longer economically viable publishing projects--or, if accepted for publication, are typically subject to lead times ranging from one to three years.

Westview Replica Editions are our practical solution to the problem. We accept a manuscript in camera-ready form, typed according to our specifications, and move it immediately into the production process. As always, the selection criteria include the importance of the subject, the work's contribution to scholarship, and its insight, originality of thought, and excellence of exposition. The responsibility for editing and proofreading lies with the author or sponsoring institution. We prepare chapter headings and display pages, file for copyright, and obtain Library of Congress Cataloging in Publication Data. A detailed manual contains simple instructions for preparing the final typescript, and our editorial staff is always available to answer questions.

The end result is a book printed on acid-free paper and bound in sturdy library-quality soft covers. We manufacture these books ourselves using equipment that does not require a lengthy make-ready process and that allows us to publish first editions of 300 to 600 copies and to reprint even smaller quantities as needed. Thus, we can produce Replica Editions quickly and can keep even very specialized books in print as long as there is a demand for them.

# Egypt's Agricultural Development, 1800–1980: Technical and Social Change

## Alan Richards

One of the principal factors underlying Anwar Sadat's willing-
ness to sign a peace treaty with Israel was the deplorable state of
the Egyptian economy. Multiplying shortages, deteriorating infra-
structures, and spiraling foreign debts fill the economic news from
Egypt. A central component of this domestic crisis is agriculture.
Agriculture forms the basis for a vast portion of the Egyptian economy, accounting for nearly half the country's employment and nearly
a third of its gross national product, as well as providing materials
for over half of its industry.

This book describes and interprets the transformation of Egyptian
agriculture from the beginning of cotton cultivation in the early
nineteenth century to the current changes under Sadat. The author
uses both microeconomic theory and social and political analysis to
show how the interaction of social classes, technical change, government policy, and the international and state systems have shaped
Egypt's agricultural development. Arguing that these forces are
bound up in a complex web of reciprocal causation, he places the
current dilemmas of Egyptian agriculture in historical perspective.

Dr. Richards is assistant professor of economics at the University of California, Santa Cruz.

# Egypt's Agricultural Development, 1800–1980

## Technical and Social Change

Alan Richards

Westview Press / Boulder, Colorado

*A Westview Replica Edition*

Published in 1982 in the United States of America by
     Westview Press, Inc.
   5500 Central Avenue
     Boulder, Colorado  80301
     Frederick A. Praeger, Publisher

Library of Congress Cataloging in Publication Data
Richards, Alan, 1946-
     Egypt's agricultural development, 1800-1980.
     (A Westview replica edition)
     Revision of thesis (Ph.D.)--University of Wisconsin--Madison, 1975.
     Bibliography: p.
     Includes index.
     1. Agriculture--Economic aspects--Egypt--History.  I. Title.  II. Series.
HD2123.R5    1981          338.1'0962          81-12919
ISBN 0-86531-099-8                            AACR2


Printed and bound in the United States of America

10  9  8  7  6  5  4  3

# Contents

# Tables

# Maps and Figures

# Egyptian Units and Equivalents

1 feddan = 1.038 acres = .42 ha.

1 qantar = 99.05 lbs = 44.928 kgs.  (cotton, sugar cane)

1 metric qantar = 346.5 lbs = 157.5 kgs.  (cotton)

1 ardeb = 5.62 bushels = 198 liters

1 ardeb (wheat)  = 330 lbs. = 150 kg.
1 ardeb (maize,
          sorghum) = 308 lbs. = 140 kg.
1 ardeb (beans)  = 341 lbs. = 155 kg.
1 ardeb (barley) = 264 lbs. = 120 kg.

1 dariba (rice) = 2,079 lbs. = 945 kg.

1 L.E. (Egyptian Pound) = 100 P.T. (Piastres Tariffe)

Until May, 1962, 1 L.E. = 1 oz 6d, Sterling.

Since then its exchange value has fluctuated widely.
In December, 1980:

Unified Exchange Rate:  1 L.E. = $.69

# Acknowledgments

I have incurred many debts in writing this book. The work began as a doctoral dissertation at the University of Wisconsin, Madison. I am indebted above all to my faculty advisor, Peter H. Lindert, now of University of California, Davis, for his probing criticism and for his unflagging, many-sided support. His help was truly invaluable. Jeffrey G. Williamson spotted various holes in the argument and encouraged me to revise the dissertation for publication. Bent Hansen read and commented upon the entire dissertation. His searching critique saved me from a variety of errors of fact and interpretation. I am very grateful to all three of these men.

I would also like to thank the following people for helpful discussions of various issues and for general encouragement: J.A. Allan, Galal Amin, Jennifer Bremer, Edmund Burke, III, Byron Cannon, Frank C. Child, Kenneth A. Cuno, James A. Fitch, Saad Gadalla, Ahmad Goueli, Walter Goldfrank, Isebill Gruhn, Nicholas Hopkins, Malcolm Kerr, Paul Lubeck, Afaf Lutfi al-Sayyid Marsot, Philip A. Martin, Samir Radwan, Vernon Ruttan, Robert Springborg, Sarah Potts Voll, John Waterbury, Marvin Weinbaum, and the members of the Santa Cruz Faculty Seminar in Comparative History. Financial support for various stages of the research came from the University of Wisconsin Graduate School, the Division of Social Science and the Faculty Senate of the University of California at Santa Cruz, and the Agricultural Development Systems Project of the University of California—Egyptian Ministry of Agriculture—U.S.A.I.D. I would like to thank Patrick O'Brien for first interesting me in Egyptian affairs. I am indebted to the University of California Press, Cambridge University Press, the University of Chicago Press, Frank Cass and Co. Ltd, MERIP, and Sage Press, for permission to draw on portions of journal articles which appeared in Agricultural History, Comparative Studies in Society and History, Economic Development and Cultural Change, Journal of Development Studies, Middle East Research and Information Project Report, Middle Eastern Studies, and Review. Thanks also to Gayla Nethercott, Alyssa Postlewait and Nikoletta Bolas for preparing the manuscript for publication on the UNIX system, to Peter Broadwell for the graphs, and to Annette Whelan for the maps and diagrams. Finally, I would like to thank my family.

Obviously, I alone am responsible for what follows.

*Alan Richards*

MEDITERRANEAN SEA

Buheirat el Idku

Buheirat el Burúllús

Abu Qîr

Rashîd

ALEXANDRIA

Buheirat Maryût

DUMYAT

Buheirat el Manzala

PORT SAID

KAFR EL SHEIKH

Kafr el Dauwâr

BEHERA

Kafr el Sheikh

El Mansûra

El Mahalla el Kubrâ

Damanhûr

DAQAHLIYYA

Kafr el Zaiyât

Tantâ

GHARBIYYA

Shibin el Kôm

SHARQIYYA

Minûf

Zagâzig

MINUFIYYA

Minya el Qamh

Ismâ'ilîya

Benha

QALUBIYYA

Delta Barrage

Qalyûb

El Gîzâ

CAIRO

GIZÂ

SUEZ

FAYYÛM

El Faiyûm

EGYPT

BENI SUEF

Beni Suef

GULF OF SUEZ

Bahr Yusef

El Minyâ

MINYÂ

Egypt – The Nile Valley

ASYÛT

Asyût

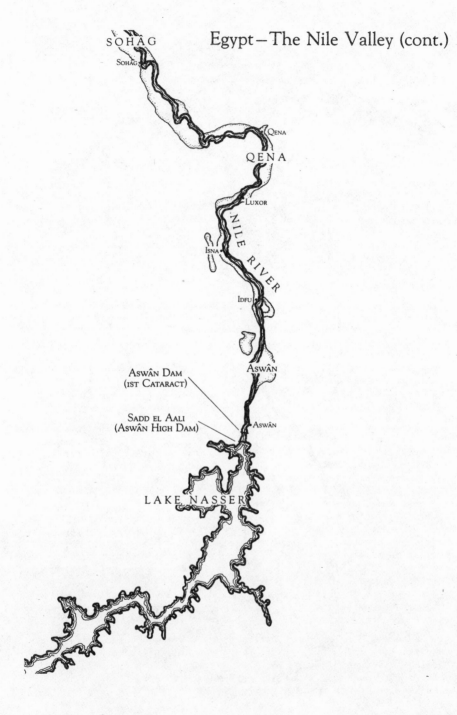

# 1

# Introduction

This book provides a survey and an interpretation of the transformation of Egyptian agriculture from 1800 to the present. It focuses on the interaction of agricultural technical change, rural social classes, and government policy. It examines the role of the international economic and state systems in generating the pattern of Egyptian agricultural development.

Numerous other works on various aspects of Egyptian economic history are available. Studies of the current state of the Egyptian economy are also legion. I have made extensive use of both sorts of work. The justification for the present book is twofold. First, at the moment no historical survey of Egyptian agricultural development exists in English which treats the whole period from the rise of cotton cultivation in the early nineteenth century to the changes of the Sadat era, such as the current spread of farm mechanization. I hope that my historical overview will be useful to those who wish to place current events in historical perspective.

Second, I believe that my approach differentiates this book from other treatments of the subject. I have relied extensively on modern micro-economic theory, that is the logic of opportunity cost, to study technical change and income distribution, without neglecting the fundamentally political aspects of agricultural change. I have drawn heavily upon Marx for an understanding of class and of labor systems, and for the emphasis upon technical change. However, my analysis of such technical change is essentially a supply-and-demand one. I employ the logic of constrained maximization through the book: actors pursue goals and are constrained by opportunities. But it must be stressed that various agents may have different goals and are most certainly constrained by their (highly unequal) resources. These resources are "given" through a political (i.e. conflictual) process. The concerns with the effects of class, government policy, land tenure systems, and the international environment upon agrarian change make this study a work of political economy.

Since this is an interpretive work, a few of the terms used require some clarification. The word "class" is used here in the "ordinary language" sense: socially distinct groups of people, with different modes of life. I have no interest in entering into the elaborate, somewhat scholastic controversies over what does, and what

1

does not, constitute a class. The basic classes which emerged in the nineteenth century and with whose fate I shall be largely concerned are the large landowners (defined as those holding more than fifty feddans of land), the small peasants (those holding less than five feddans), the rich peasants (those holding between five and fifty feddans), and the landless (those holding no land).

Although individuals on the boundaries are, of course, problematical, persons in each of these four classes differed sufficiently in life style and prospects to warrant the four fold distinction. Large landowners are the most obvious case here: wealthy, initially often non-Arabic speaking, commonly absentees, their horizons included not only Cairo and Alexandria, but also Rome, Paris and London. Their cosmopolitanism, urbanism, and very great wealth set the pashas off from the next group, the rich peasants. An important part of the story which follows is how the latter class managed to maintain its position as the dominant group of peasants resident in the villages throughout one-hundred eighty years of history. Their political position as well as their greater wealth distinguish them from their smaller neighbors. These, although poor, at least hold some land. This establishes a fundamental cleavage between them and the landless, the poorest of the poor, whose income depends entirely on selling their labor power. Peasants owning less than the subsistence minimum of land can rent supplementary parcels or maintain a buffalo or cow. The landless, who own no other assets, are insufficiently credit-worthy to be able to rent, leaving them no alternative but selling their ability to work.

Land ownership is the basis of the classification for two reasons. First, land has long constituted the principal form of rural wealth in Egypt. Second, it can stand as a proxy for overall wealth holding, for which information is not available. The dividing lines between the classes are also partly drawn for reasons of analytical convenience. It has often been observed that approximately three feddans constitutes the subsistence minimum for Egyptian peasants.[1] Arguably, therefore, three, not five feddans, should be the dividing line. However, especially for the period before World War I, such a breakdown is unavailable. Further, since most twentieth century holders of less than five feddans held considerably less than three feddans, using the "less than five feddan" boundary probably distorts the social picture relatively little.

The second category employed extensively here is "technical change." This, again, is taken in the broad sense of any change in what is produced or how this is done. It includes new crops, new crop rotations, and changes in the time pattern of inputs in addition to changes in the use of industrial inputs. Any change in which "page from the book of blueprints"[2] is used is thus considered technical change. This, of course, is entirely in line with modern economic theory.

Technical change is central to this study for two reasons. First, there is a false image of "timeless Egypt" which even now has not entirely died out. Such a view, in my opinion, results from an exclusive focus on agricultural tools. If the peasants are still using the tanbour, or Archemedian screw, the casual observer tends to conclude that Egyptian agricultural technology has been static for centuries. But such a conception of technical change is too narrow: the peasant may be irrigating heavily fertilized fields of summer cotton whose planting date was recently changed to avoid insect attacks. None of these features of the agricultural technology existed on any scale in the eighteenth century; all constitute technical change, and all had consequences for the distribution of income among rural social classes.

This leads to the second reason for emphasizing technical change: such changes were crucial for agricultural development. Technical change influenced not only the growth of aggregate production, but perhaps even more interestingly from the point of view of political economy, the position of the different classes and government policy. This is very much a two-way street. Central to my argument is the notion that class, government policy, and technical change are bound up in a web of reciprocal causation. Technical change affects social classes, and unequal access to resources moulds the pattern of technical change. Changes in technique create problems to which government policy must respond, and state decisions shape technology. And, of course, any state draws its principal social support from some social groups rather than others; it will, consequently, seek to cater to their special needs, although it must also seek to contain the inevitable conflicts with other classes. In either case, class structure shapes government policy.

Government policy, rural classes, and technical change were also strongly influenced by the international systems of trade and war. Throughout this study, I hope to stress the importance of international economic and political pressures for the actual pattern of technical and social change. Here, too, I eschew "first causes": I do not argue that the kind of agricultural development which Egypt has experienced is the "result" of international forces alone or solely derivative from changes in the "modern world-system."[3] Again, reciprocal interaction is at work: international forces and truly indigenous features of Egyptian society and economy combine to generate the type of agricultural development which she has experienced. The international forces themselves, in some cases, (e.g. the British intervention in 1882 or the oil price revolution of 1973) are partly the result of Egyptian actions. The perspective on the role of the international system then is that it shapes, but does not determine the course of technical change in Egyptian agriculture.[4]

The book is organized as follows. Chapter 2 begins with an analysis of the crisis of eighteenth century Mamluk society. I then trace the transformation of social relations and agrarian technology in Egypt from the first decade of the nineteenth century down to the British intervention. Technical change and the rise of cotton cultivation, the forcible redistribution of landownership, the development of the division of labor in agriculture, and the social background of the 'Urabi revolt are the principal themes. Chapter 3 covers the period 1880-1914. After describing the consolidation of private property rights in land and labor and their social consequences, I turn to a more extensive discussion of the modes of land and labor exploitation on large and medium-sized estates. I then describe two different crop rotations, and discuss the impact of crop rotation choice on the agroecology of Egypt and on the class distribution of cotton revenues. The question of class differences in agricultural techniques receives close scrutiny. Chapter 4 specifies the technical changes in the period between 1920 and 1940. I stress the link between the adoption of these new techniques and the earlier technical changes discussed in Chapter 3 and examine the pattern of resource use which the adoption of these new techniques implied. In Chapter 5 I turn to the picture of labor productivity in agriculture in the period 1920-1940 and discuss the probable implications of the changes described in Chapter 4 for the class distribution of income. After a brief discussion of changes between 1940 and 1952, Chapter 6 analyzes the political economy of agriculture in the Nasser and Sadat years. The focus for the period from 1952 to 1970 is upon the consequences of land reform, price policies, agricultural cooperative organization, and public investment choices for social and technical change. The section on the decade of the 1970s stresses the responses of the Sadat regime to the inherited problems of Egyptian agriculture, with particular emphasis on the spread of farm mechanization. A brief conclusion, Chapter 7, provides a summary of the principal technical and social changes, as well as a more extensive discussion of the interaction of class, technical change, government policy, and the international economy and state system in Egyptian agricultural development.

1
Footnotes

[1]     Hassan Riad, L'Egypte Nassérienne (Paris, 1964) Mahmoud Abdel
        Fadil, Development, Income Distribution, and Social Change in
        Rural Egypt: A Study in the Political Economy of Agrarian
        Transition (Cambridge, 1975).

[2]     Cf. Joan Robinson, Essays in the Theory of Economic Growth,
        (N.Y., 1968).

[3]     As a Wallersteinian might argue.  See  Immanuel  Wallerstein,
        The  Modern  World  System,  Vols  I and II.  (N.Y., 1974 and
        1979).

[4]     Cf. Peter Gourevitch, "The Second Image Reversed: The  Inter-
        national  Sources of Domestic Politics" International Organi-
        zation 32,4 (August, 1978).

# 2

# Primitive Accumulation
# in Egypt

## SOCIAL STRUCTURE UNDER THE MAMLUKS

Understanding the nature and consequences of the transformation of agricultural technology and rural social structure in the nineteenth century requires an overview of the preceding social system. After a sketch of Mamluk and Ottoman (1516-1798) social structure and trade patterns, I shall briefly describe the basin system of irrigated agriculture. I shall then trace the transformation of both in the nineteenth century before the British intervention and occupation in 1882. The focus is upon changes in the irrigation system, the emergence of private property in land, the origin of the principal social classes, and the emergent forms of land and labor exploitation. The main theme is "primitive accumulation" and its relationship to changes in agricultural technology. At the same time, I wish to demonstrate how the 'Urabi revolt of 1882 had its social roots in this transformation of hydraulic technology and social structure.

Egypt had two principal functions in the Ottoman Imperial system. First, it supplied foodstuffs to Istanbul, the Holy Places of Arabia, Rumelia, Anatolia, and Syria. Second, Egypt was the focal point of an extensive transit trade between Africa, Arabia, and the rest of Asia on the one hand, and the Mediterranean countries on the other. Revenues obtained from customs duties formed part of the annual tribute sent to the Porte. Within one hundred years of the conquest, however, both the Ottoman Imperial system as a whole and the Egyptian component of it were in decline, and by the mid-eighteenth century, Egyptian society was in full-scale crisis.

The dynamic and origins of this crisis may be sketched as follows. The decline of central authority in Istanbul and the decline in government revenues led to the creation of a system of tax-farming (iltizam) which increased the tax burdens on the peasantry. The relative decline in the transit trade in coffee after 1725 put downward pressure on government revenues; this, joined with a strong demand for revenue to finance internal and external wars of the ruling class of large tax-farmers, intensified pressures on the peasants. The heavier tax burden, the increase in Bedouin activity, and random ecological shocks (flood variations and epidemics) depopulated the countryside.

6

However, this crisis did little to alter the agrarian social structure. Certain features of that structure, primarily the organization of access to land and the lack of direct contact between the ruling class and the peasantry, inhibited the transformation of these social relations. That would only come with the resolution of the crisis by the integration of Egypt into the European systems of trade and war in the nineteenth century. The rest of this section is devoted to elaborating and documenting these assertions.

Scholars have long emphasized Egypt's function as the granary of the Ottoman Empire, supplying wheat, rice, lentils, beans and other foodstuffs to Rumelia, Anatolia, Syria, and the Hejaz.[1] The Irsaliyye, or the annual sum sent by Egypt to Istanbul, "made the austere court of the fifteenth century Sultans into the lavish splendor of the age of Sulayman and his successors."[2] The Holy Cities of the Hejaz were dependent upon Egypt for its food supplies, receiving wheat, beans, lentils, rice, oils, sugar, butter and cheese. Egypt also supported the Hajj, or annual pilgrammage to Mecca.[3]

Egypt's role in the Ottoman economy and Egyptian trade generally was not limited to that of a suppler of foodstuffs, however. First, Egypt exported raw materials for manufacturing to the rest of the Ottoman Empire, primarily flax (to Syria and the Greek Islands) and indigo (to Syria). Second, she exported her own manufactured cloth to the Maghreb, to Arabia, to Rumelia, to Anatolia and Syria, and to Darfur in the Sudan. Third and most important, Egypt was the center of a large and flourishing entrepot trade between the Orient and Africa on the one hand, and the Mediterranean on the other.[4] This transit trade underwent two transformations during the seventeenth and eighteenth centuries. First, anti-Ottoman revolts in the Yemen and renewed Ottoman-Hapsburg hostilities in the Mediterranean contributed to the demise of the transit trade in pepper and spices by disrupting the trade routes. The coming of the Dutch and the English to the Indian Ocean at the beginning of the seventeenth century accelerated this decline.[5]

However, Yemeni coffee replaced spices as the mainstay of the transit trade in the late seventeenth and early eighteenth centuries. Re-exported via Egypt to the rest of the Ottoman Empire and to Europe, the coffee trade dominated Egyptian commerce in the eighteenth century to such an extent that Andre Raymond has gone so far as to call it "the principal economic activity of Egypt."[6] It was highly profitable, bringing a net rate of return on invested capital of roughly 33%. The wealth gained in this trade underlay the power, cohesiveness, and self-confidence of the great coffee merchants in Cairo.[7] This transit trade comprised roughly 25% of all Egyptian imports and roughly 40% of all exports. It appears that roughly 28% of the re-exports of the Red Sea and Indian Ocean imports (primarily coffee and Indian cloth) went to Europe, 1% to Africa, 3% to Syria (a textile center importing its coffee directly from the Yemen), and 68% to Rumelia and Anatolia toward the end of the eighteenth century.[8] Raymond estimates that taxes on the trade amounted to roughly 50% of total government revenues, the other 50% coming from agriculture.[9]

This trade, however, also underwent a relative decline. Its zenith was roughly 1690 to 1725. The Europeans were not able to dislodge the Muslims from the coffee trade by force as they had done earlier with pepper and spices. The Ottomans held Aden until 1830 and the superior naval technology of the Europeans, so telling in the Indian Ocean, was less significant in the Red Sea.[10] Instead, the Europeans were able to set up their own coffee plantations in the Carribean. Antilles coffee first arrived in Marseilles around 1730, and by 1786-1789 some 21% of French coffee was sold in the Levant. Its price was roughly 25% lower than that of Yemeni coffee. There appears to have been an interaction among declining trade volume, increased per-unit levies on trade, and higher, less competitive coffee prices. Formally, the problem was that while the demand for Yemeni coffee was relatively elastic, ruling-class demand for tax revenues was inelastic for reasons discussed below. Therefore, any fall in the volume of revenue would lead to an increase in its "price", or per-unit taxation. When these per-unit levies were translated into higher prices, sales, and therefore revenues, also fell, and the process repeated itself. By 1764 the competition of Antilles coffee was sufficiently severe that its importation into Egypt was prohibited.[11]

The competition of European trade led not only to a decline in total trade volume and government revenue, but also to a problem in the balance of trade. Egyptian trade was composed of two sectors: trade with Africa, Arabia and the Orient on the one hand, and trade with the Mediterranean, both Christian and Muslim, on the other. The strong Egyptian deficit on the first account was covered by a strong surplus on the latter. In 1700-1730 European trade, with a balance favorable to Egypt, contributed to equilibrium in Egyptian trade. But by the end of the century, Europe was selling more to Egypt than she was buying. Raymond ascribes this to two forces: (1) the decline in the coffee and the other transit trades (the mainstay of Egyptian-European commerce), and (2) the rapid decline in Egyptian cloth exports to France after 1730.[12] The resulting imbalance contributed to the depreciation of the Egyptian currency. The roughly contemporary decline in the inflow of Sudani gold aggravated the depreciation.[13] Presumably, this depreciation of the currency raised the price of imports of European goods, largely luxury products and arms.[14] We shall see that there are reasons for supposing that the demand for luxury goods and arms by the ruling class was inelastic, and perhaps shifting out in the eighteenth century. The depreciation therefore contributed to increased demands for revenue.

Be that as it may, it would be unwise to overemphasize the role of Europe in Egyptian trade. Overall, trade with Europe accounted for about one-seventh of Egypt's total trade.[15] The trade with Istanbul alone surpassed that with Europe. Aside from the transit trade, Egypt exported some textiles to Europe (nine-tenths of these to France), and some rice and wheat to Marseilles and Livorno. Towards the end of the century, it appears that large quantities of wheat were exported to France. Girard reported that in the last three years of the eighteenth century, some 800,000 ardebs of wheat were sent to Marseilles.[16]

However, this trade suffered from several severe handicaps. Both demand and supply were highly unstable. Only in times of famine was there a demand for Egyptian grain in Europe.[17] Similarly, the vagaries of the Nile flood caused sharp fluctuations in the amount of grain which was available for export. Egyptian government policy contributed to the frailty of this trade. The export of rice and wheat to Christendom, for example, was illegal.[18] Although a clandestine trade existed and was at times openly encouraged by the Beys, European traders were exposed to extraordinary levies at virtually all times.[19]

To summarize the preceding discussion, we note that Egyptian trade played a vital role for both the Ottoman and the local economies. Egypt supplied large quantities of foodstuffs to other parts of the Ottoman Empire, as well as some textiles and raw materials for manufacturing. The transit trade in coffee was a critical component of the whole. Its decline, along with the decline in other exports, reduced government revenues, created balance of trade deficits, and contributed to the depreciation of the currency. The role of European competition, as well as the increasing demands for revenue from the ruling class, seem to have been the principal forces behind this stagnation of trade. Increased insecurity of trade routes inside Egypt and in the Eastern Mediterranean also contributed to the decline. But to understand the government's demand for revenue, we must understand the agrarian social structure.

When the Ottomans took over Egypt, they did not follow their usual practice of granting the conquered lands as timars or land grants to the principal government and military leaders. They had three reasons for not doing so. First, even in Anatolia itself, the Ottomans were trying to replace the timar holdings.[20] Second, the Ottomans feared that timar holders in Egypt might develop into a force independent of the Porte. Third, since Egypt was to be the Imperial Granary, the Sultans did not want to lose total usufruct rights over the land.[21]

Instead of creating timars, Selim tried to set up an administration in which all the lands were run by treasury bureaucrats appointed by the Porte. These men would remit all of the taxes which they collected and in turn be paid salaries. Since, however, the officials' (emins) salaries were not dependent on the amount of taxes received, they proved rather inefficient collectors.[22] High pay and/or swift removal for incompetence would have provided the necessary incentives, of course, but such measures would have required a strong central government in Istanbul. The Ottomans passed the zenith of their power some seventy-five years after the conquest of Egypt, although, of course, they continued to be formidable.

The fiscal crisis of the Ottoman state required decentralizing provincial administration. Initially, the treasury granted land to each emin, who, however, could not do all of the work himself. He therefore appointed agents or 'amils who, it happened, were drawn from the Mamluks. Early in the seventeenth century these men became the pivot of the whole tax collection system, known as iltizam or tax farming. Under this system, the Mamluks were responsible for supervising tax collection. They then paid a fixed sum to the emins (i.e., to the Ottoman Treasury) and pocketed what was left, the surplus or fa'id. In return for this privilege of tax collection, the holder paid an initial fee. As Hansen has stressed,[23] this switch to iltizam was a form of borrowing by the government, in which the latter forswore future revenues (the fa'id) in exchange for present payment. By 1671 officials sent from the Porte had little authority.[24] As the emins' importance dwindled, the Mamluks became the dominant power in the land.

By the end of the eighteenth century state lands were almost universally distributed throughout Egypt in the form of il-tizam and fell into the hands of the wealthiest and most powerful men, most of whom were Mamluks. Of 6,000 multezims, it was estimated that 300 were Mamluks who held more than 2/3 of the cultivated land in Egypt.[25]

Since iltizam was technically the property of the state, the Mamluks held only the usufruct. There were five other legal categories of land: (1) mulk, or personal property, (2) waqf or land whose revenue accrued to pious foundations, whether mosques or individuals, (3) mawat, or "dead," uncultivated land, (4) matruq, or land which was used as pasture or meadow and was held in common by the entire community, and (5) khushufiyya, plots set aside to provide revenues for the highest government officials.[26] In theory, the iltizam itself was granted for a period of years by the government bureau of land registration; of course, with the decline of the central power and the rise of the Mamluks, there was a marked tendency for the usufruct of the land to become hereditary. However, the government retained the raqaba, or "title" of the land even after the decline of its power.

The relative weakness of the central government did contribute to the indigenous evolution of private property, however. As the usufruct became hereditary, various legal fictions emerged whereby the land could be bought, sold, and mortgaged. The rising prices of agricultural products characteristic of the entire Mediterranean may have encouraged the slow spread of private property rights in land. Converting land into waqf ahli in effect placed land inside the family with very few encumbrances on its use. Waqf land as a whole may have amounted to some twenty percent of the cultivated area by 1800.[27]

The tasarruf or use of iltizam land was divided between the multezim (tax-farmer) and the peasant. The latter's land ('ard al-fellah) was reserved for their use, and they paid a variety of taxes on it to the multezim. The land reserved for the exclusive enjoyment of the multezim, the 'ard al-usya was estimated at roughly 10% of the 'ard al-fellah at the end of the eighteenth century.[28] There were three ways of exploiting this land. First, the multezim might rent out the land to the village shaykh or headman who would direct the agricultural operations of the other peasants. Second, he might exploit the lands directly using paid labor. Third, he might use corvéed labor.[29] Supervision of the agricultural activities on these lands was carried out by a variety of personnel from the Mamluk's "household." Each powerful multezim had a sort of military lieutenant (qa'imaqam), himself a Mamluk, who generally lived on the 'ard al-usya and who was in charge of supervising cultivation and taxation.[30] But he did not do this directly. Rather, the actual supervision of cultivation and other agricultural operations was the responsibility of the village shaykh and another villager, the khawli.[31] In addition, there were a host of other officials, perhaps the most prominent being the sarraf or treasurer, usually a Copt, who kept the accounts. All of these individuals received tax exemptions and payments in kind. Peasant labor was paid in cash (for cultivation), in kind (for harvest), or not at all (if corveed). Much of the labor came right from among the local villagers holding usufructory rights to the 'ard al-fellah, but there were some landless peasants. Many of these appear to have been peasants whose lands had not been watered by the annual Nile flood; they were, therefore, landless only temporarily.

The system of tax farming led to increasing burdens on the peasants. In the first place, we have already seen how changes in military technology and changes in trade constricted government revenues. In part, the Egyptian Mamluks tried to solve these problems by sending less tribute to Istanbul.[32] But this alone could not solve the problem because structural features of this society raised the Mamluks' demand for revenue. Each Mamluk "house", composed of the head of the house and his military slaves, was in severe competition with the others. Although they did form coalitions, these appear to have been unstable and no single faction was able to establish a centralized, powerful regime of its own. Doing so was especially difficult since (1) the inclusion of the offspring of Mamluks as members of the house perpetuated conflicts of revenge, and (2) the Mamluks were more formidably armed now (seventeenth century) with carbines and pistols.

> Any victory over the enemy, however decisive, could not make the life of the victorious secure as long as the defeated enemy had the slightest chance of recuperating and gathering strength. The logical conclusion was: the enemy must be wiped out.[33]

These forces, as well as the resurgence of Bedouin raids, make the political history of this period a monotonous record of feuding, raiding, assassination, and betrayal.[34]

This structure of conflict and mutual distrust meant that each Mamluk house had a large and inelastic demand for resorces in order to obtain the arms required for defense and aggression and the luxury goods needed as symbols of power to retain the loyalty of followers. But the price of these goods was pushed up by the depreciation of the currency, while revenues from trade were stagnant. The only solution was to tax the peasants more. Average real taxation per feddan (slightly more than an acre) increased by roughly 35% from the sixteenth to the end of the eighteenth century.[35] These were the official taxes, usually in kind for lands planted in wheat or beans, in cash for land planted in millet.[36] Most of the in-kind payments were transported along the Nile to Cairo, where they were stored in giant warehouses. There they were sold, or shipped directly as tribute to Istanbul or the Holy Cities of Arabia. Some of the goods were consumed in Cairo, and the rest was exported.[37]

Other extraordinary levies were common. The corvee for the 'ard al-usya has already been mentioned; in the sixteenth century, peasants had been paid for the corvee labor on irrigation works, but by the eighteenth century they were not. Other levies were imposed; the costs of feeding Mamluks whenever they arrived in the village, the costs of tax collection itself, and the cost of transporting the taxed goods to market were all borne by the peasants.[38] Shaw describes the collection of taxes at harvest time as "large scale raiding operations, with resultant depopulation, famine and devastation."[39] By the time of the French invasion, after over a century of war, bedouin raids, and oppressive taxation, the population had declined to less than four million, the lowest in its history. In Roman times Egypt had "seven and a half million (inhabitants) exclusive of Alexandria—one of the very few ancient population figures we have that is likely to be accurate."[40] At the time of the Arab conquest, the numbers may have reached twelve to fourteen million, declining to some five million by the sixteenth century.[41]

Why didn't the Mamluks realize that they were killing the goose that laid the golden egg? Hansen has argued that from the multezim's point of view there was an "optimal rate of taxation". This was the rate which would provide the maximum long-run revenue, which in turn implied taxing peasants just below the level at which they would abandon their lands and flee to the cities, to the desert and swamps, or to Syria.[42] But the structure of social relations within the ruling class largely precluded such behavior. Two Mamluks might agree that they would both be better off if they quit fighting each other, thereby reducing their demand for arms and luxury goods, which in turn would lead them to tax their peasants less heavily and so induce them to remain on the land. But if one Mamluk reduced his taxes and his acquisition of arms, but the other did not and then attacked, the result for the first Mamluk was complete catastrophe: a classic "prisoner's dilemma" game.[43] In the absence of any one ruler strong enough to impose his will on the entire country, each Mamluk house had to press for enough revenue to protect itself against attack. They were therefore driven to squeeze the peasants until, in the words of the seventeenth century peasant poet, "my testicles are like palm fiber."[44]

Nevertheless, the peasants were not entirely defenseless. The village unit afforded them some protection and, more importantly for our analysis, protected the structure of social relations within the village and between the village and the ruling class. Consequently, the dynamic of crisis outlined above would not, in itself, lead to a transformation of social relations. This protective organization had two parts. The first was the organization of peasant land. In this, as in other respects, Lower (northern) Egypt differed from Middle and Upper Egypt. In the latter region, "lands were held communally and assigned to individual cultivators annually as soon as the extent of

the Nile flood became apparent."[45] Because variations in the Nile
Flood caused considerable differences in the land available for cultivation in a given year, it was difficult to establish separate
boundaries. Land was therefore distributed by the village shaykh in
accordance with the ability of a family to cultivate the land.[46]

In Lower Egypt, where boundaries could be established more
easily, individual families cultivated a fixed area of the 'ard al-
fallah, in contrast to the communal system of the Sa'id (Upper
Egypt). These areas, called athar, were handed down from father to
son. The son had to pay the multezim an investiture tax, and then
the land, or rather, its use, was his. The usufruct could be sold,
but this seems to have mostly involved pawning the land, rather than
outright sale. The multezim could, however, remove a peasant from
this land for failure to pay his taxes. Such a peasant then either
stayed in the village as an agricultural laborer, or fled. The
actual legal protection afforded the individual peasant, then, was
not so great as might first appear. However, the peasants as a
group, the village, did not lose land abandoned for failure to pay
taxes to the multezim's 'ard al-usya. Rather, the shaykh would give
the land to another peasant.[47]

The second organizational feature protecting the peasants was
that neither the Mamluks nor the bedouin shaykhs (who held power in
Upper Egypt until 1760) had much to do with the peasants directly.
The multezim or his representatives normally dealt with the village
shaykhs, not the individual peasants. The shaykh was in charge of
distributing land in Upper Egypt, as we have seen; in Lower Egypt, he
was responsible, to some extent, for organizing work on the 'ard al-
usya. If a multezim rented his 'ard al-usya, he rented it to the
shaykh who then dealt with the problem of getting the villagers to
work it. If the multezim worked his land with a corvee, then the
shaykh supervised it. No one, however, tried to tell the peasants
which crops to raise.

les fellahs jouissent d'ailleurs de toute liberte sur le
genre de culture qu'ils veulent donner a leurs terres; ils
peuvent les ensemencer en ble, en riz, en doura, selon qu'il
leur plait.[48]

The shaykhs were responsible for both the maintenance of order and
the collection of taxes, in the sense that they were held responsible
if the tax collection came up short. The peasants did come into contact with the Mamluks at harvest and tax gathering time, but the contact was minimal, limited to a beating or two.

There may have been some direct contact between peasants and
multezims through the circulation process. If a peasant did alienate
his land temporarily, then he would have pawned it to those who had
cash. These may have been the multezims. The multezims always supplied working capital on their 'ard al-usya, but here, too, they went
through the shaykhs. Most social historians, as well as the primary
source of Lancret's testimony, emphasize the lack of direct contact
between Mamluks and peasants. The village as a whole was responsible
for taxation and for the corvee; allocation of the individual

fulfillment of these two duties, given the collective quota, was the role of the village shaykh.[49]

The principal contact that did exist between rulers and ruled was violent, usually taking the form of the Mamluks' agents administering the bastinadoe to the peasants. A fellah was considered unmanly by his fellows if he did not endure as many lashes as possible before paying his taxes.[50] Since taxes were assessed on the number of bushels harvested and not on the number of feddans sown, the peasants would simply lie about how much they had reaped and hide their surpluses. This practice of tax evasion seems to have been very ancient, and was continued into the nineteenth century. If the taxes came up short, the shaykh was also flogged, as he was if the Mamluk's agents discovered that something had been stolen from their master's property.[51]

Such minimization of contact between ruler and ruled, or between any two social groups for that matter, was a general feature of the Ottoman Empire at that time. The people were organized into corporate groups, into guilds in the cities and into villages in the countryside. After religious affiliation, this group received the main loyalties of its members.

Each population group—merchants or artisans or cultivators—was in fact, organized independently into a corporate body each with its own leadership. Allegiance was to the corporate body, rather than to the Sultan or ruling class.[52]

This isolation, and especially the organization of peasant land lent stability to rural social relations.[53] The increased tax burden, resulting from Mamluk myopia, their increased need for revenue, and the decline in trade revenues, induced peasants to abandon their lands and to flee, but it did not lead to any great change in what was produced, how it was produced, or how the labor was organized and supervised. The crisis of Mamluk society led to quantitative regression, but not to qualitative transformation.

## AGRARIAN TECHNOLOGY: THE BASIN SYSTEM OF IRRIGATION

This "corporate" social structure rested upon a particular system of irrigated agriculture. Since Egypt's integration into the world economy as a cotton producer transformed both the social relations described above and the agricultural technology, it will be helpful to outline the main features of the irrigation system and the related patterns of land use.

The basin system of irrigation had been used in Egypt since the time of the Pharaohs. The topography of the Nile Valley resembles the back of a leaf: the land slopes down gradually from the high land lying along the banks of the river and canals toward the desert. The high land along the banks of the river was flooded perhaps once every fifteen to twenty years when the annual Nile flood was unusually high.[54] The remainder of the land, comprising some seventy-five percent of the cultivated area in the Delta (Lower Egypt) and nearly ninety percent in the Valley,[55] was divided into basins by a system of dykes, some running perpendicular to the Nile, others parallel to it. Canals dug through the high land along the banks allowed the rising flood waters to flow into these basins. The water, rich with the sediment carried off from the Ethiopian highlands, was allowed to stand on the fields and soak in after being trapped there by the dykes. After standing for some forty days, the water was either allowed to drain back into the Nile, or to flow on into the next, lower basin.[56]

The irrigation system had a number of important effects upon land use patterns and the maintenance of soil fertility. First, little, if any, fertilizer was required for the basins, since they were supplied with nutrients by the flood every year. Girard asserts that fertilizer was not used on land which was "irrigated naturally," i.e., on basin lands.[57] They were used on the higher land, which makes sense since these lands were not flooded annually. Second, crops sown after the flood required relatively little land preparation: in Upper Egypt, the seeds were simply sown broadcast as the waters receded.[58]

In the Delta, the land was ploughed perhaps two times with a scratch plow and then was watered twice during the growing season of wheat (one of the principal basin crops). In Upper Egypt, no additional waterings were given. Land planted in clover (birsim) was not ploughed in either region.[59] The French savants who accompanied Napoleon noted that Egyptian basin agriculture required much less labor per acre and per-unit output than contemporary French peasant agriculture.[60] Third, the basin lands were "washed" every year and the higher lands were flooded from time to time, thus preventing the build-up of salts in the soil. Although this was helpful for the ecology of the soil, the excessive floods were very destructive of human life by destroying crops and inaugurating the usual cycle of famine and plague. The model of the decline of Mamluk society given above would be incomplete without the inclusion of such events. Fourth, these lands lay fallow for a considerable time before the arrival of the flood in late July and early August. During this period the soil heated up, cracked, and was aerated. Fifth, crops whose growing season included the dry, pre-flood months had to be grown near the river in order to obtain the necessary water. Since these lands were high, as previously noted, and since the Nile was low during this season, the water had to be raised a considerable distance to the fields. This required a great deal of labor, as did drawing and applying well water. This latter practice may have been used on as much as one-fifth of the cultivated area.[61] Some rough figures for the labor needed for the different crops are given in Table 2.1.

The cropping patterns were in large part dictated by the timing of the flood and the structure of the irrigation system. (See Figure 2.1) The basin lands were cropped once per year and followed a two-year rotation of wheat and barley alternating with birsim and beans.[62] The consumption patterns of the peasants is a bit problematical. Girard and Chabrol agree and emphasize that beans were a staple item of diet throughout Egypt. Both agree that millet was the staple cereal in Upper Egypt, despite the fact that because of its

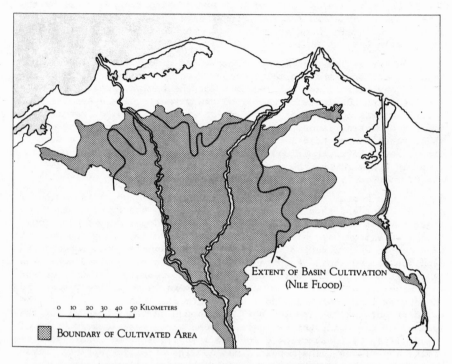

2. Extent of the Cultivated Area and of the Nile Flood, 1800
   Adapted from Jean Lozach, <u>Le Delta du Nil</u>:
   <u>Etude de Geographie Humaine</u>. (Cairo, 1935)

   (Suez Canal shown for orientation purposes.)

TABLE 2.1
LABOR INPUTS FOR CROPS UNDER THE BASIN
SYSTEM OF IRRIGATION, 1801

| Crop | man-days/feddan |
|------|------|
| Durah (millet)* | 85 |
| Maize* | 65 (Irrigation and harvest only) |
| Wheat al-bayady | 15 |
| Barley al-bayady | 7 1/2 |
| Birsim (clover) | 13-15 |
| Beans al-bayady | 20 |
| Wheat al-shitwi* | 86 |
| Barley al-shitwi* | 70 |
| Flax* | 72 (Irrigation only) |
| Cotton* | 500 plus harvest |
| Onion* | 80 |

* Indicates that the crop required labor for irrigation.

Source: Girard (1813), 565-581.

growing season it could only be cultivated on the high lands near the river. Under these conditions, it required perhaps one hundred man-days of labor per feddan. Wheat was cultivated in the basins, with much less labor, but it was used to pay taxes. Since the area was a place of refuge for exiled Mamluks "whose sole ambition was to return to Cairo," the population suffered from extremely heavy tax obligations.[63]

For the Delta, the picture is more complex. Although maize was probably introduced in the seventeenth century and had spread in some areas, there is some evidence, and an a priori reason, why maize cultivation should have been sharply limited before the advent of perennial irrigation. First, maize requires more water than either sorghum or millet.[64] This alone would have limited its cultivation. Second, because the crop needs substantial quantities of water before planting, after sowing and (especially) during the tasseling stage, maize was a flood crop (until 1965). It could not have been grown in the basins, because the flooding would have killed the plant. Growing it on the levees would have required even more work than millet, as the large quantities of water necessary had to be raised from the river at its nadir to the high ground. Maize cultivation would have been practical only on the minority of lands which had access to wells. The evidence is admittedly mixed, with some sources asserting that the peasants ate maize early in the nineteenth century. Most claim that the main food crop of the fellahin was sorghum or millet, however.[65]

18

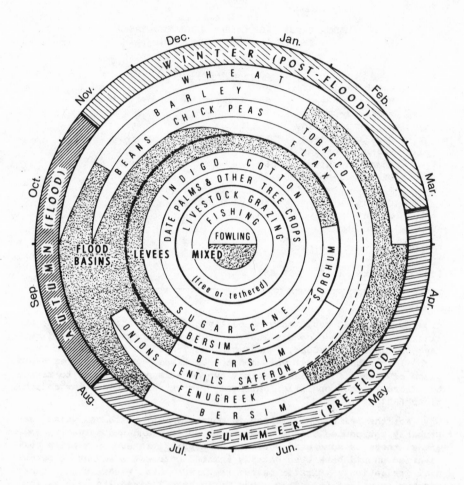

2.1  Crop Rotation, A.D. 1000–1800

Source: Karl W. Butzer, Early Hydraulic Civilization in Egypt, (1976), 49. ©1976 by the University of Chicago.

Both millet and maize were sometimes mixed with wheat in making the peasants' bread. Unmixed wheat flour was, apparently, considered a luxury, but a kind of mixed wheat and barley bread seems to have been a common article of peasant consumption.[66] On the whole, then, the consumption pattern was beans and millet bread in Upper Egypt and beans and wheat-millet, wheat-sorghum, wheat-barley, or wheat-maize bread in the Delta, with the consumption of maize spreading in the latter region. In all areas, the peasant's diet was supplemented by salt-fish, milk products, and various vegetables.[67]

In addition to these staple crops, cash crops such as sugar, indigo, and rice were grown. All of these required a great deal of labor; indigo, for example, was cultivated for four years and required nine men per feddan for eight months of the year. It was grown by "well-to-do proprietors" and/or peasant cooperatives.[68] Cotton was grown under the basin system, but it was not a particularly profitable crop.[69] It was grown as a perennial crop in Upper Egypt, although an annual variety was cultivated in the Delta. Since its growing season was from February to October, the area planted was constrained both by the need to be on high ground to protect it from the flood (although it was possible to build supplementary dykes to serve this purpose) and, more importantly, by its heavy water requirements during the summer when the Nile was at its nadir. Its watering required about four hundred and eighty man-days per feddan (two men per feddan for eight months of the year).[70] If cotton was to be grown on a large scale, the irrigation system would have to be transformed. In the process, the agrarian social structure was also altered. The process by which Egypt became a major cotton exporter can be viewed as a dual transformation of rural technology and society. Let us now turn to this transformation.

## MUHAMMAD 'ALI AND THE BEGINNINGS OF COTTON PRODUCTION

Muhammad 'Ali's massacre of the Mamluks resolved the "prisoner's dilemma" within Egypt by destroying his internal opposition. In this he had simply succeeded where others, such as 'Ali Bey Bulut Kapan, had failed.[71] But this did not lessen the tax burden on the peasants, for Muhammad 'Ali embarked on a program of "primitive accumulation." At the core of this process was the need for a large standing army, both to acquire "colonies" in the Sudan, Western Arabia, and Syria, and to defend his position in Egypt against possible local rivals and the Porte. Of course, this was very costly, especially since a modern army had to be created from scratch. His problem was acquiring the resources necessary for such a plan. Since his principal goal, independence, necessarily implied conflict with the Ottoman Sultan, and since, as we have seen, the overwhelming majority of Egypt's trade had been with the Empire, Muhammad 'Ali needed (and proceeded) to reorient Egyptian trade toward the West. As Rivlin has pointed out, it is ironical that the pursuit of independence from the Ottoman Empire led Egypt towards dependence on the West.[72] By 1823, some 76% of Egyptian exports were going to Europe.[73] The consequences of this reorientation were far reaching, as we shall see.

To facilitate revenue collection and to protect his centralized power, Muhammad 'Ali abolished the iltizam system and replaced it with the _ihtikar_, or monopoly system. This system had three essential features. First, taxes were collected directly, by government employees receiving a salary. Second, peasants delivered crops at prices fixed by Muhammad 'Ali at levels below the market price.[74] Third, he monopolized both internal and external trade. At first, this system was limited to the traditional crops of the country; it was inaugurated in Upper Egypt in 1812 when the entire grain crop was seized.[75] But increasingly he turned towards cotton as an export crop. This not only helped to integrate Egypt into the (Europocentric) world economy, but also contributed to peasant land loss.

However, the _ihtikar_ system was incompatible with the increasing contact with Europe. The usual difficulties of running a centralized bureaucracy in a pre-industrial society joined with European pressures to force Muhammad 'Ali to abolish the system and to share power. In doing so, he laid the foundations for the rise of social groups who would dominate the century, the pashas and the shaykhs. At the same time, his policies had helped to weaken, but not to destroy, the protective social relations of the village, thereby beginning their transformation into capitalist forms.

Muhammad 'Ali's reign is important for three reasons: First, he initiated the integration of Egypt into the European economy as a supplier of agricultural goods. As a part of this process, he introduced the cultivation of long-staple cotton into the country and began the transformation of the Delta from basin to perennial irrigation. Second, his policies redistributed the control over inputs into agricultural production; his policies laid the foundations of the country's class structure. Third, the peasants were brought into direct contact with the government for the first time, disrupting and partially transforming village social relations. The suffering and dislocation of peasant life caused by this contact make Egypt a classic case of the thesis that the first dealings peasants have with the central government in the modern era is military, oppressive, and violent. This dislocation was an essential part of the redistribution of land. In Egypt original accumulation was primitive indeed. For the rest of this section, I will be concerned with the changes which Muhammad 'Ali wrought in rural technology and social relations.

Since Muhammad 'Ali needed revenue, and since it had to come from Europe, he needed something to sell there. This meant an agricultural product of some sort. We have seen that in the late eighteenth century France and Italy were importing wheat from Egypt. One might suppose that the logical response to this problem, given the information costs of finding new markets, would simply be to expand these exports. But here Egypt suffered from several disadvantages. France placed a sliding-scale tariff duty on grain imports in 1819. The Restoration government raised the rates in 1821 to such an extent that imports of grain were permitted in only one month from 1821 to 1830. Tariffs remained essentially unchanged until the Second Empire in the 1850s.[76] For the period of Muhammad 'Ali's reign, wheat exports to France could not provide the kinds of revenues which he required. The same, of course, applied to Great Britain with its Corn Laws from 1815 to 1846.[77] In the rest of the

Mediterranean, Egypt faced new and formidable competition from Black Sea grain. Italy became "dependent on Russian wheat for her pasta."[78] Egyptian threshing techniques (crushing the grains with a nurag, a heavy wooden sled with iron disks drawn by cattle) contributed to its weak competitive position in the export market.[79]

On the other hand, there was a strong demand in Europe for long-staple cotton, and its price was between two and four times as high as Egyptian short-staple cotton.[80] Foreign experts were brought in and given charge over several villages in 1821 and 1822. The resulting cotton was of high quality and fetched a high price in European markets. This provided Muhammad 'Ali with the incentive to extend its cultivation.

As we have seen, however, such an extension would have been quite difficult under the basin system of irrigation. Consequently, Muhammad 'Ali embarked upon a large-scale program of public works to supply the necessary water. New canals were constructed and old ones were deepened. Deeper canals made it possible for the water in the (low) Nile of the summer months to flow out to the fields. (Although it had to be raised up to the crops). Altogether some 240 miles had been dug by 1833.[81] Saqiyas (water wheels) and shadufs (counterweight lifting devices) were also constructed. The labor for these projects was provided by the corvée, discussed below. By the early 1830's, 600,000 feddans could be placed under summer cultivation, in comparison with 250,000 in 1798.[82]

Muhammad 'Ali also embarked on large-scale land reclamation works. By 1820, he had completed the Mahmudiyya Canal, which connected the Rosetta Branch of the Nile to Alexandria, had repaired the sea dike near Alexandria, and had reclaimed the Wadi Tumayat for sericulture.[83] Land reclamation efforts in this period were directed toward the often saline, heavy clay soils of the Northern Delta, primarily along a line from Damanhur to Kafr el Shaykh to Mansurah.[84] The need to drain such soils proved quite expensive, and was increasingly given over to private interests. From 1813 to 1852 the cultivated area increased by some 600,000 feddans.[85] (See Map 3).

At first, Muhammad 'Ali tried to induce the peasants to raise the crop by offering high prices and supplying working capital. Initially he offered 175 piasters per qantar and "made generous advances...of seed, oxen, and water-raising devices."[86] Prices offered to the cultivators had declined to 100-150 piasters per qantar in 1826 and although the evidence is fragmentary, they do not seem to have risen much until 1834. Yet it appears that Muhammad 'Ali's prices were more or less in line with international prices, falling in the late 1820s and then rising again in the early 1830s.[87]

However, there is evidence of peasant resistance to cotton growing. In some areas they would surreptitiously remove the cotton seeds which they had planted "in the hope of convincing the (local) Bey that the soil was not suitable for cotton cultivation."[88] In 1834 army officers and soldiers were sent to supervise the work in the fields. Rivlin asserts that there was "a noticeable lethargy

among the fellaheen toward cotton cultivation."[89] There are perhaps
two reasons for this. First, until 1836, the peasants were not paid
in cash, but rather in tax credits.[90] Second, there were labor
power problems: cotton was a labor-using crop and the amount of labor
power which was available in a given village was declining due to
conscription, corvee, and flight, factors which are discussed below.
Consequently, peasants might well have preferred to assure their subsistence crops before undertaking the cultivation of cotton.

Muhammad 'Ali laid down strict rules on how cotton should be
grown, rules derived largely from the advice of his foreign experts
and later codified in the La'ihat zira'ah al-fellah, (Regulation of
Peasant Agriculture). These instructions were extremely precise and
included nearly all of the details of its cultivation.[91] The orders
were passed on in a chain of command starting with Muhammad 'Ali,
passing through various intermediate government officials (ma'mur,
nazir, hakim al-khatt) to the qa'imaqam and village shaykh.[92] These
latter were the direct supervisors of peasant labor: the law obliged
them to go to the fields every day to inspect the peasants' labor.
If their orders had not been carried out, and if the peasant had no
valid excuse, the peasant was to receive twenty-five blows for the
first offense, fifty for the second and one hundred for the
third.[93] A similar fate awaited disobedient or slothful shaykhs or
qa'imaqams. This same chain of command enforced central government
decisions on the crop mix for each locality and collected the taxes.
This direct interference by the government in the production process
stands in marked contrast to the Mamluk system, although the mediat-
ing role of the shaykh provided continuity with earlier practices.

THE PEASANTS UNDER MUHAMMAD 'ALI: THE BEGINNINGS OF PRIMITIVE ACCUMU-
LATION

The interference by the government in the lives of the peasants
in this period was extensive. In this section, I wish to examine the
effects of Muhammad 'Ali's policies on the peasantry and their
response to the situation. The origins of "primitive accumulation"
in the sense of peasant land loss are to be found in this conflict,
so it would be wise to look at it in some detail. First, there were
the agricultural regulations. The government told peasants what to
plant, when to plant and how to plant it. They were forced to
deliver crops at fixed prices and to be paid in tax credit, which was
evaluated in a greatly inflated paper currency. They had to bear the
costs of transporting the produce to the central government depot,
where it was evaluated by government agents who often cheated
them.[94] Formerly beaten primarily at harvest time (a practice which
continued), the peasants were now also beaten throughout the eight-
month cotton-growing season. As indicated above, the main peasant
response to such conditions was lethargy and surreptitious resistance.

But by far the greatest burden was impressment, either in the
corvée, or worse, in the military. The corvée, of course, had
existed for centuries as a means of maintaining the dykes and canals
of the basin system. However, villagers worked to maintain the irrigation works of their own village. The effects of their labor were,

therefore, clearly visible and obviously benefitted them and their families, thereby legitimizing the system. But under Muhammad 'Ali peasants were dragooned to work far from their homes. They received little, if any payment—perhaps one piaster per day for giant public works like the Mahmudiyya Canal in the Delta, nothing for other projects. 'Amr estimates that 400,000 peasants worked for four months per year on these works.[95] Often they had to supply their own food, water, and tools. Since they were working away from their village, they received few benefits from their work. Even if they were working in their own areas, they may have received few benefits, for as we shall see later, towards the end of Muhammad 'Ali's reign the lands benefitting from summer water were often appropriated by government officials. If they did have land which could be planted in summer crops, they lost this opportunity (such as it was) by being forced to work on the corvee. After the formation of large estates, the large landowners arranged to have corvéed laborers work on their estates and to get their own peasants exempted from the corvée. Under these conditions it is hardly surprising that the peasants were reluctant to be drafted into the corvée. The government, however, compensated for peasant reluctance with the usual brutality. For instance, when the Mahmudiyya Canal was built, soldiers acting as overseers rounded up the peasants and brought them to work with cords around their necks. There were many casualties; estimates range from 12,000 to 100,000 dead over a three-year period.[96]

The oppression of the corvée was certainly secondary to that of military conscription, however. Muhammad 'Ali had decided to fight his wars with native Egyptian troops, because the Porte placed an embargo against supplying him with men, because imported Mamluks would have become a threat to his power and because he realized that their brand of unruly warfare had been shattered by Napoleon's disciplined squares. To fill his military registers, he impressed peasants into the army for life. They were therefore deprived of any chance of family or clan life, were badly paid, miserably fed, and led by Turkish officers who despised them.[97] Although these conditions prompted army mutinies in 1827 and 1832, the peasants in the villages presumably knew little about army conditions and cared less—they simply regarded the idea of conscription as anathema, for it would totally disrupt their lives forever.

The peasants responded to these measures in three ways: by rebelling, by fleeing from the land, and by mutilating themselves. Revolts were fairly numerous. In 1812, when the grain crop was seized for the first time, the Upper Egyptians of the area rebelled and were massacred by Muhammad 'Ali's Albanian cavalry. In 1816 groups of peasants refused to grow the crops which they had been ordered to raise and had to be coerced before they would do so. In May, 1823, the people of Minufiyya province rebelled against conscription and high taxes. The greatest revolt of the era broke out in April, 1824, in Upper Egypt, extending from Isna to Aswan. The rebels were led by a North African shaykh, Ahmad ibn-Idris, "who declared that he had been sent by God and his Prophet to end the vexations which the Egyptian people suffered and to punish Muhammad 'Ali for introducing innovations which ran contrary to the dogma of

Islam."[98] Part of his activity consisted of dividing the contents of government warehouses (shunas) among the people.[99] At first the revolt spread, with uprisings as far north as Girga. Many of the fellahin troops sent to quell the revolt joined the rebels, until finally the Turkish mercenaries and Muhammad 'Ali's bedouin allies crushed the rebellion with the usual carnage.[100] Revolts also occurred in Sharqiyya province in the Delta.[101] Sporadic revolts continued in the 1830s, the most significant one taking place in 1838, again in Upper Egypt, and again prompted by the draft. Since these revolts were never joined by any non-peasant group in society, and since the government had an adequate and passably loyal military force in the bedouins and the cavalry, all of these revolts were crushed fairly easily.

Self-mutilation was a somewhat safer means of thwarting the government and was widespread. Blinding one eye, especially the right one, cutting off the right index finger, and pulling the front teeth seem to have been the favored forms.[102] This worked so well that in Girga, a province of ninety-six villages, there were only seven suitable recruits. The practice of blinding seems to have fallen off somewhat when Muhammad 'Ali formed a one-eyed regiment, but there is no evidence to suggest that his order to throw those who deliberately maimed themselves into the galleys as slaves had any effect.

Finally, the peasants simply fled. About two thousand families went to Syria with their flocks where they were granted refuge by Muhammad 'Ali's enemy, 'Abd-allah of 'Akkah. They also fled to distant villages, to swamps, to bedouin tribes, and to the larger towns and cities, especially Alexandria. One major round-up netted between six and seven thousand refugees in that city. By 1831 twenty-five percent of the cultivable land in Upper Egypt was lying fallow because of the labor shortage,[103] despite frequent government attempts to catch the refugees and forcibly return them to the land.

THE RISE OF THE PASHA AND SHAYKH CLASSES UNDER MUHAMMAD 'ALI

Thus we see how Muhammad 'Ali began the transformation of the agricultural technology in Egypt and in the process brought the peasants into direct contact with the central government for the first time. This contact and the peasant response to it deprived many peasants of decision-making power over the land. Muhammad 'Ali's policies also created a new class of large landholders and strengthened a middle group, the village shaykhs. In effect, his policies radically redistributed land. Let us see how these changes came about.

After the massacre of the Mamluks in 1811, Muhammad 'Ali tried to make all of the important final decisions. This attempt to concentrate all of the power in his own hands quite naturally failed, especially since the bureaucracy was trying to run virtually the entire economy. So much land had been abandoned by the peasants that the government revenues noticeably declined; he had suffered heavy losses during the world economic crisis of 1836; and he was under pressure from the foreign powers, notably England, to accept the

Anglo—Turkish Convention of 1838 and abolish his monopolies. Consequently, Muhammad 'Ali made increasing numbers of grants of land to
military officers, foreigners, members of his own family, and miscellaneous flatterers, favorites, and the like.

These grants formed the basis for the creation of large estates.
They were of three kinds: (1) 'uhdah, (2) ib'adiyyah, and (3) chiflik. 'Uhdah grants resembled the old iltizam in that the fellahin
paid their taxes to the muta'ahhid ('uhdah holder) instead of to the
government. Indeed, such grants were primarily made because of
peasant tax arrears. The grant holder could not impose a tax burden
on the peasants higher than the official government tax rate. Like
multezims, part of the land grant was reserved for his own use, for
which he could use the unpaid labor of fellahin and upon which he
paid no taxes. Muhammad 'Ali still issued orders on land use, however. The muta'ahhid supervised cultivation to ensure that these
orders were carried out and provided working capital. He also "exercised judicial and executive powers previously discharged by government servants."[104] Muhammad 'Ali gave some 327,762 feddans as
'uhdah in the Delta and the Fayyum to members of his own family in
1842.[105] The second category, ib'adiyyah, were grants of uncultivated lands. The recipient of such land paid no taxes on it if he
brought it into cultivation. In 1842 "Muhammad 'Ali was
compelled...to grant almost complete rights of ownership (to
ib'adiyyah), including the right of sale and transfer,"[106] this
being the first sort of land to gain such status. Originally, most
of these grants were in Middle Egypt. The recipients were largely
high officials, who held some 164,960 feddans of such land in
1848.[107]

Finally, chiflik, "the most important factor in the formation of
large estates in the last century,"[108] were grants to Muhammad 'Ali
himself and to the royal family. From 1838 to 1846, some 334,286
feddans were so granted. There were other grants of land and the
three main categories sometimes overlapped. The origin of chifliks
is of interest here: "the greater part of them consisted of villages
abandoned because of the heavy tax burden and transferred to the
royal family."[109] Ninety-five percent of the chifliks were in the
Delta.[110] 'Uhdah were also often obtained by a grantee's paying off
the accumulated tax arrears of a village.[111] Rivlin summarizes the
tenure changes as follows:

> In a twenty—three year period (1821—1844) the best Delta land
> most suited to cotton cultivation had passed into the hands
> of large landholders. The peasants...held land which did not
> benefit from the program of irrigation works...Furthermore,
> the status of peasant held land remained unchanged; it was
> state land burdened with the kharaj (land tax) and to which
> the peasants had rights to usufruct only.[112]

Although the statistics of the period are very rough, it is possible to get a general idea of the changes in landholding wrought by
Muhammad 'Ali. (see Table 2.2)[113] 'Amr asserts that in the early
years of Muhammad 'Ali's reign (c.1813—1818) the total cultivable

TABLE 2.2
LAND TENURE UNDER MUHAMMAD 'ALI

|  | Peasant Land (feddans) | | Land held by Estate |
| --- | --- | --- | --- |
|  | 1820 | 1844 | (feddans) holders 1844 |
| Lower Egypt | 1,003,866 | 674,914 | 1,464,559 |
| Upper & Middle Egypt | 952,774 | 1,339,000 | 112,000 |
| Total | 1,956,640 | 2,013,914 | 1,576,559 |

Source: Rivlin, (1961), 73.

land of some two million feddans was divided as follows:[114]

1. Ib'adiyyah and chiflik            200,000 feddans
2. usya (lands of old,
   presumably non-Mamluk
   iltizam holders)                  100,000 feddans
3. 'arad al-mashayikh
   (shaykhs' lands)                  154,000 feddans
4. 'arad al-rizqah
   (lands given to foreign
   experts and advisors)              6,000 feddans
5. 'arad al-athri
   (peasant land)                   (no numbers)
6. 'arad al-'arabin
   (bedouin land)                   (no numbers)

Grant holders of all types, including shaykhs, held roughly 28% of the cultivated area in about 1820, but nearly 44% of it by 1844. These numbers are, of course, quite rough. Nevertheless, they do provide an indicator of the very large increase in estate lands during this period. The grantee holders did not wield much independent power in the late 1830s and 1840s, for the state was still sufficiently powerful to thwart any possible attempts to challenge the ruler. But these land grants went far toward creating a ruling class of landlords.

This new Turko-Egyptian landlord class was supplemented by some native Egyptians, at least in the bureaucracy, although the highest officials remained Turks. Muhammad 'Ali needed manpower to fill the posts which the expansion of the civil and military bureaucracy created. Since Muhammad 'Ali was suspicious of his Turkish subordinates, he needed a group of officials who would carry out his orders more effectively, or at least, upon whom he thought he could rely. The native Egyptians could fulfill both needs because they were newcomers to power, raised up by Muhammad 'Ali and therefore more dependent on him than were the Turks. The appointment of Egyptian officials had an additional advantage for Muhammad 'Ali. Since they were less secure in their tenure than the Turks, he could afford to pay them less, thereby lowering government expenditure. The new Egyptian bureaucrats made up for this loss by extorting the difference between their pay and that of the Turks from the peasants. They were uniquely qualified to do this because many of the appointees were village shaykhs or their sons or relatives.[115]

This brings us to the "middle" group whose power increased during Muhammad 'Ali's reign, the village shaykhs. Like the Egyptian bureaucrats with whom they were closely associated (indeed, they were often the same person), their power only began its growth in this period, during the second half of Muhammad 'Ali's reign. Initially, his policies may have weakened them. The abolition of _iltizam_ "stopped up the source of most of the payments from which the village shaykhs had benefitted."[116] However, the shaykhs still functioned as the government's administrator, being paid by a grant of land (_masmuh_) which amounted to between 5 and 10% of the village areas.[117] Further, they acquired a good deal of the land which was abandoned by peasants fleeing conscription. They also managed to turn their powers of tax collection to their own advantage, as did everyone in the administration, where corruption was rife.[118] No doubt the army officers and bureaucrats in Cairo took the lion's share, but the shaykhs could endure the bastinado as well as anyone and their physical endurance quite probably enriched them. Finally, their position as directors and supervisors of agricultural labor was extended considerably. As we shall see, these shaykhs, as a "middle" group in the countryside, play a central role throughout the modern history of rural Egypt.

In summary, although Muhammad 'Ali did not introduce capitalist social relations on a full scale, defined here as private property in land and the creation of a landless class and a free labor market, he did contribute toward their later rise. First, peasants were brought into direct contact with the central government for the first time, a contact which caused major dislocations in peasant life. Muhammad 'Ali weakened the protective organization of the village and its internal solidarity by his land grants, but by retaining collective tax obligations, including the labor tax of the corvée, he did not destroy it. Second, such dislocations contributed to the foundation of a new class of large landowners and to the strengthening of a middle group, the shaykhs. In short, the outlines of the class structure of nineteenth-century Egypt began to emerge during his reign and as a result of his policies. Third, he made important changes in the agricultural technology by beginning perennial irrigation, by reclaiming Delta lands, and by introducing the cultivation of long-staple cotton. Finally, he acquainted Europeans with the possibility of Egypt as a source of supply of an important raw material for their factories. The Muhammad 'Ali years were a time of beginnings.

THE TRANSFORMATION OF AGRARIAN SOCIAL RELATIONS, 1848-1882: LAND TENURE

The transformation of both rural social structure and technology
continued and intensified from the death of Muhammad 'Ali in 1848 to
the British intervention of 1882. The period saw the consolidation
of the large landowning class of Turko-Egyptians, the further rise in
the power of the village shaykhs, and the dispossession of large
numbers of the peasantry. The irrigation system and the transportation network were developed further, cotton became the main cash
crop, and foreigners came to occupy an important role in the economy.
The stresses which this dual transformation produced culminated in
the 'Urabi revolt, which in its early phases pitted shaykhs, native
Egyptian government officials, and numerous peasants against the
Turko-Egyptian landlords and foreigners. Given Britain's economic
interest in India, and Egypt's strategic position vis-a-vis the
latter, Britain intervened. For the rest of this chapter I will try
to trace the acceleration of the transformation of rural technology
and social relations, to see how it affected different social groups,
and what their responses to these changes were.

The transformation of tenurial institutions took place in
several stages. It was a complex affair, and I shall only sketch it
here.[119] There are two issues to be discussed: one, the change in
the definitions of the various legal categories of land, and two, the
growth of private property rights for these various categories. What
happened to the 'uhdah holders is by no means clear. However, it
appears that 'Abbas (r. 1848-54) confiscated many of them. This did
not, apparently, help the peasants much. (See below) It seems to
have meant a return to more direct taxation. Further, Abbas seems to
have granted full ownership rights to some 'uhdah holders. Some new
'uhdahs were created under Isma'il (r. 1863-1879). He then abolished
them between 1866 and 1868 by granting ownership rights to their
holders in return for payment of tax arrears. I shall have more to
say about the relationship between taxes and the rise of private property in land later. Ib'adiyya and chiflik seem to have continued as
before, although both were expanded by Isma'il. He made extensive
grants of both kinds between 1863 and 1870. The former type was
especially prevalent in the northern Delta provinces of Behera and
Gharbiyya.[120]

These categories were more or less subsumed by two other
categories which were created by Sa'id (r. 1854-1863) in his Land Law
of 1854. The measures taken by Sa'id and Isma'il greatly increased
the security of both sorts of tenure, although they favored one over
the other. The first sort was known as 'ushuriyya, since in 1854 a
tax of ten percent was imposed on them ('ushr = "tenth" in Arabic).
The second category was kharajiyya or athariyya, which included all
peasant holdings. 'Ushuriyya included chiflik, some ib'adiyya land,
and some other sorts of land, including, later, the 'uhdahs granted
by Isma'il.[121] Such lands form a subset of all land held in large
grants: ib'addiyya grants under Abbas fell outside the category, as
did 'uhdahs granted before Abbas. Nevertheless, the growth of this
type of land can be taken as a lower-bound estimate of the growth of
large-landholding at the expense of the peasants. I shall examine
reasons for supposing that the actual land loss of the peasants
exceeded this lower bound later on. As can be seen from Table 2.3,

the peasant holders of kharajiyya land lost some 300,000 feddans to
the large 'ashariyya holders from 1863 to 1880. All 'ushuriyya land
was declared to be full private property in the Land Law of 1858.
This law also provided for foreign ownership of land, although some
restrictions remained here until 1873.[122]

The second category of land, kharajiyya or athariyya, presents a
somewhat more complicated case. In 1846 the peasants were allowed to
mortgage the usufruct; at the same time, the seizure of the usufruct
for failure to pay taxes was formally recognized. In 1858 they were
allowed to inherit the usufruct, and Sa'id's Land law of 1858 further
secured this inheritance.[123] However, the peasants still held and
inherited only the usufruct since the government could confiscate the
land without indemnity. An exception to this right of confiscation
was that anyone who erected buildings, constructed a saqiya, or
planted trees on kharajiyya land became the full owner. Also, kharajiyya could not be endowed as waqf without the express permission of
the ruler.

Under Isma'il the government's need for cash gave impetus to the
creation of private property from kharajiyya land. In 1871 the Muqabala Law "freed from one-half his tax liability anyone who paid six
year's taxes in advance and stated that the difference would never be
claimed."[124] They would also become full owners of the land. At
first people held back, but as the government made payments compulsory in 1874, those who could muster the capital came through. By
1881, private ownership of land was the general rule.

## THE TRANSFORMATION OF AGRARIAN TECHNOLOGY, 1848-1882: PUBLIC WORKS AND THE COTTON BOOM

The impetus behind these changes was the government's increasing
need for revenue, revenue which was used, in part, for large-scale
public works. As a result, the infrastructure of Egypt was considerably improved. The length of railroad track tripled; between 1851
and 1856 the Alexandria-Cairo railroad was built; in 1856-1857 the
Cairo-Suez line was finished; side lines to Samanud and Zagazig in
the Delta were constructed. By 1877, Egypt had 1,519 kilometers of

TABLE 2.3
LAND TENURE, 1863-1880

| Year | 'ushuriyya Area (feddans) | % | Kharijiyya Area (feddans) | % |
|------|---------------------------|------|---------------------------|------|
| 1863 | 636,177 | 14.5 | 3,759,125 | 85.5 |
| 1875 | 1,194,288 | 26.0 | 3,509,168 | 74.0 |
| 1880 | 1,294,343 | 27.4 | 3,425,555 | 72.6 |

Source: Baer (1962), 20.

standard gauge railroad. This made marketing for export much easier
and extended the area where export crops could be grown profitably.
Other public works were the improvement of the harbor facilities at
Alexandria, and a large program of irrigation works. Isma'il had
8,400 miles of canals dug, the largest being the Ibrahimiyya Canal in
Middle Egypt, thus bringing perennial irrigation to a portion of that
area for the first time.[125] Isma'il's spending on the country's
infrastructure is summarized in Table 2.4. The problem was, of
course, that such spending could only be financed through increased
taxation and foreign borrowing, particularly when one adds the spending on luxuries for the court and military adventures in the Sudan.
By 1876 Isma'il was in debt some nine and one half million
pounds.[126]

The change in tenurial relations, the extension of irrigation,
and the improvement in transportation facilities laid the groundwork
for the spread of cotton cultivation in Egypt. Two factors stimulated the growth of this crop. First, although there is some doubt
here, there were government attempts to increase the amount of taxes
collected in cash. We don't know how successful this was; we do know
that 'ushuriyya landholders were explicitly allowed to pay either in
cash or in kind by the law of 1854. Owen doubts that there was
enough money in circulation at the time to allow all peasants to pay

TABLE 2.4
PUBLIC WORKS SPENDING OF ISMA'IL

| Work | Cost ('000 L.E.) | Remarks |
|---|---|---|
| Suez Canal | 12,000 | After deduction of L.E. 4 million for shares sold |
| Canals | 12,600 | L.E. 1,500 per mile |
| Bridges | 2,150 | 430 bridges |
| Sugar Mills | 6,100 | 64 mills with machinery |
| Alexandria harbor | 2,542 | |
| Port of Suez | 1,400 | |
| Alexandria waterworks | 200 | |
| Railways | 13,361 | 910 miles |
| Telegraphs | 853 | 5,200 miles |
| Lighthouses | 188 | 15 in Mediterranean and Red Sea |
| Total | 51,394 | |

Source:

Abbas, H.H. The Role of Banking in the Economic Development of
Egypt., Ph.D. Dissertation, University of Wisconsin—Madison,
1954.

in cash. Nevertheless, the attempts to increase cash payment of taxes probably did induce some peasants to try to raise a marketable crop.[127] In the 1850's this may have been wheat; in 1856 the U.S. Consul was reporting that wheat was more profitable than cotton. The repeal of the Corn Laws in England, the liberalization of French tariff policy under Napoleon III, and the Crimean War all helped to stimulate production in Egypt. Whereas average annual wheat exports were 485,021 ardebs in the period 1840-1843, and 527,065 ardebs in 1848-1850 for the 1850s (1852-1858), average annual exports climbed to 1,143,770 ardebs, reaching a maximum of 1,674,852 ardebs in 1855 and a minimum of 1,522,572 in 1857.[128] Wallace presents some peasant testimony that, at least in some areas, the fellahin abandoned the cultivation of cotton upon the death of Muhammad 'Ali in 1848.[129] However, by 1860 Egypt was already the sixth most important supplier of the British cotton market, with Britain taking 65% of the crop. When the U.S. Civil War broke out and the Union navy blockaded or occupied Southern ports, British textile manufacturers turned to Egypt as a source of raw material, a move which doubled the value of Egyptian exports. Area figures are not available for the period, but the growth of cotton cultivation can be traced in the export figures, since very little cotton was retained for home manufacture. Although the cotton boom collapsed, with social consequences discussed below, cotton exports remained well above their pre-1860 level (see Table 2.5).

THE POSITION OF THE SOCIAL CLASSES, 1848-1882: THE PASHAS

Thus, in general, the period 1848 to 1880 saw the extension of capitalist institutions (private property in land) and the development of Egypt as an exporter of cotton. The question we now ask is how the different groups in the countryside were affected by these changes. Who gained ownership of what sort of land? How was this reflected in the methods of cultivation? How was agricultural labor organized? How did the landless class emerge? What social forces lay behind the Urabi revolt of 1882?

First, the class of Turko-Egyptians consolidated their position as large landowners during this period. As we have seen, they had been given extensive grants of land under Muhammad 'Ali and other Viceroys, and their private property rights over these lands had been extended during the period under consideration. We have seen how their holdings expanded at the expense of peasant lands and how they acquired tracts of uncultivated land. They also enjoyed lower tax rates than the peasants. (See Table 2.6) The rates on 'ushuriyya lands rose during the period, but the differential remained: in 1877 kharajiyya land tax rates in the Delta were between 120 and 170 piasters per feddan, and sometimes went as high as 200 per feddan. The national average on this sort of land was 116.2 per feddan. 'Ushuriyya land holders, on the other hand, paid an average rate of 30.3 piastres per feddan.[130] When under the Muqabala law of 1871 kharajiyya lands became private property if six years' taxes were paid in advance, the Turko-Egyptians, along with the shaykhs and foreigners, must have been the main beneficiaries. The way in which land became private property favored the Turko-Egyptian "pashas."

TABLE 2.5
VOLUME AND PRICE OF EGYPTIAN COTTON EXPORTS

| Year | Volume (cantars) | Price (rials/cantars = 1/5 L.E.) |
|------|------------------|----------------------------------|
| 1858 | 502,645 | 12.75 |
| 1860 | 501,415 | 12.00 |
| 1861 | 596,200 | 12.00 |
| 1862 | 721,052 | 13.0 |
| 1863 | 1,181,888 | 23 |
| 1864 | 1,718,791 | 36.25 |
| 1865 | 2,001,169 | 45 |
| 1866 | 1,288,762 | 21.25 |
| 1867 | 1,260,946 | 35.25 |
| 1868 | 1,253,455 | 22.25 |
| 1869 | 1,289,714 | 19 |
| 1870 | 1,351,797 | 22.5 |
| 1871 | 1,966,215 | 19.5 |
| 1872 | 2,108,500 | 15.75 |
| 1873 | 2,013,433 | 21 |
| 1874 | 2,575,648 | 19 |
| 1875 | 2,206,443 | 19.5 |
| 1876 | 3,007,719 | 15.5 |
| 1877 | 2,439,157 | 13.12 |
| 1878 | 2,583,610 | 13 |
| 1879 | 1,680,595 | 16.38 |
| 1880 | 3,000,000 | 14.52 |
| 1881 | 2,510,000 | 13.83 |
| 1882 | 2,811,000 | 14.24 |
| 1883 | 2,140,000 | 14.71 |
| 1884 | 2,565,000 | 13.52 |
| 1885 | 3,540,000 | 12.37 |
| 1886 | 2,788,000 | 11.71 |
| 1887 | 2,864,000 | 12.37 |
| 1888 | 2,964,000 | 12.3 |
| 1889 | 2,780,000 | 13.27 |
| 1890 | 3,203,000 | 13.4 |
| 1891 | 4,054,000 | 11.52 |
| 1892 | 4,662,000 | 9.06 |
| 1893 | 5,117,000 | 9.3 |
| 1894 | 5,073,000 | 8.49 |
| 1895 | 4,840,000 | 8.46 |
| 1896 | 5,220,000 | 10.03 |
| 1897 | 5,756,000 | 8.68 |
| 1898 | 6,399,000 | 7.18 |
| 1899 | 5,604,000 | 7.90 |
| 1900 | 6,512,000 | 10.84 |
| 1901 | 5,391,000 | 10.87 |
| 1902 | 5,526,000 | 9.81 |
| 1903 | 5,860,000 | 13.65 |

| 1904 | 6,147,000 | 14.41 |
| 1905 | 6,376,000 | 12.18 |
| 1906 | 6,033,000 | 15.11 |
| 1907 | 6,977,000 | 16.87 |
| 1908 | 6,913,000 | 14.42 |
| 1909 | 6,813,000 | 13.44 |
| 1910 | 5,046,000 | 21.49 |
| 1911 | 7,477,000 | 17.6 |
| 1912 | 7,367,000 | 17.25 |
| 1913 | 7,375,000 | 18.28 |
| 1914 | 7,369,000 | 19.02 |

TABLE 2.6
TAX RATES BY LAND CATEGORIES (PIASTERS/FEDDAN)

| | 'ushuriyya 1854 | 1864 | Kharajiyya 1856 | |
|---|---|---|---|---|
| Lower Egypt | 26 | 35 | 100 | Lower and Upper |
| | 18 | 25 | 90 | Egypt |
| | 10 | 18 | | |
| Upper Egypt | 20 | 31 | | |
| | 14 | 21 | | |
| | 8 | 14 | | |

Source: Baer, (1962), 18.

Furthermore, they had the best and the most fertile tracts lying along the summer canals, lands which were the most suitable for cotton cultivation. This had been true since Muhammad 'Ali's original grants. "The Englishman, Thomas Clegg, who made a tour of the Mediterranean cotton areas in the mid-1850s reported that something like 3/8 of the total Egyptian crop came from the estates of 'Abbas and the family of Ibrahim alone."[131] They were able to use their power to get water, to appropriate animals, and to corvée fellahin to work their land. They could afford to erect saqiyas more easily. The lower tax rates on their lands and their greater wealth presumably gave them a longer time-horizon than the smaller cultivators.

All of these advantages were reflected in their superior techniques of cultivation. They planted their lands in cotton "only once every four or five years, thus preserving the quality of the soils"[132] in the 1850s. They paid greater attention to seed selection and spaced their rows of cotton three feet apart.[133] The result of these techniques was higher yields:

By the end of the 1850s a marked difference was being observed in the crops produced by the two groups—that from the

large estates being known as..."princes'" cotton and generally enjoying a premium of 1 1/2 to 2 dollars a cantar on account of its greater cleanliness and length of staple...
Balli (peasant) cotton rarely produced more than two cantars
a feddan, whereas those with money to invest in the oxen and
saqiyas necessary to provide adequate watering could provide
one to one and a half cantars more.[134]

In short, there were few advantages in agricultural production which
the pashas did not have: higher quality land, more of it, more money,
longer time-horizons, lower tax rates, and the ability to dragoon the
one input which they did not own outright, labor power. Clearly, a
bimodal land tenure system had emerged in Egypt.[135]

It is unclear precisely how the pashas exploited their estates.
However, if we look at the period 1848-1880 as a whole, we can draw a
few tentative conclusions. Central to the discussion here is the
probable emergence of the 'ezbah system at roughly the time of Muhammad 'Ali's moves toward decentralization of agricultural administration.[136] As the word implies, 'ezbahs were hamlets, established by
the landlord at some distance from the local village.[137] The
essence of the system was the granting of parcels of land to peasants
to grow their subsistence crops (maize and beans) and fodder for the
animals (birsim) in exchange for labor services in the landlord's
cotton fields. These subsistence plots were rotated in accordance
with the general rotation of the fields among crops and fallow.
Agricultural operations as a whole were supervised by the owner or
his agent, especially the cultivation and harvest of cotton, questions of irrigation and drainage, and crop rotation. Supervision of
the labor for food and fodder crops was unnecessary, since the
peasants had every incentive to produce as much as possible. Admittedly, this sketch summarizes what we know about the system from
information starting in the 1880s and 1890s.[138] But it seems likely
that some such system was practiced on the large estates earlier.
First, we have the assertions of Lozach and Hug, and Ayrout on the
date of origin of the system. Second, Bayle St. John, travelling in
Egypt in the 1850s, reported that on estates, "almost everywhere the
fellah is allowed to possess a small allotment, which he cultivates
when he is able."[139] He also noted that the estate holder could use
the labor of those who worked on it, and often paid no wages to the
fellahin. Third, the evidence presented in the preceding paragraph
on techniques of cotton cultivation on the estates indicates that it
was probably a closely supervised crop in the 1850s. The peasants
had much less incentive for working hard in the cotton fields, work
for which they received little, if any, direct remuneration, than
they had for carefully tending the maize, bean, and clover patches.
Yet the cotton grown on the estates produced higher yields of cotton
than did small farms and used different techniques. Cotton laborers
must, therefore, have been supervised.

Despite their considerable power, several factors mitigated the
pashas' power and the security of their wealth. Muslim inheritance
law requires that land be divided proportionately among all heirs.
This produced a tendency toward fragmentation, and many of the

families of high officials lost their estates in this way.[140] However, the result seems to have been that those who continued in power or continued to have money simply acquired other lands, with the result that some large landowners held lands in different and dispersed areas, rather than in contiguous plots forming vast estates. If the land units became small enough, it seems likely that share-cropping (metayage) was the principal mode of land and labor exploitation. Sharecropping was the principal mode of exploitation for medium properties in the 1880s and later, and Gali asserted in 1889 that the system was less widespread then than it had been in the past.[141] This is what one would expect on a priori grounds, since, (1) the supervision costs of the 'ezbah system would rise sharply with fragmentation, and (2) the poverty of the fellahin would make it difficult for them to pay cash rents or to give guarantees. Sharecropping would then be the logical alternative. Gali notes that the risks of Nile flood variations also contributed to the choice of sharecropping.[142] For the period 1850-1880, the available evidence is simply too scanty to permit any conclusions about how much estate land was cultivated under a metayage system and how much under the 'ezbah system. Such an assessment is made particularly difficult by the fact that within the 'ezbah system, the subsistence plots were sometimes cultivated on a share basis.[143] One can note, however, that the inheritance laws, and therefore one of the pressures toward choosing a metayage system, could be and were evaded by turning land into waqf.[144]

In addition to fragmentation, the power of the central government made the tenure of property insecure for the Turko-Egyptians. Because officials got their land through politics and the favor of the Viceroy, they could lose them in the same way. Sa'id destroyed some of the fortunes of the grantees who had been favored by 'Abbas. When evidence that an official had fallen from favor appeared;

the peasants plunder him with impunity. Old claims are raked up against him; new ones are forged; his crops are destroyed; his cattle are stolen; he becomes..."a thing to be eaten."[145]

It was really only with the British occupation and administration that all land could be considered entirely free from confiscation.

Nevertheless, Egypt in the nineteenth century was far from being an Oriental despotism a la Wittfogel. Large landowners were fairly well protected by the 1858 law, and there were other mechanisms for circumventing the ruler's power, such as turning land into waqf. The Viceroy also needed the cooperation of his fellow Turko-Egyptians, so he couldn't antagonize them too much. There is no question that he exercised a great deal of arbitrary power, but not so much that he prevented the Turko-Egyptians from being the wealthiest and most powerful social group in Egypt.

THE POSITION OF THE SOCIAL CLASSES, 1848-1882: THE VILLAGE SHAYKHS

They were not the only powerful group, however. Just as Turko-Egyptians turned office holding into land holding, so did the native Egyptian officials. During Sa'id's rule, and especially during that

of Isma'il, for example, a fellah, Hamid Abu-Satit, became <u>mudir</u> of
Girga and then Qena provinces. During his time in office he acquired
seven thousand feddans. 'Urabi Pasha, leader of the revolt in 1882,
held eight and a half feddans from his father, a village shaykh, but
managed to expand his holdings to 800 feddans by the end of his army
career. It seems, however, that most such officials who owned land
had acquired it <u>before</u> they became officials, usually by virtue of
being village shaykhs.[146]

Indeed, the village shaykhs showed the most notable rise of any
group during this time. Their wealth and power derived directly from
their political position as intermediaries between the government and
the peasants. By exploiting this key role, by taking advantage of
economic trends, and by joining the civil service, some of them rose
to positions comparable to the Turks, as in the examples given above.
Of course, the shaykhs had always linked the peasants to the government. What happened between 1850 and 1880 was that the functions of
the shaykh expanded as the state performed new tasks and undertook
new projects. The spread of private property rights in land allowed
them to convert such power into land, always at the expense of the
peasants.

Their rise began during Mahammad 'Ali's reign, as shown above,
when they were responsible for taxation. He established manpower
quotas from each village for the army and the corvee, and made the
shaykhs responsible for filling them.[147] This put them in an ideal
position to extract bribes from the peasants in exchange for exemptions from the draft or corvee. This was not as important during the
reign of 'Abbas as it would have been under Muhammad 'Ali, since
there were no foreign wars to fight and no major public works projects were undertaken. Still, the corvee and the draft continued,
and the shaykhs continued to get land which the peasants abandoned
when they fled from these two scourges. The shaykhs also used their
power to supervise the transfer of land when a fellah died without
heirs to defraud peasants and to increase their own fortunes.[148]

Sa'id attempted to weaken the shaykhs' power, although with limited success, by reducing their authority in land distribution and in
taxation. In 1855 land transfers were taken out of their hands and
put into the hands of the <u>mudir</u>. The Land Law of 1858 which secured
inheritance of usufructory rights on peasant land undermined the
shaykhs' power to redistribute land. The abolition of collective
taxation (except for the corvee) reduced their power in tax affairs.
Finally, Sa'id extended conscription to include the shaykhs' sons,
who had previously been exempt.[149]

Such a weakening was only temporary and was limited because the
shaykhs remained the government's representatives and agents in the
villages.

They had a large and sometimes decisive say in fixing tax assessments; they determined virtually alone the classification
of land for purposes of taxation; and they decided which land
should be classed as "unproductive" and therefore tax exempt.
It was on the strength of what the <u>umdah</u> told him that the

mudir decided which land should be expropriated for public use.[150]

The shaykh remained "the Pasha in miniature."[151]

If Sa'id left the shaykhs with considerable power, Isma'il raised them to their apogee. They continued to be responsible for the corvee and to reclassify land for taxation. The weakening of the central government toward the end of his reign only strengthened them by removing any checks on their arbitrary exercise of power, so that by the time of the British intervention, "the general impression of contemporary observers was that the village shaykhs were practically uncontrolled, virtually masters of the country."[152] We shall see, however, that they had competitors for this position in the countryside.

As a result of their power, the shaykhs accumulated considerable wealth and lands, although not so much as the Turko-Egyptians. Baer estimates that the higher Turkish officials held estates of between 1,500 and 2,500 feddans apiece, whereas large shaykh estates were between 800 and 1,000 feddans.[153] Some owned as much as 3,000 feddans and more, but these seem to have been exceptional. Their lands came chiefly from former peasant holdings, for they seem to have been the main beneficiaries of peasant flight from the land.[154] It is likely that the richer shaykhs exploited their lands using the 'ezbah system, whereas the more modest ones used sharecropping. Wallace in 1876 speaks of landless peasants "associating themselves with richer neighbors," and Villiers Stuart in 1883 reported conversations with shaykhs resident on their 'ezbahs. [155] Their long experience as supervisors of agricultural operations must have served them well in either case.

THE POSITION OF THE SOCIAL CLASSES, 1848-1882: THE PEASANTS AND LAND LOSS

Finally, there are the fellahin. The ways that they lost their land, alluded to at various times above, can be summarized here: (1) outright seizure—the land grants of Muhammad 'Ali; (2) flight from the land to escape corvee, conscription, or taxes; (3) failure to pay taxes; and (4) foreclosures for nonpayment of private debts. The steady rise in the land tax rates and their heavy incidence on the fellahin have been noted above. Although 'Abbas reduced the size of the army, conscription continued with the result that whole villages sometimes left and fled to the hills, as in the area near Luxor mentioned by Bayle St. John.[156] When 'Abbas abolished the 'uhdahs, he demanded tax payment from the peasants; a "large-scale exodus" ensued.[157] Sa'id's tax increases also led to mass flights. He then limited the peasant's right of return to the land and sold the raqaba or title of the land to officials and to foreigners.[158] Under Isma'il, not only the increase, but also the timing of these increases contributed to peasant land loss: as the cotton boom collapsed, taxes were increased. In 1868, the British consul at Alexandria reported that the peasants were paying seventy percent more in taxes than in 1865. Widespread flight from the land resulted.[159]

Taxation contributed to peasant land loss in another way, through debt and foreclosure. Its rise seems to have begun during the cotton boom. We have seen how cotton prices rose in the 1860s and how output responded. It should be noted, first, that most cotton grown before 1860 was produced on the large estates. Estimates vary; the best available study, that of Owen, asserts that "only a fraction" of the total cotton crop was grown on peasant-held land in this period.[160] But after the precipitous price climb (see Table 2.5) they did, increasingly, plant the crop. In part, this was no doubt a response to the ordinary sort of price incentives. There is much evidence, however, that the heavy tax rates played a role as well, leading in some cases to overcropping.

> ...the fellahin being hard pressed for money, took to rais-
> ing as much of it (cotton) as possible, regardless of the
> fact that the overcropping might in a few years impoverish
> the soil...as the overcropping did not suffice to provide all
> the money which the peasants required for the payment of
> their taxes and other purposes, and as the influx of Europe-
> ans and European institutions furnished new means of easily
> borrowing money on land and other securities, the peasants
> gradually sank into a state of indebtedness from which a very
> large section of them can never hope to extricate them-
> selves.[161]

If the peasants were growing cotton, this rise is hardly surprising. Although peasants' techniques (the balli method) differed sharply from those of the landlords, using less labor, land, and seed than the methods of the pashas, increased cotton production would have increased their need for working capital, inducing them to borrow. However, most contemporary observers (e.g., Wallace, Villiers Stuart, and Lady Duff Gordon) emphasized the increased tax burden as the cause of most peasant debt. These moneylenders loaned at rates which varied from one to five percent per month.[162] Such loans, incurred during the boom, proved burdensome later: (1) the price of cotton fell sharply (see Table 2.5), (2) their cattle were wiped out by a plague in 1863, and (3) tax rates rose after the boom collapsed, as noted above. Bankruptcies and land sales followed.[163] This form of land loss continued and accelerated during the 1870s, especially after the establishment of Mixed Courts in 1875.

CONCLUSION

In summary, Egyptian agricultural development from 1800 to 1882 was a contradictory process. On the one hand, land, labor, and output expanded very rapidly. The transformation of the irrigation system advanced considerably. Perennial irrigation had become the rule throughout the Delta, as well as in parts of Middle Egypt. This change made possible the dramatic diffusion of "new" crops, especially cotton and maize, whose output expanded at the annual compound rate of roughly 7% and 3.5% per year, respectively, between 1821 and 1872-78. Sugar output grew at just under 8% per year, wheat at 2%, and beans at 1.4%. Rice production, however, declined, presumably replaced by cotton in some northern Delta areas. The cultivated area more than doubled (from 2,032,000 to 4,742,000 feddans) during the same period as a result of land reclamation. (See Map) The population also grew, although at a much slower rate: from just under 4.5 million to just over seven million from 1821 to 1876.[164] As a result, O'brien estimates that per capita farm output increased nearly six fold during the period. One should note, however, that these figures are measured from the early nineteenth century, perhaps the nadir of Egyptian agriculture and rural history for several millenia. (See above, e.g. on population). Further, when considering the increase in output per head of the rural population, one should recall that the peasants were also very probably working more days per year than before, tending the labor-intensive crops produced during a (formerly) "dead" season. Nevertheless, the aggregate growth figures are very impressive.

And yet there were aspects of rural development during the period which boded ill for the future.[165] First, the creation of a landless class was an integral part of Egypt's incorporation into the world market as a cotton exporter. We have seen how the rise in private property in land, the increase in taxation to finance irrigation works and the other infrastructural investments which cotton production required contributed to peasant land loss. Summer canals and transport facilities required labor and capital. The corvée provided the former, while increased taxation and borrowing abroad supplied the latter. As the size of the foreign debt mounted, the government increased taxes further, mainly striving to meet its obligations. As a result of such fiscal pressures, peasants fled from their lands, forfeited them for failure to pay taxes, and faced foreclosure for private debt. Their lands then accrued to pashas, shaykhs, and village moneylenders. We shall see in subsequent chapters that the problem of the landless class in rural Egypt which this process created even now remains unsolved.

The same change which created a landless class also generated a bimodal land tenure system. Landholders were divided into different classes, with sharply unequal access to productive resources. By 1888 less than 1% of landholders held more than 40% of the land on large estates over fifty feddans. At the other pole of land distribution, over 80% of landowners held only some 20% of land on small minifundia holdings. Egypt provides no grounds for dissent from the current consensus that bimodal land tenure systems retard agricultural development, as we shall see in the next chapter. There I will examine social structure under the British administration, the consequences of such a structure for the choice of agricultural technology, and the resulting consequences for income distribution and the productivity of the land itself.

APPENDIX: THE 'URABI REVOLT OF 1882

Such large scale social transformations usually do not occur without resistance and upheaval, and Egypt was no exception. All of the social developments described above contributed to the 'Urabi revolt of 1882. Although the causes of the actual military uprising are most complex, involving a great deal of intrigue, false starts,

40

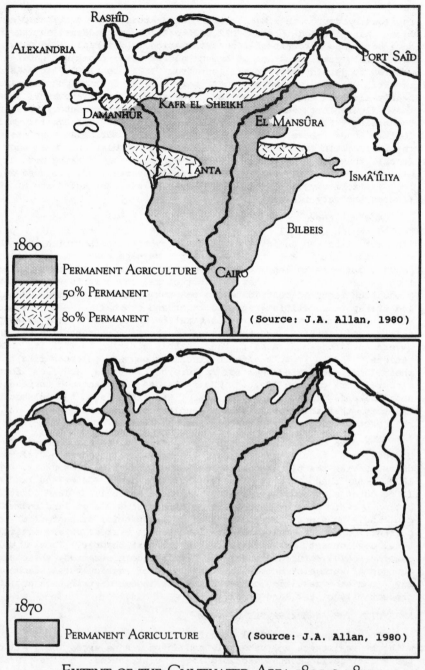

EXTENT OF THE CULTIVATED AREA 1800 & 1870

misunderstandings and the like by Egyptian political figures and the foreign ministries in London and Paris, a number of important facts about the social composition of the rebels, their targets, and sources of support up to the intervention of the British seem critical. I am not trying to offer a detailed explanation for all of the maneuvers which occurred; I merely wish to place the revolt within the context of the rural transformation outlined so far.[166] As mentioned above, Urabi Pasha belonged by birth and career to the class of shaykhs and native government officials. Men such as this were also his principal co-plotters; it was a revolt of the junior officers, a product of the recruitment of native Egyptians.[167] These men were angry that the Turko-Egyptians blocked their advancement by monopolizing the highest ranks. Furthermore, there were "cutbacks in every branch of administration under Anglo-French control from 1876-1882."[168] All that was needed to trigger a revolt was a collapse of legitimacy, which had in fact happened when Isma'il placed tax collection under the aegis of the Caisse de la Dette Publique, (the corporation of aggrieved foreign creditors) and abdicated in 1879.

Initially at least, the shaykhs sided with 'Urabi. We have seen how their power grew throughout this period. Although their wealth was still less than that of the Turko-Egyptians, they had become increasingly insubordinate to the pashas and the higher (i.e. Turkish) authorities.[169]

> The backbone of the Arabist party, in so far as that party represented a national movement and not a military movement, was to be found amongst this class. The greater part of the yeomanry of the country were sympathetic with Arabi; he was their kith and kin; they looked to him to deliver them from the usurer and the pasha.[170]

The shaykhs had a specific grievance: the Caisse de la Dette Publique was proposing to abolish the Muqabala Law as part of its program of fiscal reform. Since the notables had benefitted from this law, they protested its repeal.[171] They also called for the reform of tax responsibility and for the abolition of the tax on animals.[172] The initial split was between the native Egyptian bureaucrats and their allies (often their blood relatives), the shaykhs, on one side, and the Khedival house, the Turko-Egyptian landlords, and foreigners (whether Greek moneylenders or French financiers), on the other.

It was not obvious where the peasants would fit in. They had lost land to pashas, shaykhs, and moneylenders; both pashas and shaykhs oppressed them with tax collection and corvees. If anything, one might expect the peasants to prefer the pashas. There is some evidence that they regarded the shaykhs as rather more oppressive than the pashas, possibly because the peasants had so little personal contact with the latter.[173]

However, although the shaykhs made loans to the peasants, most peasant land loss from debt was to foreigners, especially Greeks, Syrians, and Lebanese Christians, not to shaykhs.[174] The moneylenders exploited their new legal position under the Mixed Courts and the peasants' need for cash for taxes with the usual ruthless tactics. For instance, peasants lost land worth L.E. 50 per feddan for a L.E. 10 debt. The overall result was a tidal wave of peasant debt:

> Between 1876 and 1882 legal mortgages alone rose according to Lord Dufferin, from five-hundred-thousand to seven million pounds (approximately), of which five million pounds were village mortgages. To this an estimated three or four million pounds worth of debts to moneylenders should be added. By 1882 the Mixed Courts had handled foreclosures to the value of L.E. 24,000.[175]

The situation of the peasants was becoming desperate.

In this situation the government's inability or unwillingness to protect them was critical. The government acquiesced (or was compelled to acquiesce), in the financial rule of the very foreigners who managed and profited by the Mixed Court's mortgage proceedings. In addition, the fact that the local moneylenders were Christians delegitimized them in the eyes of Muslim peasants. 'Urabi's promise to cancel peasant debt and to "banish the usurers" tipped the balance: Villiers Stuart, a Member of Parliament on a fact-finding mission for that body after the occupation, repeatedly reported land loss to Greek and Levantine moneylenders through the mechanism of the Mixed Courts as the cause of peasant discontent.[176] Their support for 'Urabi was largely due to his promise to "banish the usurers,"[177] although his promises of tax reduction must also have been popular.[178] Wallace reported that students from al-Azhar (many of whom were the sons of shaykhs) appeared in the villages, agitating against the foreign infidels, including, apparently, the moneylenders.[179] The peasants did not actually join 'Urabi's forces, but they did become increasingly insubordinate from 1876 on. In 1879 armed bands were formed against tax collectors in the area between Suhag and Girga. Revolts occurred against the corvée in the rice-growing areas in 1880. "In 1882 the British consuls reported a general state of rebelliousness among the peasants."[180]

There were, then, all the makings of a full-scale revolution: a discontented class of rising, well-to-do but politically subordinate men both in the countryside and in the cities, closely linked with each other; a weakened and discredited central government; a peasantry in the throes of change from one set of social relations of production to another, and who had a specific, acute grievance which was due to new and unheard-of innovations (mortgage foreclosure). Finally, these peasant grievances were appealed to by the leaders of the discontented "middle" group.

But, of course, the revolt failed, going the way of so many indigenous uprisings against foreign penetration and the dislocations which such foreign, capitalist penetration fostered. The British intervened when riots in Alexandria gave them their excuse to crush the revolt. The Turko-Egyptians retained their hold on the land, although they had to cede much political power to the British. The failure of the revolt seems to have been the result of this intervention, although the incompetence of the leadership undoubtedly helped.

Some insight into the reasons for the failure of the revolt can be gained by comparing the Egyptian case with other countries which experienced peasant revolts as a result of the transitionist agriculture. Eric Wolf[181] demonstrates that the middle peasants, defined as those peasants who have at least enough land on which to support a family and hence do not have to enter the market for either labor or land (although they may buy and sell other inputs and outputs), played a crucial role in all of the major peasant revolutions of this century: Mexican, Russian, Chinese, Vietnamese, Algerian, and Cuban. Their importance derives from two facts: (1) they have a limited amount of leverage against more powerful groups because of their (potential) economic self-sufficiency, and (2) they are vulnerable to economic changes which will weaken them, separate them from their land, and reduce them to dependent relations.

His is a balancing act in which his balance is continuously threatened by population growth, by the encroachment of rival landlords; by the loss of rights to grazing, forest, and water; by falling prices and unfavorable conditions of the market; by interest payments and foreclosures.[182]

The Egyptian case presents some important differences with those cases which Wolf studied. We have seen that the shaykhs were supporters of 'Urabi in the early stages of his political struggle with the Khedive and the Pashas. Clearly, these men are better viewed as "rich peasants" than as "middle peasants" as Wolf defines them. They were beneficiaries, not victims, of the dissolution of the old, precapitalist village community. They were, to a lesser degree, part of the power system which was being used to transform the countryside, rather than outside of it as were Wolf's middle peasants. Unlike the middle peasants, the political status of the shaykhs rose as market relations spread.

It should be remembered, as well, that the 'Urabi revolt was a revolt by military men. The shaykhs' support was mostly verbal. There is a resemblance here to the behavior of rich peasants during the Chinese revolution: "Only when an external force, such as the Chinese Red Army, proves capable of destroying these other, superior power domains, will the rich peasant lend his support to an uprising."[183] It would appear that the existence of this "external force" was, in fact, crucial; it seems that the mere appearance of the British forces dissolved the support of the provincial notables for 'Urabi.[184] Most of all, the shaykhs wanted to be on the winning side.

It is also necessary to emphasize that the issue of legitimacy, couched in more-or-less traditional Islamic terms, played a very important role in the rebellion. Foreclosures were contrary to Muslim law. The accession to power of European Christians over state finances would naturally have been anathema to the 'ulema. These men were the "city cousins" of the village shaykhs. As in Algeria, there were linkages between an urban elite and peasants. As in Algeria, some form of Islam provided the two groups with a common sense of "us" as opposed to "them". But, unlike Algeria, the links were

between an urban elite and <u>rich</u>, not middle, peasants. In the face
of superior force, the 'ulema would back the winners, supplying a
theological reason for their actions.[185] These social factors, plus
the obvious differences in terrain (there are no inhabited "peri-
pheral areas" in Egypt like the Kayble mountains of Algeria) may help
to account for the failure of the 'Urabi revolt.

## Footnotes

[1]   Shaw, Stanford J., Ottoman Egypt in the Age of the French Revolution, (Cambridge, Mass., 1964), 6, 129; Andre Raymond, Artisans et commerçants au Caire au dix-huitième siècle, (Damascus, 1973), 188, 190. Syria received primarily rice, getting wheat only in times of famine. P.S. Girard, "Mémoire sur l'agriculture, l'industrie, et le commerce de l'Egypte", in Description de l'Egypte, Etat Moderne, Vol. 2, Part 1, (Paris, 1813), 661.

[2]   Shaw, (1964), 6.

[3]   Raymond, (1973), 129-131; ibid., 6.

[4]   Shaw, (1964), 129, 135; Girard, (1813), 542.

[5]   Braudel, Fernand. The Mediterranean and the Mediterranean World in the Age of Phillip the Second. (N.Y. 1974) Vol. 1, Part II, Chap. III, 1, "The Pepper Trade"; J.H. Parry, "Transport and Trade Routes," Cambridge Economic History of Europe, Vol. IV, (Cambridge, 1967), 165-167.

[6]   Raymond, (1973), 80. This, of course, is hyperbole, since at least 80% of the population was engaged in agricultural activities.

[7]   Ibid., 412-414.

[8]   Calculated from ibid., 196.

[9]   Ibid., 652. This is a dramatic revision of Shaw, (1964), 151, who reports that urban tax farms accounted for 19% and 14% of government revenues for 1585/6 and 1795/6, respectively.

[10]  Cipolla, Carlo, Guns, Sails, and Empires, (N.Y. 1965), 102-103.

[11]  Raymond, (1973), 149, 156-157, 178-179.

[12]  Ibid., 180-184.

[13]  Ibid., 196.

[14]  Shaw, (1964), 127.

[15]  Raymond, (1973), 194.

[16]  Girard, (1813), 676.

[17]  Ibid., p. 671.

[18]  Raymond, (1973), 180.

[19]  See, e.g., Paul Masson, Histoire du commerce francais dans le Levant au xviii siècle, (Paris, 1911), 462ff.

[20]  A timar grant gave full, temporary possession of the usufruct of the land in return for rendering services to the government, usually furnishing light cavalry troops (sipahis) for

campaigns. Although these troops were very formidable in the fourteenth and fifteenth centuries, early in the sixteenth century Austrian firearms and tactics were rendering them obsolete. Timars were, consequently, also less useful as a fiscal measure. See Shaw, Stanford J. The Financial and Administrative Organization and Development of Ottoman Egypt, 1517-1798, (Princeton, 1962), 26; Cipolla, (1965), 91; Bernard Lewis, The Emergence of Modern Turkey, (London, 1961), 30; Halil Inalcik, "The Heyday and Decline of the Ottoman Empire," Cambridge History of Islam, Vol. 1, (Cambridge, 1970), 345.

[21]    Shaw, (1962), 29-30.

[22]    Shaw, Stanford J. "Landholding and Land Tax Revenues in Ottoman Egypt," in P.M. Holt, ed., Political and Social Change in Modern Egypt, (London, 1964).

[23]    Hansen, Bent. "An Economic Model for Ottoman Egypt: The Economics of Collective Tax Responsibility," unpublished paper presented to the Conference on the economic history of the Near East at Princeton University, June 16-20, 1974, 42-43.

[24]    Shaw, "Landholding...."

[25]    Rivlin, Helen A.B. The Agricultural Policy of Muhammad 'Ali in Egypt, (Cambridge, Mass., 1961), Ch. 2.

[26]    Ibid.

[27]    Kenneth M. Cuno, "The Origins of Private Ownership of Land in Egypt: A Reappraisal"", International Journal of Middle Eastern Studies, 12,3 (November 1980).

[28]    Lancret, Michel-Ange. "Mémoire sur le système d'imposition territoriale et sur l'administration des provinces de l'Egypte dans les dernières années du gouvernement des mamlouks,"" in Description de l'Egypte, Etat Moderne, Vol. 1, (Paris, 1809), 236. 'Abd al Rahim challenges this figure however, claiming that the 'ard al-usya was sometimes as high as 50% of the total. Since he cites examples ranging from 53% to 0% of the village land held by the multezimin, it is evident that there were large variations. 'Abd al Rahim 'Abd al Rahim and Wataru Miki. Village Life in Ottoman Egypt and Tokugawa Japan: A Comparative Study, (Tokyo, 1977) 16. See also 'Abd al Rahim 'Abd al Rahim Al-Rif Al-Masri Fi qarn al-thamin 'ashr, (Cairo, 1974). (The Egyptian Countryside in the Eighteenth Century)

[29]    Esteve, Comte. "Mémoire sur les finances de l'Egypte depuis sa conquête par le sultan Selym Ier, jusqu'à celle du général en chef Bonaparte," in Description de l'Egypte, Etat Moderne, Vol. 1, (1809), 243.

[30]    Lancret, (1809), 249.

[31]   The remainder of this paragraph is based on  Esteve,  (1809),
       311; Rivlin, (1961), chap. 2; and Lancret, (1809), 243–244.

[32]   Shaw estimates that Egyptian expenditures for the Porte   fell
       87% from 1596 to 1796 in nominal terms.  Shaw, Ottoman
       Egypt..., 151.  Hansen estimates that the share of the  Porte
       in  land taxes fell from roughly 25% in the sixteenth century
       to 15% in 1798.  Hansen, (1974), 10.

[33]   Ayalon, D. "Studies in al–Jabarti." Journal of the Economic
       and Social History of the Orient, 3, 306ff.

[34]   See, e.g., P.M. Holt, Egypt and the Fertile  Crescent,  1516–
       1922, (London, 1966), passim.

[35]   Hansen, (1974), 16.

[36]   Girard, (1813), 529.

[37]   Shaw, (1962), 22–23; Girard, (1813), 529.

[38]   Shaw, Ottoman Egypt..., 133, 136–139; Lancret,  (1809),  250–
       251.

[39]   Shaw, "Landholding...."

[40]   Finley, M.I., The Ancient Economy, (Berkeley, 1973) 31.

[41]   Justin A. McCarthy, "Nineteenth–Century Egyptian Population",
       in  Elie  Kedourie, ed., The Middle Eastern Economy:  Studies
       in Economics and Economic History, (London, 1976), 1–40.

[42]   Hansen (1974), 40ff.

[43]   The "Prisoner's Dilemma," is a non–zero–sum game in which the
       pursuit  of individual ends leads to a sub–optimal result for
       both players.  See Donald Luce and Howard Raiffa,  Games  and
       Decisions, (N.Y. 1957), 94ff.

[44]   'Abd al–Rahim, 'Abd al–Rahim "Hazz al–Quhuf: A New Source for
       the  Study of the Fallahin of Egypt in the XVIIth and XVIIIth
       Centuries." Journal of the Economic and Social History of the
       Orient, 18, (1975), 260.

[45]   Shaw, "Landholding...."

[46]   Rivlin, (1961), chap. 2.

[47]   Ibid.; Baer, Gabriel, A History of  Landownership  in  Modern
       Egypt, (London, 1962), 6.

[48]   Lancret, (1809).

[49]   Owen, E.R.J., Cotton and  the  Egyptian Economy,  1820–1914,
       (London,  1969),  71; Gabriel Baer, "The Dissolution of the
       Village Community," in his Studies in the Social  History  of
       Modern Egypt, (Chicago, 1969), 17–29.

[50]   Volney, C.F.  Travels Through Syria and Egypt  in  the  Years
       1783, 1784, and 1785.  Vol. 1, (Dublin, 1973), 203.

[51]   Shaw, "Landholding...."; Earl of Cromer, Modern  Egypt,  Vol.
       2,  (London,  1908),  192–193; Crouchley, A.E., The Economic

Development of Modern Egypt, (London, 1938), 47.

[52]   Gibb, H.A.R. and H. Bowen, Islamic Society and the West I:
       Islamic Society in the Eighteenth Century, Vol. I, (London,
       1950), 213.

[53]   Ibid.

[54]   Willcocks, William, Egyptian Irrigation, 2nd. ed., (London,
       1913), 38-39.

[55]   Girard, (1813), 557. 564.

[56]   Ibid., 497; Willcocks, (1913), 36-38.

[57]   Girard, (1813), 562.

[58]   Willcocks, (1913), 42.

[59]   Girard, (1813), 515-517, 568-569.

[60]   de Chabrol de Volvic, "Essai sur les moeurs des habitants
       modernes de l'Egypte," in Description de l'Egypte, Etat
       Moderne, Vol. 2, Part 2, (Paris, 1822), 511.

[61]   John Waterbury, Hydropolitics of the Nile Valley, (Syracuse,
       N.Y., 1979), 29-30.

[62]   Girard, (1813), 563.

[63]   Ibid., 511-519.

[64]   Hugh Doggett, D.L. Curtis, F.X. Laubscher, O.J. Webster,
       "Sorghum in Africa," in Joseph S. Wall and William M. Ross,
       eds., Sorghum Production and Utilization, (London, 1970),
       291.

[65]   Among modern historians, Roger Owen has argued that maize was
       "spreading" in the eighteenth century. "Introduction:
       Resources, Population, and Wealth", in Thomas Naff and Roger
       Owen, eds. Studies in Eighteenth Century Islamic History,
       (Carbondale, Ill., 1978), 146. Those arguing that maize's
       widespread diffusion is primarily a nineteenth century
       phenomenon include Gamal Hamdan, "Evolution of Irrigation
       Agriculture in Egypt," in L. Dudley Stampp, ed, A History of
       Land use in Arid Regions, (Paris, 1961), 129; Waterbury,
       (1979), 32; Abdal Rahim and Wataru Miki, (1977), 50; Hussein
       Kamel Selim, Twenty Years of Agricultural Development in
       Egypt, (Cairo, 1940), 145; Henry Habib Ayrout, The Egyptian
       Peasant, J.A. Williams, trans., (Boston, 1963), 50. Girard
       (1813), 517-520 noted that although dhurra beladi, (itoleus
       sorghum) was the principal peasant food stuff, dhurra shami
       (zea mays) had begun to replace it in certain Delta areas,
       especially around Tanta and Samanoud. Dhurra for both
       plants, as well as their physical similarity before tassel-
       ing, may account for the confusion.

[66]   Girard, (1813), 563-564.

[67]   Chabrol, (1822), 409.

[68]     Girard, (1813), 572, 545.

[69]     Ibid., 581.

[70]     Ibid., 708-709.

[71]     Holt, P.M., "The Later Ottoman Empire in Egypt and the Fer-
         tile Crescent," in The Cambridge History of Islam, Vol. I,
         (1970), 381.

[72]     Rivlin, (1961), 253.

[73]     Owen, (1969), 69.

[74]     This was also a Mamluk practice.  See Shaw, Ottoman Egypt...,
         121.  A comparison of the ratio of prices paid to market
         prices or selling prices given in Shaw for wheat and rice
         with those of the same crops for Muhammad 'Ali given by 'Amr
         indicates that the difference between the two prices was
         smaller in the case of Muhammad 'Ali:

| Crop  | P selling | P purchase    |
|-------|-----------|---------------|
|       | Mamluks   | Muhammad 'Ali |
| Wheat | 3.33      | 1.85          |
| Rice  | 1.90      | 1.56          |

         This helps to explain 'Amr's assertion that the peasants'
         position had advanced (tuqaddamat) under Muhammad 'Ali rela-
         tive to Mamluk times.  As we shall see, on other grounds pre-
         cisely the opposite conclusion seems more likely.  Ibrahim
         'Amr, Al-Ard w'al-fallah: al-mas'ilah az-zira'iyya fi misr,
         (The Land and the Peasant: The Agricultural Problem in
         Egypt), (Cairo, 1958), 81.

[75]     Rivlin, (1961), chap. 10.

[76]     Clough, S.B., France:  A History of National Economics,
         (N.Y., 1939), 98-100, 130.  By contrast, wheat exports had
         been prohibited before the Revolution.

[77]     Chambers, J.D. and G.E. Mingay, The Agriculture Revolution,
         1750-1880, (London, 1966), chap. 6.

[78]     Herlihy, Patricia, "Odessa and Europe's Grain Trade in the
         First half of the Nineteenth Century," unpub. m.s., (1971),
         24.

[79]     Owen, (1969), 29.  However, Muhammad 'Ali's first financial
         success was selling wheat to the British for use in the pen-
         insular wars.  But such trade was not sustainable for the
         reasons outlined.

[80]     Ibid.  Owen also gives the fact that peasants couldn't cheat
         on the monopoly system by eating cotton or otherwise consum-
         ing it locally as a reason for the choice of cotton versus
         wheat.  This argument essentially posits lower enforcement
         costs to government revenue collection for cotton as  opposed

to wheat. Now, it is true that although peasants in various districts had spun cotton, weaving was performed in larger urban areas where it would have been more difficult for "cheaters" to escape detection by Muhammad 'Ali's officials. (See Girard, 1813). However, given that the production of cotton on a large scale required huge corvees for irrigation works and considerably increased field, labor it is not clear that such reasoning is valid.

[81]  Ibid., 47.

[82]  Ibid., 48-49. He notes, however, that it is not clear that the units of measure are the same in the two cases.

[83]  Cuno, (1980)

[84]  J.A. Allan, "Some Phases in Extending the Cultivated Area in the Nineteenth and Twentieth Centuries in Egypt," paper presented to Middle East Studies Association of north America Meetings, Washington D.C., November 6-9, 1980.

[85]  Patrick O'Brien, "The Long-Term Growth of Agricultural Production in Egypt: 1821:1962", in P.M. Holt, ed., (1964), 172.

[86]  Owen, (1969), 140.

[87]  Ibid., 34.

[88]  Wallace, D. Mackenzie, Egypt and the Egyptian Question, (London, 1833), 264.

[89]  Rivlin, (1961), 141.

[90]  Owen, (1969), 36.

[91]  Rivlin, (1961), 138-139.

[92]  Al-Hitta, Ahmad Ahmad, Ta'rikh az-zira'iyya al-misriyya fi 'ahd Muhammad 'Ali al-kabir. (The History of Egyptian Agriculture in the Age of Muhammad 'Ali the Great), (Cairo, 1950), 124.

[93]  Ibid., 126.

[94]  Rivlin, (1961), 139.

[95]  'Amr, (1958), 81. Unfortunately, it is unclear from the context whether he means every year or only in certain years, although the latter is almost certainly the case. See also Rivlin, (1961), 244-245.

[96]  Rivlin, (1961), chap. 12.

[97]  Ibid., chap. 11.

[98]  Ibid., 201-202.

[99]  Berque, Jacques, Egypt: Imperialism and Revolution, Jean Stewart trans., (London, 1972), 137.

[100] Rivlin, (1961), chap. 11, and Bayle St. John, Egypt and Nubia, (London, 1852).

[101]   Baer, Gabriel, "Submissiveness and Revolt of the Fellah," (1969), Cuno, (1980) notes that the areas of revolt were often regions where the Mamluk presence had been felt only lightly.

[102]   Rivlin, (1961), chap. 11.

[103]   Ibid., chap. 12.

[104]   Owen, (1969), 60.

[105]   Ali Barakat, <u>Tatawwur al-milkiyyah az-zira'iyyah fi misr w'atharuh 'ala al-harakah as-siyasiyyah 1813-1914</u>. (The Development of Agricultural Landownership in Egypt, and Its Influence Upon Political Development, 1813-1914), (Cairo, 1977), 103.

[106]   Baer, (1962), 17.

[107]   Barakat, (1977), 33-34, 470.

[108]   Baer, (1962), 17.

[109]   Ibid., 18, fn. 5.

[110]   Barakat, (1977), 493.

[111]   Baer, (1962), 14.

[112]   Rivlin, (1961), 73.

[113]   These figures are extremely rough, have somewhat peculiar definitions, include double counting, etc. See Rivlin, (1961), Appendix II for a discussion of these numbers. She notes that due to these numerical problems and after many complicated adjustments, the "increase in peasant holdings was minute." Rivlin, (1961) 322, fn. 32.

[114]   'Amr, (1958), 78-79.

[115]   Rivlin, (1961), chap. 6.

[116]   Baer, Gabriel, "The Village Shaykh...." in (1969).

[117]   Cuno, (1980).

[118]   Rivlin, (1969), 107-110. The following discussion relies primarily upon Cuno, Rivlin, and Baer, (1962).

[119]   More detailed discussions may be found in Cuno and Barakat.

[120]   Baer, (1962), 14, 17.

[121]   Ibid., 15-18.

[122]   Ibid., 6-8.

[123]   Ibid., 9, 'Amr, (1958), 83.

[124]   Owen, (1969), 92-93; See also 'Amr, (1958), 86-87.

[125]   Crouchley, (1938), 117.

[126]   Ibid.

[127]  Owen, (1969), 69.

[128]  Owen, (1969), 80.

[129]  Wallace, (1883), 265-266.

[130]  Baer, (1962), 31.

[131]  Owen, (1969), 75.

[132]  Ibid., 75.

[133]  Ibid., 76, 104.

[134]  Ibid., 76-77.

[135]  The term "bimodal" is used here, "in a generic sense to dep-
       ict  a  situation where a small subsector of large farm units
       cultivates a large part of the farm land, while  the  greater
       part  of the farm population is confined to very small, semi-
       subsistence holdings," Bruce F.  Johnston  and  Peter  Kilby,
       Agriculture  and  Structural  Transformation:  Economic Stra-
       tegies in Late-Developing Countries, (N.Y., 1975), 32.

[136]  Lozach and Hug, writing in 1928, asserted that the system was
       "about  one hundred years old"; Ayrout, writing in 1938, says
       (p. 111) the same thing.  Lozach and Hug, L' Habitat Rural en
       Egypte  (Cairo,  1930);  Ayrout, (1963), 18.  This is roughly
       contemporaneous  with  Muhammad  'Ali's  decentralization  of
       agricultural  administration, which began in 1837.  See Owen,
       (1969) 58ff.  A full discussion of the 'ezbah system, and its
       economic rationale and implications, is postponed to chap. 3,
       covering the period of  British  administration  (1880-1914),
       when somewhat better documentation is available.

[137]  The word derives from the root "'azaba," "to be distant."

[138]  Documentation for these assertions can be found in  chap.  3.
       It should be emphasized that the 'ezbah system described here
       is an "ideal type," since the  evidence  indicates  the
       existence of many variations.

[139]  St. John, (London 1852), 190.

[140]  Baer, (1962), chap. 2.

[141]  Gali, Kamel, Essai sur l'agriculture de l'Egypte, (Paris,
       1889),  140;  Muhammad  Saleh,  La Petite propriete rurale en
       Egypte, (Grenoble, 1922), 35.

[142]  Ibid., 138.

[143]  Ibid., 136-137.

[144]  Crouchley, (1938).  On economies of scale in supervision, see
       J.F.  Nahas,  Situation economique et sociale du fellah egyp-
       tien, (Paris, 1901), 141.

[145]  Senior, Nassau, Conversations and Journals in Egypt and
       Malta, 1855-6, (London, 1882), 175.

[146]   Baer, (1962), chap. 2. See also Ra'uf 'Abbas Hamid, Al-Nizam
        al-Ijtima'i Fi Misr Fi Dhill al-milkiyyat al-zira'iyya al-
        kabira, 1837-1914 (Cairo, 1973). (The Social System in Egypt
        under the Influence of Large Agricultural Landownership,
        1837-1914).

[147]   Baer, "The Village Shaykh....", in (1969).

[148]   Crouchley, (1938), chap. 3.

[149]   Baer, "The Village Shaykh....", in (1969).

[150]   Baer, (1962), 52.

[151]   Senior, (1882), 279.

[152]   Baer, (1962), chap. 2.

[153]   Ibid.

[154]   Ibid.

[155]   Wallace, (1883), 363; Villiers, Stuart, Egypt Since the War,
        (London, 1883), 29.

[156]   St. John, Bayle, Village Life in Egypt, Vol. 2, (London
        1852), 60ff.

[157]   Baer, (1962), 29.

[158]   Ibid.

[159]   Owen, (1968), 144-145; Baer, (1962), 30.

[160]   Owen, (1969), 93, fn.

[161]   Wallace, (1883), 337.

[162]   Lady Duff Gordon, Letters from Egypt, G. Waterfield, ed.,
        (London, 1969), 182; Owen, (1969), 105.

[163]   Baer, (1962), 35. One might object that if yields were ris-
        ing, then perhaps the incidence of taxation was steady or
        even declining. This was almost certainly not the case.
        Even if we consider average yields (a misleading indicator,
        as I will argue in a moment), the incidence of taxation rose
        sharply in certain critical years: from 1865/66 to 1869 there
        is no evidence of any change even in the average yields of
        cotton. Yet during those years, taxes rose by 70%. As I
        mentioned in the text, this increase coincided with a fall in
        the price of cotton of 50% (see Table 2.5). At the same
        time, a cattle plague decimated the peasants' work animals.
        The peasants were also subjected to a host of other taxes:
        salt, poll, date palm, not to mention the corvee.

        Owen does mention a rise in average yields from two to two-
        and-one-half cantars per feddan in 1869-1871, at the same
        time as the 25% surcharge was levied. But this is really
        beside the point for two reasons: First, average yields are
        deceptive. We know that there was a marked difference
        between the techniques of cultivation of pashas and peasants:

we also know that the latter group was losing land to the former. Peasants continued to employ the balli method into the 1880s, so it seems very likely that yields on peasant lands changed very little, if at all. There is certainly no testimony that peasant yields were rising, nor any reason to suppose that they might have been. Second, the numbers on the tax rates give a most inadequate picture of the incidence of taxation, which was highly capricious, and assessed with no regard for ability to pay. See Owen, (1969), 146-147.

[164]   O'Brien, (1964); McCarthy, (1976), 38.

[165]   Cf. Charles Issawi, "Egypt: A Study in Lop-sided Development", Journal of Economic History, 21, 1 (March 1961).

[166]   For a detailed interpretation of the revolt on roughly these lines, see Alexander Schoelsch, Aegypt den Aegyptern. Die politische und gesellschaftliche Krise der Jahre 1878-1882 in Aegypten. (Zurich, 1972).

[167]   Although apparently a faction of Turko-Egyptians also supported 'Urabi. See Schoelch, (1972).

[168]   Tignor, R.L., Modernization and British Colonial Rule in Egypt. (Princeton, 1966), Ch. 1.

[169]   Baer, "The Village Shaykh...." in (1969).

[170]   Earl of Cromer, (1908), 187. See also Schoelch, (1972), 243.

[171]   Hourani, in Schoelch, (1972), 11; 'Amr, (1958), 87.

[172]   'Amr, (1958), 87.

[173]   See, for example, Earl of Cromer, (1908), 187; the report from the British Consul, Borg, quoted in Baer, (1962), 53; Wallace, (1883), 222-228. Schoelch points out, however, that the evidence of the first two men may have been intentionally distorted to improve the "moral position" of the British intervention against 'Urabi.

[174]   Baer, (1962), 36.

[175]   Ibid.

[176]   Stuart, Villiers. Egypt Since the War, (London, 1883), 17-21; 39-40; 54-59; 70; 100; 137.

[177]   Ibid., 17-21; 47; 143.

[178]   'Amr, (1958), 87.

[179]   Wallace, (1883), 290-291.

[180]   Baer, "Submissiveness and Revolt..." in (1969).

[181]   Wolf, Eric R., Peasant Wars of the Twentieth Century, (N.Y., 1969).

[182]   Ibid., 292.

[183]   Ibid., 291.

[184]   Schoelch, (1972), 268.

[185]   Ibid.

# 3 | Technical and Social Change, 1890–1914

## INTRODUCTION

In the last chapter we saw how the integration of Egypt into the international division of labor in the nineteenth century as a cotton exporter destroyed the old quasi-communal forms of land tenure, broke up the protective web of village social relations, replaced them with private property in land and individual tax responsibility, and helped to create four classes: large landowners or pashas (a group of some 12,000 men owning about 40–45% of the cultivated area), rich peasants or village notables (about 10–20% of the landowners holding roughly 35% of the land in farms of 5–50 feddans), small peasant landowners (owning plots of less than five feddans, comprising 80–90% of the landowners), and a landless class. We also saw that the creation of these classes was an integral part of the process of the government's attempts to foster the export of cotton. The British occupation completed the transformation of rural social structure and technology in Egypt by (1) finally removing all impediments to full private ownership of land, (2) by abolishing the corvée (thereby creating a free labor market and abolishing the last legal vestige of the solidarity of the village unit), and (3) by greatly extending the irrigation system, converting all of the Delta to the perennial system.

In this chapter I will first sketch the consequences of British changes in administration and law for the positions of the shaykhs and pashas and then turn to a more extensive discussion of the organization of land and labor on large estates and medium properties. We will then turn to a discussion of how these classes fared during a period of rapid growth of exports, the period of the British occupation and administration, 1882–1914. The focus here will be on the nature of the technical changes in agriculture, the ecological consequences of these changes, the reasons for the adoption of a new technique, and the impact of its adoption on the short-run class distribution of revenue. All of these themes have received much attention in recent studies of the "Green Revolution," and this chapter may shed further light on the nature of the interaction of technical changes in agriculture with the rural class structure in Third World countries.

BRITISH RULE AND RURAL SOCIAL RELATIONS: PASHAS AND SHAYKHS

Although there could be no question of a return to the old order after the failure of the 'Urabi revolt, the British occupation quite definitely strengthened the old ruling class of Turko-Egyptians. Since the British were limited by lack of manpower, and since at first the occupation was supposed to be a temporary expedient to reestablish security, to ensure debt collection, and to protect their route to India, they were obliged to rule with the help of indigenous groups. The attack on the intervention by the left wing of Gladstone's cabinet further limited the size of the troop commitment which the British government could make.[1] The British would rule not by total conquest and absorption as the French attempted in Algeria and elsewhere, but by "treaties, informal persuasion, and the threat of force."[2] Some of the implications of the nature of the occupation for agricultural technology and production will be examined below.

The logical group through which to rule was the old landowning class of Turko-Egyptians. Needless to say, there was no love lost between them and the British; it was, rather, a case of mutual need. The old aristocracy was discredited among large sections of the population; the British military presence ensured that the events of 1882 would not be repeated. The Turko-Egyptians were able to use their political leverage with the British to strengthen or defend their economic interests. For example, the landowners were able to scuttle the British plans in the 1880s to change the tax rates such that 'ushuriyya land would be taxed more and kharajiyya less.[3] Although the alliance became increasingly strained, especially under 'Abbas II (r. 1892-1914), it was never broken; the British simply increased their own power without doing away with their partners altogether.

The British made some significant legal changes in accordance with their liberal ideology. First, all land became full private property. This is perhaps the least important of their legal changes, since only a few types of land were not held in full private ownership when the British arrived. The main such category was kharajiyya land on which the Muqabala had not been paid. Such land became full private property in 1891.[4] Kharajiyya land could be endowed as waqf after 1893 without the permission of the Khedive.[5] The only difference between the two types of land that remained was the tax rate; for all other purposes, full private property was the rule.

This final consolidation of private property rights under the British was accompanied by something similar to what Marx described in Tudor England as "bloody legislation against the expropriated."[6] It would appear that although the 'Urabi revolt was easily defeated, the peasant bands which had sprung up in 1879-1882 were not so easily suppressed. A Khedival decree of October 14, 1884 established a Brigandage Commission to deal with this crime, defined as "attacks by armed bands on a house or village at night."[7] Until the abolition of this commission in May 15, 1889, it inflicted severe repression on the rural offenders; for example, one fellah received life imprisonment for stealing two camels. The criminal registers were filled with cases of brigandage, but it was often difficult to convict the offenders for lack of evidence and witnesses.[8] Not only did the peasants refuse to testify against the brigands, but they also helped the bands elude capture.  It appears that the bulk of this "primitive rebellion"[9] was directed at landlords and their estates.[10] By the first decade of the twentieth century brigandage had become infrequent, although in 1902 and again in 1907 the police succeeded in breaking up gangs in the sugar cane districts of the Sa'id and in Behera in the Delta.[11] By then peasant resistance had assumed more individualistic forms, such as arson and cattle-poisoning.[12] More violent acts of peasant resistance to pashas persisted, however.  Sir Thomas Russell, Commandant of Police in Cairo from 1913 to 1946, reported a conversation with a rich pasha in which the former mused on the pleasure of reading on the veranda of a country estate house in the cool of the evening.

> My friend said at once: "You don't really think that a land-
> lord in the districts could sit out on the veranda after
> dinner, with a bright light over his head, do you, and not
> get shot?"[13]

In addition to consolidating private property rights in land and dealing with its consequences, the British introduced a number of tax reforms.  Although they could not abolish the Kharajiyya-'ushuriyya distinction at first, they did reduce tax rates, bringing them more into line with cultivators' ability to pay (see Table 3.10). Finally, after a cadastral survey in 1898, new tax rates, the same for both categories, were introduced, thereby removing the last legal distinction between peasant-held land and lands held by the rich.[14]

British tax reforms, as well as other policies, reduced the power of the shaykhs.  Tax assessment, which had been such an important source of their power and aggrandizement earlier, was removed from their hands, as was the use of the kirbaj, or whip, their primary means of coercion.[15] The law of March 16, 1895, removed disputes involving immovable property from their jurisdiction. The British weakened the shaykhs further by applying Muslim inheritance law to kharajiyya land, the bulk of their property. This led to fragmentation of their holdings.  Such a policy, of course, had similar consequences for the peasants.  The shaykhs were further weakened when the British assigned them new, onerous duties.  They were now in charge of the census and of the supervision of canals and were required to report on a wide variety of matters.  As Cromer put it:

> The omdeh has a great deal more to do than heretofore.  He is
> at the beck and call of the inspectors of every department.
> He is responsible for the execution of a number of regula-
> tions, which he often fails to understand thoroughly.[16]

However, the shaykhs retained some power in the villages. First, they had jurisdiction in suits of up to L.E.1, the equivalent of about one month's wages for an agricultural laborer.[17] They also retained some penal jurisdiction and could imprison suspects for twenty-four hours.[18] Second, they apparently consistently exceeded their authority; in 1902, for example, "nearly one-third of the village omdehs received punishment of one sort or another"[19] for various excesses. Given the peasants' reluctance to provide evidence to the authorities, the actual number of abuses was probably considerably larger. Third, they continued to play an important role in the rural economy as supervisors of agricultural labor. They exploited their own lands using sharecropping or, if their estates were large enough, with the 'ezbah system. In the later cases, they themselves resided on the 'ezbah. [20] Sometimes they were hired as supervisors on the pasha's 'ezbahs. [21] Finally, when large landowners rented out their estates, they often leased them for cash to the shaykhs and 'umdahs, who then sublet the lands using sharecropping.[22] It was perhaps for these reasons that a certain Johnson Pasha, reporting on his inspection of the judicial work of the 'umdahs, asserted that;

> there is just as much difference between the head of an Egyptian village of, say, 5,000 inhabitants and the ordinary fellah of his village as between an English country gentleman and the labourers on his own and surrounding properties.[23]

The shaykhs were not able to use these powers to gain a larger portion of the land area, however. Their share of land (measured as land holdings between five and fifty feddans) fell from 37.7% in 1894 to 30.5% in 1907, showing stability from then until 1914.[24] The pashas, on the other hand, managed to use their wealth and influence to gain land during the period. It is true that some members of the royal family lost land early in the period; however, they were able to buy a lot of it back through a variety of maneuvers.[25] Grants of uncultivated areas to encourage land improvement increased the holdings of this class.[26] Finally, the State Lands were auctioned off during this period. The purpose of those sales which took place early in the period was to pay off the State Debt. The managers of the Caisse de la Dette Publique insisted that the lands be sold in large lots, ensuring their purchase by large landowners.[27] The average lot sold appears to have been between thirty and forty feddans.[28] Land sales in 1905-1907 similarly benefitted the large landowners; the percentage of land held in lots of over fifty feddans rose from 42.5 in 1894 to 45.1 in 1907, largely as a result of these sales.[29] I shall return to this increase in pasha land holdings in the section on income distribution.

LAND AND LABOR ON LARGER COTTON FARMS

The systems of land/labor control employed in Egyptian agriculture during this period were quite varied, ranging from the exclusive use of cash wage-labor to cash renting. Nevertheless, certain features of the division of labor in agriculture for the period 1890-1914 stand out. First, large estates were exploited either by the 'ezbah system or by cash renting. If the latter was done, it is likely that the proprietor either rented out part of his estate while exploiting another part directly[30] or leased his land to intermediaries (often village notables) who then sublet using sharecropping.[31] Second, although it is impossible to come to any firm conclusions here, it seems probable that the 'ezbah system predominated on the large estates. Third, there is evidence of a shift away from sharecropping (métayage) toward cash renting over the two-decade period.[32] Given the role of intermediaries, however, the importance of this shift is problematical. Fourth, there is evidence that the landlord specified methods of crop-rotation, irrigation, and drainage regardless of the mode of land exploitation employed. Some proprietors neglected these critical areas, but the evidence suggests that they were a minority. Fifth, medium properties apparently continued to use sharecropping in this period. Finally, it bears repeating that the highly fragmentary and often contradictory nature of the evidence renders all of these conclusions provisional and tentative.

Let us look at the 'ezbah system first. There can be little question that this system was widespread. Gali, writing in 1889, noted that unlike the State Domain and Daira Saniah lands, which were rented out, lands belonging to land companies and to "large proprietors" were exploited by agents (regies). [33] Nahas refers to "bon nombre" of large estates being exploited "a la journee," and Willcocks asserted that "if the land is good," the large proprietors generally worked the estates themselves.[34] In 1929, at least 60% of the villages in every Delta province had 'izab.

Government statistical sources provide some support for the contention that the 'ezbah system was common. Land area figures for different tenure systems are not available until 1929 with the publication of the First Agricultural Census. In 1929, 22% of the total land area was leased either for cash or on shares, the rest being exploited directly. In 1937, 21% of the total land area was leased in some way.[35] The same percentage of land held in estates of fifty feddans and over was leased at that time.[36] The census of 1917 attempted to gather information on land tenure. "Census Form Number 2" asked respondents to specify the amount of land owned, the land area cultivated directly, and the area held "as a tenant or on the

---

TABLE 3.1

PERCENTAGE OF VILLAGES IN DELTA PROVINCES WITH 'IZAB, 1929

| Province | Percent of Villages with 'izab |
|----------|-------------------------------|
| Buhaira | c. 100% |
| Daqahiliyya | 60% |
| Gharbiyya | 78% |
| Giza | 65% |
| Minufiyya | 70% |
| Qalyubiyya | 88% |
| Sharquiyya | 81% |

Source:
Lozach and Hug, L'Habitat Rural en Egypte, (Cairo, 1930).

crop-sharing system (métayage)."[37] However, the census administration added that "all the arrangements in the schedule, the card, and the tabulation concerning this subject have proved ineffective,"[38] and were, therefore, not published. It seems likely that the mounting antagonism of Egyptians of all classes toward the British administration and its World War I policies contributed to the difficulties of securing such information.

Population census data of 1907 is of some help, but here, too, there are problems. (See Table 3.2)

One cannot assume that all of the "fermiers et cultivateurs" are renters. In the first place, the term is sufficiently vague to include small peasant proprietors or 'ezbah workers. However, it seems likely that this is not the case in view of the other two categories. There is, however, a more serious problem. Subsequent censuses include an additional category of agricultural workers, "agricultural workers assisting on relatives' land." We do not know how many of the "fermiers et cultivateurs" may have fallen into this category. Comparison with the 1917 census is difficult here because in that year two categories of such family labor were given: "agricultural laborers on family land" (612,473) and "agricultural laborers on family land (inferred from the schedules)" (1,545,278). This second category looks suspiciously like a "residual," in which the census takers put their mistakes, double-counts, etc. Its inclusion also brings the sum of the four categories of "agricultural laborers (wage earning)," "laborers cultivating their own land," "cultivators of land taken on lease," and "agricultural laborers on family land" of both types to 3,927,819.[39] This is implausible, since the population in these four categories in 1927 was 3,292,625 and rural

---

TABLE 3.2
PRINCIPAL OCCUPATIONS OF THE RURAL WORKFORCE, 1907

| Occupation | Number | Percent of sum of the three categories* |
|---|---|---|
| "Proprietaires cultivant eux-memes leurs terres" | 519,693 | 23% |
| "Fermiers et cultivateurs | 920,435 | 40% |
| "Ouvriers et domestiques de ferme" | 832,785 | 37% |

Source:
The Census of Egypt taken in 1907, 279.

* These three categories comprised 98% of all workers listed as employed in agriculture.

---

population increased by some 7% during the previous decade.[40]  Hansen and Marzouk give 2.82 million as the population occupied in agriculture in 1917.[41] Since in that year 506,181 workers were listed as cultivators of land on lease (both cash and share), this would be roughly 18% of the agricultural population. Since there is little evidence that the percentage of "renters" so defined changed from 1907 to 1917, we might conclude that something like the picture in Table 3.3 would be appropriate for 1907.[42] Finally, note that some 37% of the agricultural workforce in these four categories is classed as laborers. Large estates occupied nearly 45% of the privately owned land in 1907.[43] Since we shall see that they employed a crop rotation which used less labor per feddan than either the small peasant proprietors or the rich peasants, we would expect that their share of the labor force would be somewhat below their share of the land area.

Indeed, suppose that we make the following assumptions:

(1)  small peasants hired no labor outside
of their families;
(2)  all "fermiers et cultivateurs" were renters,
either for shares or for cost;
(3)  pashas and middle landlords used the same
amount of labor per feddan.

Then, the agricultural work force hiring itself out (excluding small peasants) would have the following composition: renters 52%, paid laborers 48%. Then, since pashas owned 45% and rich peasants roughly 30% of the land area, the latter owned 40% of the sum of pasha and rich peasant land. If, under our assumptions, we assume that middle proprietors used only share or cash renting, they would have employed

---

TABLE 3.3
"GUESSTIMATE" OF THE DISTRIBUTION OF OCCUPATIONS
OF THE AGRICULTURAL WORKFORCE, 1907

| Occupation | Percent of each category in sum of the four categories |
|---|---|
| Small Peasant Proprietors | 23% |
| Renters (cash and share leases) | 18% |
| Laborers on Family Land | 22% |
| Paid Laborers | 37% |

Source:
    The Census of Egypt Taken in 1907, 279, and the above calculations.

---

77% of the renters. Since assumption (3) is in fact false (see below on agricultural techniques), the rich peasants' share of the labor force should have been greater than their share of the land area. Then they could have accounted for nearly all of the renters. Of course, assuming that medium proprietors used only cash or share rents is too extreme an assumption; they, too, employed paid agricultural laborers. But we shall see evidence that they primarily used sharecropping or cash rental systems. The (admittedly scanty) evidence would still be consistent with the assumption that at least a majority of large estates were exploited using agricultural laborers under the 'ezbah system.

Labor on such estates was of two sorts, those "attached to the domain" or tamaliyya, and daily wage laborers, often migrants (tarahil). The former were usually hired by the year, and received both money wages and payment in kind. The latter was sometimes a share of the subsistence crops, such as maize or millet, at other times the granting of a small parcel of land. Sometimes this plot of land was given in exchange for a reduced rent,[44] but at other times the fellah had to pay only the land tax on the parcel.[45] In either case he was obliged to furnish labor services to the landlord for which he received a wage below that of ordinary day-laborers.[46] The time spent working for the landlord may have amounted to as much as twenty-five days per month.[47] Such labor was allocated to a variety of tasks, but it was especially employed in cotton cultivation.[48] The workers were housed on the 'ezbah in mud huts constructed by the proprietor. Writers often commented on the poor ventilation and sanitary conditions of these dwellings, but it is not obvious that small peasant proprietors living in the villages were much better off in this respect. Apparently, the proprietor often furnished the working capital for the subsistence plot; this frequently led to the peasants' falling into debt to the landlord. Consequently, most of the cash wage for cotton labor was absorbed by the landlord as payment for debts.[49] His labor in the cotton fields thereby became labor for which he received no payment. Here, too, resident 'ezbah workers resembled small peasant landowners, as we shall see (see below, on small peasant cultivators' debt). Finally, Nahas noted that proprietors exercised "quasi-feudal" privileges, such as administering justice and settling disputes, over the resident workers, despite the fact that such authority was not legally sanctioned.[50] It was, no doubt, conditions such as these which led the Nationalists in 1919 to demand the abolition of the 'ezbah system and which led 'Amr to describe the system as being not very different from the "feudal system" (nizam al-iqta'). [51]

These resident workers were supplemented by laborers hired by the day for certain tasks, notably the wheat and especially the cotton harvest, emergency labor such as fighting pests, and the digging and cleaning of irrigation canals and drains.[52] Sometimes this labor was supplied by land-poor peasants from the neighboring villages,[53] but often it was supplied by migrant workers, usually from the Sa'id. Several social and technological features of that region induced Sa'idis to seek this kind of work in the Delta. Most areas of Upper Egypt retained the basin system of irrigation, with its long "dead" season in the summer, well into the twentieth century. Not only did this lead to a high degree of seasonal unemployment there, but it also gave large proprietors in that region little reason to hire resident workers.[54] The population density was higher there: 2.28 persons per cultivated feddan compared with 1.95 in the Delta in 1913.[55] Finally, the percentage of landless families was higher in Upper Egypt. All of these factors increased the Sa'idis' need to seek wage labor and made it relatively easy for labor contractors to recruit them.[56] Sa'idis came north to labor on the public works of those years.[57] Indeed, the system appears to have originated at the time of the abolition of the corvée and as a result of the existence of large numbers of peasants dispossessed in the 1860s and 1870s.[58] This seems logical, since, as we have seen, before the British occupation, peak season labor demand on estates could be obtained through the corvée.

All authorities agree that these tarahil laborers were the most miserable of the fellahin. The contractor took roughly 10% of their wages, they were either not housed at all or housed in tents (even in winter), and brought their food (large sacks of bread and onions) with them.[59] They received higher daily cash wages than did tamaliyya workers, which is logical, given that the latter also received subsistence plots and housing. Tarahil also had to bear the costs of transportation to the job.[60] The cash wage differential can be partly described as an "unemployment insurance premium" paid by the tamaliyya worker. Such workers were employed by the year and were assured of their subsistence. Tarahil, by contrast, were unemployed for part of the year. Those from Upper Egypt were apparently able to find work from between two to six months of the year.[61] If they were unable to find enough work to assure subsistence, they avoided starvation by being fed by relatives who were better off.[62] Tarahil workers were of all ages, children being used especially for the cotton harvest and to combat insect attacks by stripping off the leaves of the plant. Regardless of age, when working as harvest laborers, they were supervised both by their contractor or ra'is and by the supervisory personnel of the 'ezbah. [63]

Indeed, both kinds of 'ezbah labor was closely supervised. Nahas speaks of the employment of "brigades" of supervisory personnel. The landlord, his son, or most often, his agent watched over the cotton harvest with a stick ready to punish any laxity.[64] The supervisor paid close attention to the threshing of wheat and regulated questions of irrigation and drainage, such as the cleaning of canals and the connecting of the estate's canals with those of the government.[65] The important question of crop rotation was also decided by the proprietor.[66] Indeed, Nahas describes a kind of "agricultural Taylorism" on the estates, in which "le paron fixe le nombre d'ouvriers et de journees necessaires a l'execution de chaque travail agricole."[67]

We may now summarize the discussion of the 'ezbah system and draw a few conclusions about its economic rationale and its implications for the political economy of rural Egypt. In addition to such sine qua nons of the system as the existence of large estates, of a large landless class, and of private property rights in land and labor, the system had two bases: the crop rotation of cotton, wheat, maize, beans, and clover (see below), and the time pattern of labor inputs in cotton. The proprietor was primarily interested in cotton, for which the best organized markets existed, and, to a lesser extent, in wheat, which was sold for consumption in the cities. Cotton required a good deal of labor throughout the eight-month growing season, both for the regular waterings (a cycle of twelve days of watering, followed by six dry days was the best practice in 1914), and for hoeing (requiring one to two workers per feddan per day).[68] However, the cotton harvest required a considerably larger work force, roughly twice as many workers as during the rest of the year.[69] The proprietor thus needed a year-round work force, but he also required a supplementary one for the harvests. The year-round force was also required to tend the clover, beans, and maize which had to be grown in rotation with the cotton to ensure high yields of the latter and to preserve soil quality. But the owner had less interest in supervising the growing of fodder and food (except wheat), since he was not as likely to market these crops. By granting small, rotating[70] parcels of land for such crops, the costs of supervising the daily labor in these fields could be reduced considerably. The peasant producing crops which he himself would eat required little supervision. This would free the overseers to devote their full attention to cotton in the spring, summer, and early fall (also the maize season), and to wheat and irrigation works in the late fall and winter (also the bean and clover season).[71]

Under the 'ezbah system, then, the tamaliyya workers received a guaranteed subsistence in return for working year-round for the proprietor, both in cotton labor and at other tasks. The landlord could pay such workers a lower wage for cotton labor because of these subsistence plots. Since the crop rotation included maize, beans, and clover, and since the proprietor was less likely to market such crops, he lost relatively little by granting these parcels of land to the peasants. Indeed, he gained by reducing the amount of supervisory labor which was necessary, allowing his overseer to concentrate his energies on more profitable crops and activities.

Of course, the proprietor did give up the cash which could have been obtained by marketing the maize, beans, and clover in exchange for this reduced cost of supervision. This helps to explain why tarahil or other day laborers were used. There was a cost to retaining workers on the farm for the whole year, and no rational proprietor would want to keep a work force twice as large as necessary for eight or ten months of the year just in order to have enough labor for the peak season demands of the cotton harvest, irrigation work, etc. The tarahil system solved this problem for the proprietor. As a "reserve army of the (seasonally) unemployed," the tarahil workers also helped to keep wages down for the resident workers, since it seems likely that a tarahil worker would have been eager to take the place of the more secure tamaliyya worker. The existence of such a pool of day laborers also increased the power of the overseers, who could enforce their decisions and impose discipline not merely with the kirbaj (reported still in use on the estates in the late 1930s), but also through fines charged and deducted from cotton wages, and with the ultimate threat of dismissal.[72]

In general, this description seems to bear out those writers who characterize the political economy of Egyptian agriculture as a form of "backward colonial capitalism."[73] As noted above, 'Amr describes the 'ezbah system as being similar to "feudalism." Yet, the same author emphasizes the capitalist nature of Egyptian agriculture, with its "free" land and labor markets. Other "non-feudal" features of the land tenure system include the pasha's treating labor as a cost, their (increasingly) consolidated private property rights in land, their close supervision of the labor force, and their attention to agricultural techniques.

A central feature of the 'ezbah system was the "two-tiered" labor force. As argued above, the origins of such a system lay with the seasonal fluctuations in the demand for labor. Tarahil was straight-forward wage labor, but the case of the tamaliyya workers was more complex. It is true that they enjoyed the double freedom of the capitalist worker, freedom to sell his labor power in the market, and "freedom" from any other salable commodity except his labor power. (Although 'ezbah workers often owned their own hoe and some- times a water-buffalo). It is also true that workers who fell into debt because of inability to repay consumption or working capital loans on their subsistence plots labored in the pasha's cotton fields without receiving the usual money wage. However, in general, they do not seem to have been forced to remain on the estate: labor mobility was, therefore, much less inhibited than under European feudalism.

Yet one cannot simply describe the land tenure system as capi- talist, tout courte. Here it is useful to separate production from consumption. The market structure of the estates is quite clear. They produced for the market, and that market did not include their own labor force. Because of the low level of cash payments, the low levels of income, and the existence of the subsistence plots, there was no local demand for the 'ezbah's produce. Indeed, there was a sharply limited local market demand for either food crops or simple industrial goods. To use the structuralist term, the economy was "disarticulated."[74]

The form of payment of tamaliyya workers may also have had qual- itative consequences for rural social relations. As Marx stressed, capitalist social relations take the form of commodity relations. Not only is the worker free to sell his labor power, not only is this the only commodity which he has to sell, but also the only way in which he subsists is by selling his labor power. Not only are his labor power and his product commodities, but so also are the goods which he consumes. Marx argued that a consequence of turning all goods needed for social reproduction into commodities was that exploitative social relations were "concealed." The worker did not know when he was working for himself (reproducing his labor power) and when he was working for the owner (producing a surplus appropri- ated by the dominant class). He simply received a money wage, the same for each hour (abstracting, of course, from overtime).[75] But the tamaliyya worker did not obtain his food on the market; rather, he produced it himself with his own labor. When working in the maize and bean patches, he knew that he was working to reproduce himself and his family. In contrast, his work in the cotton fields was more clearly work "for the boss". This would have been especially true if the money wage received for such work was absorbed as debt-payment to the landlord. Under such conditions, the resident 'ezbah worker was, indeed, similar to a worker on the multezim's 'ard al-usya in Mamluk times.

For all that, there remained important differences between the 'ezbah worker and the eighteenth-century peasant. Unlike the latter, the tamaliyya worker held no rights of any kind to his subsistence plot. He could not inherit, lease, sell, or dispose of the plot in any way. The size of the plot itself was changed by the landlord in accordance with changes in the market prices of food crops.[76] By contrast, peasants who paid their taxes in the eighteenth century were ensured of their usufructory rights to the 'ard al-fellah. A second difference between peasants in Mamluk times and tamaliyya laborers lay in the degree of supervision. The landlord supervised all of the cotton labor, decided the crop mix and crop rotation, regulated questions of water, etc. This was a far cry from the administration of the 'ard al-fellah. Of course, labor on the 'ard al-usya was supervised and was often unpaid. However, since the area of such land was perhaps between 10 and 30% of the total area, and since primarily the same crops were grown on both peasant and al-usya lands, we might conclude that peasants divided their time in roughly the same proportions. By contrast, the 'ezbah worker spent twenty-five days per month working under supervision for the landlord. It seems safe to conclude that the 'ezbah worker was more closely supervised and had less decision-making power over his day-to-day labor than did the Mamluk peasant. Whereas in Mamluk times the ruling class had been content to skim off the surplus product of the worker, under the 'ezbah system, as in the Manchester mills spinning and weaving Egyptian cotton into cloth, the owners entered directly into the process of production itself.[77]

The 'ezbah system was not the only organizational method used on large estates. Gali noted that 78% of the cultivated land of the Daira Saniah and 37% of the State Domain lands were rented.[78] Most of these were leased to entrepreneurs, often village notables, who then sublet the land on a sharecropping system.[79] There is some evidence of small peasants renting for cash.[80] In such cases the landlord was often paid in installments, usually one-third after the winter harvest and the remainder after the cotton harvest.[81] Under such conditions, we would expect to find fairly close supervision of such tenants, which helps to explain why landlords often combined such direct cash renting with the 'ezbah system.[82] Whether leased directly to agricultural workers or indirectly to intermediaries, leases were short, never longer than three years, usually for one year, and sometimes for the duration of a single crop.[83]

The degree of supervision of lands rented for cash is uncertain. Willcocks noted that after poor lands improved, they were often let for three years, with the stipulation that cotton be grown only once during that time. But he also noted that "many" proprietors rented "good and bad land together on three year leases," and that such tenants "do as they please," i.e., grew cotton as often as possible.[84] Van Vloten noted that when lands of 100–200 feddans were sublet cotton was overplanted, and Martin and Levy asserted that when peasants rented land, they followed an "irrational" rotation.[85] On the other hand, Lecarpentier asserted that when larger proprietors rented out their land, they imposed "very strict" conditions on the tenant and reserved the right to have their agents inspect the properties at any time.[86] This was especially the case for those combining cash renting with the 'ezbah system. The examples of small peasant renters in the Report to Parliament of 1903 show them following the three-year crop rotation (cotton once every three years, the "best practice" technique. (See below on crop rotations).

As we shall see, there is considerable testimony that large proprietors retained the three-year crop rotation, unlike the small peasants who planted cotton every two years. This might be taken as evidence supporting the hypothesis that cash renters were closely supervised, or merely as supporting the view that large landlords by and large used the 'ezbah system rather than cash rents. In any case, in the interwar period the sources agree that cash renters were closely supervised, with the landlord specifying the crop rotation.[87] The available evidence precludes any definite conclusions. We can only note that some, probably a minority, of large landowners rented out their land for cash, and that a subset of those did not supervise their tenants adequately, with the result that the tenants naturally sought to plant cotton as often as possible (i.e., use the two-year crop rotation).

Sharecropping systems in which the cropper received a portion of the cotton crop were also used. This system seems to have been common among rich peasants on medium-sized properties. Gali in 1889 spoke of sharecropping as being the most common mode of exploitation on medium properties; Saleh says the same thing in 1922.[88] However, there is also evidence of both the 'ezbah system and cash renting being used on such lands. Lozach and Hug described the common case of a middle proprietor, resident in the countryside, living in a house next to his 'ezbah which permitted him "de connaitre directement ses gens, leurs besoins, leurs gouts, leurs habitudes...."[89] Guimei, speaking specifically of those holding estates of from twenty to forty feddans says that some are "successful fellahin," whose occupation is directing the agricultural work.[90] The fact that in 1913 a law was passed prohibiting the construction of 'ezbahs on estates smaller than fifty feddans provides evidence that such a practice existed.[91] Guimei also noted that other holders of 20-40 feddan estates were professionals or merchants living in urban areas who rented out their lands for cash.[92]

Although the evidence is fragmentary, it is likely that the following description of the modes of exploitation of medium-sized properties is appropriate. First, if the land was owned by a rich peasant, he would exploit his land using the 'ezbah system if his estate and the proprietor's means were above a certain (unknown) "threshold" size.[93] Below that size, he generally would have used sharecropping.[94] Since such landowners were already resident in the countryside, the costs of supervision were relatively lower for them. Cash renting would have been correspondingly less attractive. Further, there is testimony that the fellahin preferred sharecropping to cash renting because of the risk-sharing inherent in sharecropping contracts.[95] Gali listed lack of resources by the proprietor as one of the reasons for using sharecropping;[96] this would be consistent with a picture of "rich peasants" using the 'ezbah system and "middle peasants" using sharecropping. If the proprietor was an urban resident, he apparently rented his lands for cash. This, of course, is logical, given that his not residing in the countryside precluded his supervising agricultural labor. Nor, apparently, would an estate of this size warrant the employment of an overseer. In addition to testimony on economies of scale in supervision, there is the indirect, negative evidence that all of the descriptions of 'ezbahs on medium estates which I have seen indicate that the owner was resident and was himself the director of the agricultural operations. Under such circumstances, it is logical that urban holders of medium-sized properties rented for cash.

The terms of sharecropping contracts were highly variable, with the share of cotton and other crops received varying directly with the share of inputs supplied. Since the peasants often had little to supply but their labor power, it is not surprising that they usually received only one-fourth or one-fifth of the cotton harvest.[97] There were numerous variations and special arrangements, such as sharecropping for one crop only, or the granting of a parcel of land to be planted in birsim to feed the animals.[98] From the latter example one can see how the métayage system could grade off into the 'ezbah system. Indeed, the care of animals on 'ezbahs was usually arranged on a share basis (shirk), in which the landlord purchased the animal, then entrusted its care to an 'ezbah worker who fed the animal with the maize stalks and birsim from his subsistence plots, and consumed the animal's milk himself. The landlord had the use of the animal whenever necessary, and when the animal was sold for butchering, the proceeds were divided between the 'ezbah worker and the landlord. This system was described as "spreading" in 1911, presumably to spread the risks of animal plague, endemic until after 1913, and of cattle poisoning, a common crime in the first decade of the twentieth century.[99] Other specialized agricultural operations were let on a share basis. Rice lands were often let on a 1/2:1/2 basis, to "provide the incentives for the careful work which this plant requires."[100] The sources agree that cotton sharecroppers were closely supervised, although we would expect such workers to require less supervision than the 'ezbah workers.

We may now summarize the broad outlines of agrarian social rela-tions for the period 1890-1914. (1) Large estates were primarily exploited by the 'ezbah system, using resident and temporary workers, but (2) some large estates were exploited by a combination of the 'ezbah system with cash renting while (3) other estates were leased for cash to intermediaries, often village notables, who then sublet them using a sharecropping system. Rich peasants used either (4) the 'ezbah system if their resources permitted or (5) sharecropping, or some combination of the two. Those using sharecropping were often themselves working in the fields.[101] (6) Labor supervision appears fairly close on all lands, with the possible exception of some lands leased for cash. This was, perhaps, especially the case on (7) medium properties owned by urban residents, who usually leased their lands for cash. Finally, (8) some 20-25% of the land area was exploited by small peasant landholders working the land themselves and with family labor. Like the resident 'ezbah workers, many of these peasants were in debt. The feedback of these social relations on the choice of technique in agriculture will be our next concern. First, however, we must look more closely at the ecology of these techniques.

AGRICULTURAL TECHNOLOGY:  IRRIGATION

The British completed the transformation of the irrigation system which had begun under Muhammad 'Ali and had been continued by his successors. Perennial irrigation required the construction of dikes, dams and canals in order to raise the level of water in the river and in the canals in those dry months. The completion of this process under the British occurred in several stages. First, they repaired and made functional the barrage at the fork of the Nile just below Cairo which Muhammad 'Ali had begun but had been forced to abandon. This work, completed in 1891, raised the level of the water in the summer canals. This had two effects: (1) it extended the area over which cotton could be grown, and (2) it reduced the amount of labor required to put a given quantity of water on the fields in summer by reducing the vertical distance from the water to the fields.[102] The barrage made perennial irrigation and cotton cultivation possible over the entire cultivated area of the Delta. Another barrage, at Asyut, was completed in 1902; the subsequent construction of feeder canals converted all of Middle Egypt to perennial irrigation by 1909. The completion of the Aswan dam in 1902 increased the supply of summer water throughout the country, including Upper Egypt, allowing some areas in the region to switch from basin to perennial irrigation by 1909.[103] By 1914, perennial irrigation extended as far south as Deirut, in northern Asyut province.[104]

All of these changes made possible a rapid expansion in cotton production, as a glance at Table 3.4 shows. This came about because of the increase in the supply of summer water and also because of the higher water level in the canals. It should be noted that the increased water did not necessarily imply a fall in the labor input into cotton. This may have been the case for the landlords; however, I doubt that it was true for the peasants for two reasons: first, low yields under the old peasant technique was attributed in part to the lack of water. Now that water could be obtained more easily (defined as less labor time per watering), peasants probably watered more times than previously.[105] Second, a large number of peasants were able to grow cotton more often than before. Growing the crop every two years instead of every three became the norm for peasant cultivation during this period.

AGRICULTURAL TECHNOLOGY: CROP ROTATION SYSTEMS

The more intensive use of land, the switch from a three- to a two-year crop rotation, is the second technical change of importance in this period, complementary to the extension of perennial irrigation. Although there are numerous, slightly differing descriptions of the rotations, the following would be roughly accurate:[106]

Three Year:
        First year:    <u>Birsim</u> (Nov.-March)
                        Cotton (April-Oct.)
        Second year:  Beans (Nov.-May)
                        <u>Sharaqi</u> or fallow (June-Oct.)
        Third year:   Wheat or Barley (Nov.-May)
                        Maize (June-Oct.)

Two Year:
        First year:   Cotton (April-Nov.)
                        Wheat, barley, or beans (Nov.-May)
        Second year:  Fallow (May-July)
                        Maize (July-Nov.)
                        <u>Birsim</u> (Nov.-March)

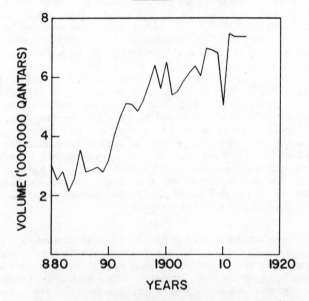

Fig. 3.1 Volume of Egyptian Cotton Exports 1880-1914 (Source: Table 2.5)

TABLE 3.4
VOLUME OF EGYPTIAN COTTON EXPORTS, 1880-1914

| Year | Volume (cantars) | Year | Volume (cantars) |
|------|------------------|------|------------------|
| 1880 | 3,000,000 | 1900 | 6,512,000 |
| 1881 | 2,510,000 | 1901 | 5,391,000 |
| 1882 | 2,811,000 | 1902 | 5,526,000 |
| 1883 | 2,140,000 | 1903 | 5,860,000 |
| 1884 | 2,565,000 | 1904 | 6,147,000 |
| 1885 | 3,540,000 | 1905 | 6,376,000 |
| 1886 | 2,788,000 | 1906 | 6,033,000 |
| 1887 | 2,864,000 | 1907 | 6,977,000 |
| 1888 | 2,964,000 | 1908 | 6,913,000 |
| 1889 | 2,780,000 | 1909 | 6,813,000 |
| | | | |
| 1890 | 3,203,000 | 1910 | 5,046,000 |
| 1891 | 4,054,000 | 1911 | 7,477,000 |
| 1892 | 4,662,000 | 1912 | 7,367,000 |
| 1893 | 5,117,000 | 1913 | 7,375,000 |
| 1894 | 5,073,000 | 1914 | 7,369,000 |
| 1895 | 4,840,000 | | |
| 1896 | 5,220,000 | | |
| 1897 | 5,756,000 | | |
| 1898 | 6,399,000 | | |
| 1899 | 5,604,000 | | |

Source: Crouchley, (1939), 263-265.

Only the three-year rotation was used in the late 1880s, but by 1900, 40% of the cultivated area was on the two-year system and by 1908, this figure had grown to 58%[107] (see Table 3.5). The main features of this shift were: (1) an increase in the intensity of land use, measured by the amount of time per year that a given piece of land was cropped (and, therefore, a reduction of the fallow); (2) an increase in the amount of land planted in cotton per unit of time; and (3) a reduction in bean production.[108] Proxies for the first and second feature are provided by the ratio of cropped area to cultivated area, and by the ratio of cotton area to cultivated area,

TABLE 3.5
ROTATION OF CROPS IN LOWER EGYPT, 1894-1909

| Year | Two-year Area | Two-year Percent | Three-year Area | Three-year Percent | Total |
|------|------|---------|------|---------|-------|
| 1894 | 159 | 17 | 753 | 83 | 912 |
| 1895 | 177 | 18 | 750 | 82 | 927 |
| 1896 | 211 | 21 | 757 | 79 | 968 |
| 1897 | 281 | 27 | 748 | 73 | 1,028 |
| 1898 | 248 | 24 | 781 | 76 | 1,029 |
| 1899 | 286 | 27 | 779 | 73 | 1,065 |
| 1900 | 455 | 40 | 689 | 60 | 1,143 |
| 1901 | 443 | 38 | 700 | 72 | 1,143 |
| 1902 | 460 | 40 | 710 | 60 | 1,170 |
| 1903 | 462 | 40 | 709 | 60 | 1,171 |
| 1904 | 526 | 44 | 668 | 56 | 1,194 |
| 1905 | 629 | 50 | 627 | 50 | 1,256 |
| 1906 | 668 | 53 | 592 | 47 | 1,260 |
| 1907 | 722 | 56 | 568 | 44 | 1,290 |
| 1908 | 726 | 56 | 573 | 44 | 1,299 |

Source:
J.I. Craig, "Notes on the Cotton Statistics of Egypt," L'Egypte Contemporaine (March, 1911), 177.

respectively. These are given in Table 3.6 and 3.7. They are generally in line with Craig's estimates.

Both the first and the second features of the new rotation imply that the rotation switch was labor-using and land-saving in the short run. However, the long-run effects of the two-year rotation are best described as "land-destroying." Average cotton yields declined starting at the turn of the century, after a sharp upward rise in the 1890s. Output per worker in agriculture followed a similar pattern (see Table 3.8).

Contemporary observers stressed four sources of the fall in cotton yields: (1) a rise in the water table, (2) soil deterioration due to shortening the fallow, (3) insect attacks, and (4) deterioration in the quality of the seed. The agricultural intensification outlined above contributed to the first three of these problems.[109] The rise in the water table suffocated the deep roots of the cotton plant and promoted soil salination through capillary action and evaporation. Under the basin system, the level of the Nile when not in flood was six to eight meters below the surface level of most of the land. Since the canals were without water during that period, they acted as drains. At first, this was not a problem, since due to a limited supply of summer water, only the southern Delta was so irrigated. This area was high enough to allow free flow drainage. But

TABLE 3.6
RATIO OF CROPPED AREA TO CULTIVATED AREA,
1897/1898, 1912/1913

| Year | Cropped Area/Cultivated Area |
|------|------------------------------|
| 1897/98 | 1.346 |
| 1898/99 | 1.356 |
| 1899/00 | 1.369 |
| 1900/01 | 1.384 |
| 1901/02 | 1.393 |
| 1902/03 | 1.405 |
| 1903/04 | 1.410 |
| 1904/05 | 1.400 |
| 1905/06 | 1.401 |
| 1906/07 | 1.418 |
| 1907/08 | 1.426 |
| 1909/09 | 1.427 |
| 1909/10 | 1.443 |
| 1910/11 | 1.433 |
| 1911/12 | 1.453 |
| 1912/13 | 1.460 |

Source:
Department of Statistics. Annuaire Statistique, 1914, 322.

TABLE 3.7
RATIO OF COTTON AREA TO CULTIVATED AREA
1897/1898, 1912/1913

| Year | Cotton Area/Cultivated Area |
|------|------------------------------|
| 1897/98 | .2204 |
| 1898/99 | .2224 |
| 1899/00 | .2352 |
| 1900/01 | .2373 |
| 1901/02 | .2391 |
| 1902/03 | .2550 |
| 1903/04 | .2672 |
| 1904/05 | .2899 |
| 1905/06 | .2821 |
| 1906/07 | .2967 |
| 1907/08 | .3080 |
| 1908/09 | .2972 |
| 1909/10 | .3073 |
| 1910/11 | .3251 |
| 1911/12 | .3258 |
| 1912/13 | .3262 |

Source:

Egypt, Ministry of Finance, Department of Statistics, Annuaire Statistique, 1914, 322.

with the completion of the Aswan dam, more water was used, and summer cultivation was extended toward the north. This "choked back the flow of drainage from the south and center of the Delta."[110] In the absence of an alternative drainage system, a lack discussed below, the switch to perennial irrigation created severe drainage problems.

At the beginning of the nineteenth century the subsoil water in the middle of the Delta during the summer was six or eight metres below the level of the land; in 1886-7 it had risen to three or four metres below the surface; in 1908 it was only one metre.[111]

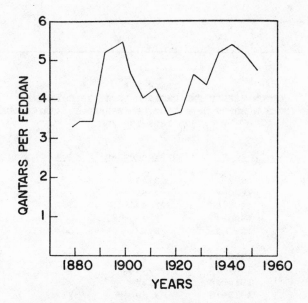

**Fig. 3.2** Cotton Yields per Cropped Area
1879-1950/54 (Source: Table 3.8)

The rotation shift contributed to the problem in a different way, for the more intensive cropping led to more water being on the soil per unit time—not only because fallow time was shortened but also because cotton was a water-intensive crop (see Table 3.9). The relatively saline summer water evaporated rapidly under the high temperatures, depositing salt on the soil.[112] Since the lands no longer received the periodic washings of the basin system, this became a problem.

Shortening the fallow created an additional set of soil problems. During the long fallow period of basin irrigation or of the three-year crop rotation under perennial irrigation,[113] the land heated up, dried out, and cracked. This had three beneficial effects on the soil (1) Physically, the soil was aerated, thus increasing its permeability and porosity, which, in turn, increased its capacity to hold air and water in the pores of the soil where the plant roots could utilize them. (2) Chemically, it broke up the colloids which bound with the minerals in the soil water. This freed these minerals for use by the plants. (3) Biologically, it promoted the proliferation of nitrifying bacteria, partly by killing their predators, a species of protozoa. The switch to the two-year rotation with its much shorter fallow period deprived the soil of these beneficial effects.[114]

The problem of _pests_ was also, in part, the result of the two technical changes. Serious attacks occurred every year after 1904.

---

**TABLE 3.8**
**COTTON YIELDS PER CROPPED AREA, 1879–1950/54**
**AND INDEX NUMBERS OF OUTPUT PER WORKER IN AGRICULTURE,**
**1895/1899–1950/54**

| Year | Cantars/Feddan | Index |
|------|----------------|-------|
| 1879 | 3.29 | |
| 1880–4 | 3.43 | |
| 1885–9 | 3.43 | |
| 1890–4 | 5.21 | |
| 1895–9 | 5.47 | 100 |
| | | |
| 1900–04 | 4.67 | 102 |
| 1905–09 | 4.03 | 98 |
| 1910–14 | 4.27 | 96 |
| 1915–19 | 3.58 | 76 |
| | | |
| 1920–4 | 3.67 | 74 |
| 1925–9 | 4.63 | 79 |
| 1930–4 | 4.36 | 73 |
| 1935–9 | 5.2 | 74 |
| | | |
| 1940–4 | 5.39 | 64 |
| 1945–9 | 5.14 | 69 |
| 1950–4 | 4.73 | 73 |

Sources:
    Cotton Yields: "Agricultural Production in Historical Per-
    sepctive," in Egypt Since the Revolution, P. Vatikiotis ed. (New
    York: Praeger, 1968), 53.

    Index Nos. of Q/L in Agriculture: Patrick O'Brien, "The Long
    Term Growth of Agricultural Production in Egypt," in P.M. Holt,
    ed., Political Social Change in Modern Egypt (1968), 191.

---

TABLE 3.9
WATER REQUIREMENTS PER CROP
($M^3$/feddan), 1920

| Crop | Lower Egypt | Upper Egypt |
|------|-------------|-------------|
| Wheat | 1,140 | 1,510 |
| Barley | 1,030 | 1,575 |
| Maize | 2,670 | 3,400 |
| Millet | 2,670 | 2,510 |
| Rice | 10,000-16,000 | ---- |
| Cotton | 3,740 | 4,640 |
| Birsim | 2,630 | 3,560 |

Source:
M. Macdonald, Nile Control (1920), reprinted in W.B. Fisher, The Middle East: A Physical, Social, and Regional Geography, 5th ed. (London: Methuen and Co., 1963), 500.

In part, this was simply a question of increased water, which favored the egg laying of cotton worms.[115] More intensive cropping also favored the pests.

As a result of the extra water, maize could now be sown in...July...whereas previously it had been necessary to wait for the arrival of the flood. This in turn allowed some proprietors to harvest their maize a month or so earlier than before and thus fit in a crop of birsim before they were required to prepare the land for cotton in the following February or March.[116]

Such a practice multiplied the number of host plants on which the pests could feed, with the moths eating the clover flowers and the worms the leaves. Further, the small cultivator who used the more intensive rotation system faced a dilemma. If, as was most common, he took two, three, or more cuttings of birsim to ensure fodder for his animals, he delayed the planting of cotton. This meant later ripening, at a time when it was more subject to attacks of the cotton boll worm.[117] This was the most common small peasant practice,[118] for they needed the fodder and/or short-term cash which the additional cuttings of birsim afforded them. The final possible cause for the deterioration of yields, decline in seed quality, does not seem to have been connected with either perennial irrigation or the two-year rotation, but rather with the mixing of seed in ginning establishments and in the fields.[119]

THE POLITICAL ECONOMY OF DRAINAGE

Contemporary observers felt that the principal remedy for at least some of these problems was improved drainage. This would have prevented a rise in the water table, improved soil aeration, and reduced collidity. The Domains' Commission conducted experiments as

early as 1903 which clearly showed the relationship between good
drainage and cotton yields. But the British failed to provide adequate public drains. This failure was not the result of ignorance.
They had encountered similar problems in India before,[120] and
seemed to have been quite aware of the problem. Consider the statements of Scott-Moncrieff, the chief British engineer in Egypt and
head of the Irrigation Department from 1883 to 1892:

> He (Nubar Pasha) was a great agriculturalist, and therefore
> keenly interested in irrigation. A chance remark to him one
> day, that it would be more difficult to drain the water off
> the land than to pour the water over the land, seemed to
> strike him as original and true....

and

> Irrigation without drainage always tends to injure the
> soil...drainage has been quite neglected....[121]

The statements of Professor Wallace, who accompanied Lt. Col. Justin
Ross, Inspector General of the Irrigation Department, on a tour of
the country, are even more explicit:

> Of all the schemes for the regulation of water in Egypt, and
> they are many, there is nothing nearly so urgent as drainage,
> and it may be safely assumed that any material addition to
> the amount of water without extensive drainage works would
> probably on the whole be more of a curse than a blessing to
> the country.[122]

Armed with such knowledge, the British proceeded to remodel old
drains and to construct new ones. From 1885 to 1905 some L.E. 1.4
million were spent on drainage.[123] However, the British seem to
have concentrated a good deal of their energy on renovating old
drains; Eldon Gorst (Cromer's successor) wrote in 1910 that the
"efforts (of the Irrigation Department) have been largely concentrated on the improvement of the present system by means of a sys-
tematic overhauling and remodeling of the existing drains."[124]
Kitchener's report of the same year asserted that "for some years
minor works of partially deepening drains in (badly drained) areas
were taken in hand. It has been seen, however, that if real improvement was to be made, much larger areas must be dealt with at one
time...."[125] Gorst added that the solution to the "admittedly most
difficult" problem of drainage could only be gradual, primarily
because "simple improvement alone" would absorb "all the money that
can possibly be afforded by the country for the purpose for some
years to come."[126] He pointed out that the drainage problem had
been dealt with "to the fullest extent of the funds available."[127]

Here we find a key to the problem. The British were aware of
the need for drainage, made considerable efforts to improve and
extend the drainage system, yet in the end and by their own admis-
sion, had failed.[128] In part, this failure may have stemmed from
thinking on too small a scale; that is, they knew drainage was important, but they miscalculated.[129] Beyond that, however, the British
faced budgetary constraints which prevented them from undertaking the

truly large-scale spending which would have been necessary. These constraints, in turn, derived from the nature of the British occupation. Unlike, say, the French occupiers of Algeria, contemporary international rivalries sharply circumscribed the decision-making powers and political maneuvering room of the British administration in Egypt. Since the British occupation of Egypt worsened French security in the Mediterranean, France did everything she could to obstruct the British there. The Caisse de la Dette Publique, which represented European bondholders, including many Frenchmen, sought to extract the maximum possible percentage of Egyptian government revenues for debt payments. The "need" to spend heavily on military forces to fight the Mahdi in the Sudan was an additional pressing claim on the budget.[130]

Furthermore, the administration in Egypt had to finance itself--no money would be forthcoming from the home government to support a venture which was opposed by a substantial faction within the Liberal Party and a considerable section of the Foreign Office.[131] These last constraints faded away in the 1890s, but French opposition continued until the Entente of 1904. Even then, however, the Egyptian government could not borrow more than one million Egyptian pounds annually without the permission of the Ottoman Sultan. Since by that time enmity between Great Britain and the Porte was growing, it is not surprising that the latter refused to grant such permission.[132] As the following two quotes show, financial constraints remained binding throughout the period:

By the extension and maintenance of a well conceived drainage system, we hope to avert the excessive salting of the soil. Very great efforts have been made to better drainage by very small sums of money available, but the conditions of Finance as laid down by the "Caisse de la dette Publique" hamper us terribly.

and

I have seen canals as well as drains, especially the latter, which have been allowed to get overgrown with weeds, and this naturally reduces their efficiency. I was told in the Ministry of Public Works, under which the Irrigation Department is, that until recently there was not sufficient money at their disposal to keep all the drains clear....[133]

Nor could the British rely on taxation to raise revenue for drainage. Recall that Isma'il's increases in the land tax had precipitated widespread peasant indebtedness and land loss. This in turn, had fueled peasant discontent and had made them receptive to 'Urabist propaganda. Reduction of land taxes was part of British "pacification" policy.[134] The British wished not only to forestall peasant unrest, but also to secure the cooperation of their chief Egyptian political allies, the large landowners. Accordingly, land tax rates were lowered, and the burden of taxation fell (see Table 3.10). The priorities of the British administration of this period are best summarized by Lord Cromer himself:

...each zealous official, anxious to improve the administration of his own Department, hurried in demands for money on a
poverty-stricken Treasury...The engineer showed that it was
false economy not to extend the system of irrigation, to
drain the fields.... It was thought that...it was essential
to alleviate the burdens which weighted on the mass of the
population. Fiscal relief had a prior claim to administrative reform. I was, therefore, decided that, whilst penuri-
ously doling out grants to the spending Departments, the
principal efforts of the Government should be devoted to
devising means for the relief of taxation.[135]

In summary, the British failed to invest in drainage because of
miscalculation, bureaucratic inertia, revenue constraints, and the
desire to have quick results at low cost to stifle opposition to the
occupation at home, abroad, and in Egypt itself. We have seen how
the interplay of domestic and international forces generated this
outcome. Both the British occupation and the economic and political
constraints on their administration derived from the interaction of
the European balance of power (the international state-system) with
Egyptian socio-political forces. As argued in Chapter 2, these
latter forces were shaped by the spread of cotton cultivation (the
international trade system). We shall see in Chapter 4 that the
disequilibria which this failure engendered led to further technical
change in agriculture in the interwar period, when the damage had to
be repaired while the population continued to grow. We shall also
see, in Chapter 6, that despite a wholly different regime and a
transformed international system, the same failure to invest in
drainage occurred under Nasser. It is a sad irony of Egyptian agricultural history that the same, very costly mistake could have been
made both by the colonial government and by a fervently nationalist
regime.

SOCIAL RELATIONS AND THE CHOICE OF CROP ROTATION

The serious long run ecological consequences of the intensification of agriculture lead us to ask why farmers switched to the two-
year crop rotation. We have seen evidence that some estate holders
rented out their lands and that some of these apparently failed to
specify carefully the crop rotation which the tenant was to follow,
with the result that the tenants employed the more intensive two-year
system. But since it is unlikely that poorly supervised cash rent
farms constituted a large proportion of pasha lands, this must have
made only a marginal contribution to the spread of the two-year system. Indeed, there is considerable contemporary testimony that
larger proprietors retained the three-year rotation while small
peasants overwhelmingly adopted the two-year system.[136] However,
since small peasant holdings were between 22-24% of the agricultural
land, whereas the two-year system was practiced on between 40-55% of
the fields (see above, Table 3.5), it appears that most medium
proprietors and rich peasants must have adopted the two-year rotation. If the pashas refrained from switching in order to avoid soil
deterioration, why did the others switch?

### TABLE 3.10
### BRITISH LAND TAX RATES, 1881-1913

| Year | Land Tax Rates (L.E./ feddan) | Tax Rate/ Maize Price (ardebs feddan) | Tax Rate/ Cotton Price (cantars feddan) | Tax as % of cotton value per feddan[a] |
|---|---|---|---|---|
| 1881 | 1.035 | | | |
| 1882 | 1.045 | | | |
| 1883 | 1.005 | | | |
| 1884 | 1.046 | | | |
| 1885 | 1.027 | | | |
| 1886 | 1.030 | | | |
| 1887 | 1.034 | (No data) | (No data) | (No data) |
| 1888 | 1.018 | | | |
| 1889 | 0.979 | | | |
| 1890 | 1.036 | | | |
| 1891 | 1.006 | | | |
| 1892 | 1.005 | | | |
| 1893 | 0.980 | | | |
| 1894 | 0.851 | | | |
| 1895 | 0.922 | 1.99 | 0.46 | 8.7% |
| 1896 | 0.891 | 1.70 | 0.42 | |
| 1897 | 0.903 | 1.61 | 0.52 | |
| 1898 | 0.842 | 1.19 | 0.54 | |
| 1899 | 0.836 | 1.28 | 0.43 | |
| 1900 | 0.798 | 1.12 | 0.30 | 6.7% |
| 1901 | 0.829 | 1.17 | 0.38 | |
| 1902 | 0.836 | 1.34 | 0.36 | |
| 1903 | 0.815 | 1.27 | 0.26 | |
| 1904 | 0.840 | 1.30 | 0.27 | |
| 1905 | 0.851 | 1.13 | 0.32 | 7.9% |
| 1906 | 0.858 | 0.97 | 0.25 | |
| 1907 | 0.876 | 1.10 | 0.23 | |
| 1908 | 0.880 | 0.99 | 0.29 | |
| 1909 | 0.891 | 0.86 | 0.26 | |
| 1910 | 0.904 | 1.22 | 0.20 | 6.7% |
| 1911 | 0.895 | 1.12 | 0.24 | |
| 1912 | 0.909 | 1.07 | 0.27 | |
| 1913 | 0.916 | 0.90 | 0.25 | |

Source:
   Egypt, Ministry of Finance, Department of Statistics,
   Annuaire Statistique,
   1914.
   Tax Rates, 414; maize, cotton prices, 387.
a. Average for each five-year period.
   See Table 3.8 for source.

The task of explaining the difference in technical choice is eased only partly by considering class differences in access to agronomical information. The "largest landowners in all parts of the country" were members of the Khedival Agricultural Society (founded 1898), which strongly advocated the use of the three-year rotation, as did government agronomists.[137] The pashas, as rich city dwellers, had much better access to this sort of information than did the small peasants. There was no extension service to compensate for this lack: although an Agricultural School was founded by the Ministry of Public Works in 1890, in 1910 it had only 196 students.[138] Demonstration farms, when established at all, were set up on the estates of large landowners.[139] Furthermore, it is likely that peasants continued to use the same techniques of watering which they had always used, not realizing that by doing so they would contribute to the deterioration of the soil. The fund of peasant knowledge about cotton cultivation which had been accumulated before 1890 stressed the importance of getting <u>enough</u> water for the plants. This, indeed, had been their main problem here until the irrigation changes of the early 1890s. Acting on this inherited information after 1890 would then have caused trouble. While there is no reason to suppose that the peasants could not perceive the decline in their yields, they may not have perceived the cause-and-effect relationships between crop rotation, overwatering, and the fall in cotton yields. The pashas, on the other hand, had the benefit of agronomical advice.

However, a certain skepticism is in order here. Soil science was only about one generation old at the time,[140] and casual empiricism might have been sufficient to understand the real problem. Further, government circulars warning of the dangers of overwatering and of the two-year rotation were distributed to the Mudirs and read in the village squares.[141] It seems quite possible that the peasants knew of the consequences of the two-year rotation, but chose it anyway.

It is natural to ask whether differences in the relative input prices facing the two classes could explain the difference in their choice of crop rotation.[142] The fact that the two-year rotation was labor-intensive suggests that perhaps the peasants faced a lower price of labor (in terms of cotton and maize) and therefore chose the two-year rotation. This is plausible, but again, some skepticism is in order. Hiring was apparently common, even the hiring of children. We know that small peasants hired themselves out in the 1960s, which suggests that the opportunity cost of family labor used on the farm was roughly in line with market wage rates.[143] Evidence from the pre-World War I era is scanty. Ayrout, writing in the 1930s, mentions that families hired out their own children to work on the pashas' estates at cotton-picking time. Schanz, in 1913, implied that this was the case at that time, as well.[144] This would be consistent with the evidence that land-poor peasants often hired themselves or their families to 'ezbah proprietors for the cotton harvest.[145]

Further, Robert Mabro[146] argues that, indeed, the opportunity cost of hiring out family labor may exceed the market wage rate. This is the result of the interaction of factors of risk and the timing of agricultural operations on the one hand, and of the existence of a large number of landless laborers on the other. Anyone hiring labor for, say, the cotton harvest, would want the wage contract to cover a long enough time period to ensure than an adequate work force would be on hand to meet any unexpected contingency, such as a dust storm or an insect attack which might arise at harvest time. The peasant family, for its part, would want to be sure that its family members would be available to meet such emergencies. Therefore, the supply price of labor from peasant landowning families to outside employers would be higher than the supply price of labor of landless peasants, who had no such alternative use for their own labor. If the landless class formed a relatively large percentage of those offering their labor for hire, the market wage rate would be below the opportunity cost of hiring out family labor for the small landholding peasant. Such a description seems to fit the Egyptian case for the 1880-1914 period, as well as for the post-World War II era for which Mabro developed his argument. In this case a lower opportunity cost of labor to the peasants could not explain the observed class differences in choice of crop rotation.

A different approach would be to posit that peasant choices can be described by a lexicographic preference ordering, in which peasants place top priority on assuring themselves a food supply and secondary priority to maximizing gains by choosing a seemingly profitable mixture of cash crops. This ordering has been referred to as a "safety-first" decision rule.[147] In its present form it might also be thought of as a food-first rule, and it implies that low-income peasants will tend to favor devoting a large share of their land to growing crops for their own consumption.

On balance it would appear that the shift to the two-year rotation jeopardized peasant food supplies and nutritional intake. Peasants had to assure themselves an adequate supply of two principal food stuffs, maize and beans. Although maize was grown during the same season as cotton, the above rotation description would indicate that maize was not squeezed out by the shift. Such an impression is borne out by the available data: neither the total area nor the percentage of land planted in maize declined during the first decade of the twentieth century (see Table 3.11). It is likely that the maize area expanded at the expense of the fallow. This would imply that the maize was being watered during the pre-flood months in the summer; the government forbade the watering of maize from May 15 to July 31 in 1903 to keep the water for the cotton crop. This regulation remained on the books until 1913.[148] It seems likely, however, that this regulation was evaded. Schanz, writing in 1913, said that maize was mostly planted in June, an indication of widespread evasion.[149] Furthermore, the increased supply of water permitted a dramatic rise in yields: for the State Domain lands in 1879, average yields were 3.64 ardabs per feddan whereas the national average in 1913 was 6.33 ardabs per feddan.[150]

TABLE 3.11
AREA PLANTED IN MAIZE
1898/1899-1912/1913

| | Maize Area (feddans) | | | Maize Area as Percent of Total Cultivated Areas |
|---|---|---|---|---|
| Year | Lower Egypt | Upper Egypt | Total | |
| 1898-99 | 1,069,059 | 490,600 | 1,559,659 | 22.18 |
| 1899-00 | 1,103,941 | 519,984 | 1,623,925 | 22.68 |
| 1900-01 | 1,165,027 | 552,576 | 1,717,603 | 23.55 |
| 1901-02 | 1,195,385 | 579,510 | 1,774,895 | 23.90 |
| 1902-03 | 1,173,624 | 566,487 | 1,740,111 | 23.70 |
| 1903-04 | 1,190,848 | 585,139 | 1,775,987 | 23.43 |
| 1904-05 | 1,163,425 | 579,532 | 1,742,957 | 23.04 |
| 1905-06 | 1,164,413 | 605,758 | 1,770,171 | 23.66 |
| 1906-07 | 1,167,089 | 632,192 | 1,799,281 | 23.48 |
| 1907-08 | 1,169,059 | 630,646 | 1,799,705 | 23.69 |
| 1908-09 | 1,119,398 | 677,347 | 1,796,745 | 23.42 |
| 1909-10 | 1,140,487 | 700,166 | 1,840,653 | 23.87 |
| 1910-11 | 1,119,632 | 653,054 | 1,772,686 | 23.49 |
| 1911-12 | 1,170,665 | 662,403 | 1,833,068 | 23.86 |
| 1912-13 | 1,174,086 | 678,674 | 1,852,760 | 24.02 |

Source:
Department of Statistics, Annuaire Statistique, 1914, 326-327.

The area planted in beans, on the other hand, did decline (see Table 3.12). The fall in area was not, apparently, offset by a rise in yields. The only available series on yields before 1911 is from the State Domains. Using this as a proxy for national yields, and the figures on area, we find a decline of bean production of over 700,000 ardabs from 1899 to 1914.[151] One might argue that the peasants simply substituted maize for beans in their diet. They may well have done so. However, the two are poor substitutes, since maize is deficient in the amino acids isoleucine and lysine, which are found in much greater quantities in broad beans. Since the human body uses the eight essential amino acids, including these two, in fixed proportions, a fall in bean consumption would have to be compensated for by a very large increase in maize consumption or by alternative protein sources, if the same level of nutrition were to be maintained.[152] Until 1907 such an alternative supply may have been provided by imports of butter, margarine, and cheese. Imports rose at 8.5% per year from 1897 to 1907[153] (see Table 3.13). Only a part of this, for the first decade of the twentieth century, can have been to replace losses due to a decline in the number of cattle and buffaloes[154] (see Table 3.14). However, most of this was probably consumed by city dwellers or rich peasants, rather than small peasants. There is the further problem that such a protein source had to be obtained in the market; the usual assumption in a "safety-first" framework is that peasants will always seek to produce their own food to avoid the risks of relying on the market.[155] It is at least doubtful that peasants who adopted the two-year rotation achieved their subsistence goal (so defined) before 1907.

Furthermore, from 1907 to 1914, it seems quite clear that the goal was not met. Imports of margarine, cheese and butter fell at the rate of 4.45% per year during that time (see Table 3.13). The bean area continued to decline at a rate of 2.8% per year, and the number of cattle and buffaloes fell at the rate of 3% per year. If we assume that the population was growing at 1.3% per year (the average growth rate for 1907-1917), there is clear evidence of a decline in per capita protein supply during those seven years. The maize

---

TABLE 3.12
AREA PLANTED IN BEANS
1898/1899-1912/1913

| | Beans Area (feddans) | | | Beans Area as Percent of Total Cultivated |
|---|---|---|---|---|
| Year | Lower Egypt | Upper Egypt | Total | Areas |
| 1898/99 | 165,644 | 472,108 | 637,752 | 9.07 |
| 1899/00 | 170,067 | 480,283 | 650,350 | 9.08 |
| 1900/01 | 169,114 | 490,665 | 659,779 | 9.05 |
| 1901/02 | 162,321 | 471,532 | 633,853 | 8.53 |
| 1902/03 | 157,614 | 461,294 | 618,908 | 8.43 |
| 1903/04 | 146,388 | 527,185 | 673,573 | 8.88 |
| 1904/05 | 124,686 | 460,531 | 585,217 | 7.74 |
| 1905/06 | 118,856 | 463,352 | 582,208 | 7.78 |
| 1906/07 | 124,648 | 471,035 | 595,683 | 7.78 |
| 1907/08 | 112,827 | 428,258 | 541,085 | 7.12 |
| 1908/09 | 100,718 | 465,970 | 566,688 | 7.39 |
| 1909/10 | 93,117 | 467,518 | 560,635 | 7.27 |
| 1910/11 | 73,434 | 467,971 | 541,405 | 7.18 |
| 1911/12 | 69,197 | 448,622 | 517,819 | 6.74 |
| 1912/13 | 63,540 | 414,647 | 478,187 | 6.26 |

Source:
Egypt, Ministry of Finance, Department of Statistics, Annuaire Statistique, 1914, 326-327.

area was expanding, but only at the rate of .42% per year.  Insofar as the increase in yields was due to increased amounts of water, the main gains in yields presumably occurred well before 1907.  In view of the protein deficiencies of maize noted above, it probably did not compensate for the other declines.

We should note that the reduction in per capita protein supply was _not_ accompanied by a switch out of the two-year rotation back to the more "food-intensive" three-year system.  Intensification, measured by the ratio of cropped area to cultivated area or by the ratio of cotton area to cultivated area advanced from 1907 to 1914,  albeit at  a  slower rate.  (See Tables 3.5 and 3.6) The _switch_ of rotations appears to have been made during the preceding, relatively more prosperous  decade, but the decline in the per capita food supply did not induce any "reswitching." It would seem that under  the  usual  "food first"  strategy the peasants should have rejected the two-year rotation.

---

**TABLE 3.13**

**IMPORTS OF BUTTER, MARGARINE, AND CHEESE,**

**1885-1913**

| Year | Imports of Butter, Margarine & Cheese (kg) | Year | Imports of Butter, Margarine & Cheese (kg) |
|------|------|------|------|
| 1885 | 2,440 | 1900 | 4,509 |
| 1886 | 2,508 | 1901 | 4,179 |
| 1887 | 2,423 | 1902 | 4,149 |
| 1888 | 2,860 | 1903 | 4,225 |
| 1889 | 2,318 | 1904 | 5,272 |
| 1890 | 2,088 | 1905 | 6,058 |
| 1891 | 2,271 | 1906 | 6,254 |
| 1892 | 2,604 | 1907 | 5,838 |
| 1893 | 2,383 | 1908 | 5,739 |
| 1894 | 2,357 | 1909 | 5,423 |
| 1895 | 2,817 | 1910 | 5,711 |
| 1896 | 2,995 | 1911 | 5,244 |
| 1897 | 3,143 | 1912 | 4,567 |
| 1898 | 2,595 | 1913 | 4,020 |
| 1899 | 3,036 | | |

Source:

Egypt, Ministry of Finance, Department of  Statistics,  _Annuaire Statistique_, 1914, 300.

---

TABLE 3.14
CENSUS OF BUFFALOS AND CATTLE, ALL EGYPT,
1903-1914

| Year | Cattle | Buffalos | Total |
|------|--------|----------|-------|
| 1903 | 859,669 | 718,023 | 1,677,692 |
| 1904 | 605,022 | 645,796 | 1,250,818 |
| 1905 | 655,156 | 708,233 | 1,363,389 |
| 1906 | 732,537 | 775,149 | 1,507,686 |
| 1907 | 778,896 | 761,486 | 1,540,382 |
| 1908 | 737,732 | 750,548 | 1,488,280 |
| 1909 | 725,116 | 728,284 | 1,553,400 |
| 1910 | 762,091 | 765,392 | 1,527,483 |
| 1911 | 656,166 | 657,406 | 1,313,572 |
| 1912 | 619,540 | 652,186 | 1,271,626 |
| 1913 | 637,098 | 632,725 | 1,269,823 |
| 1914 | 601,136 | 568,388 | 1,169,524 |

Source:
   H.A. Ballou, "Notes on the Cotton Crop," Agricultural Journal of
   Egypt, 9, 42.

To resolve the puzzle of the peasant rotation shift, we must focus on another feature of peasant behavior that divided them sharply from the pashas. The peasants faced objective reasons for maximizing shorter-run goals than those pursued by larger proprietors. This myopia was the direct result of the social and economic inequality which placed peasant families near the border of subsistence. Even if they were aware of the link between the two-year crop rotation and the subsequent gradual declines in yields, peasants were compelled to opt for the higher short-run revenues offered by the two-year rotation. The very poorest peasants had little choice but to adopt the two-year rotation. Once a family's land holding was below the minimum size needed for subsistence,[156] they had to rely on the market for food. In that case, they had every reason to switch to the two-year system, which would generate more cash and therefore more food. Population growth, the highly unequal land distribution which emerged over the nineteenth century, and Muslim inheritance law interacted to force the smallest peasants to choose the two-year rotation.

This same social and economic inequality forced even small peasants who held enough land for their subsistence to choose the two-year rotation with its higher short-run revenues. A wealthy pasha could obtain the mortgages or other credit that farsightedness often requires by borrowing from banks at relatively low interest rates.[157] Peasants, on the other hand, had reason to shun the banks. First, they would have been charged more due to the higher transaction costs to the banks of small loans. Second, the peasants feared the banks: government-sponsored loans were collected by tax collectors, and the peasants had not forgotten their suffering at the hands of such men during earlier decades.[158] Consequently, the peasants borrowed from village moneylenders, at rates ranging from 25 to 50% per year.[159] Contemporary sources agree that this was the principal source of credit for the peasants.[160] A government inquiry in 1913 provided further evidence of the extent of peasant debt. Since local creditors no doubt encouraged evasion of this inquiry, and probably found a receptive ear in the peasants who feared and shunned government officials of all types, the estimates are presumably below the actual extent of indebtedness. The commission found that over 50% of the small peasant land area in the Delta was owned by indebted peasants (see Table 3.15). Peasants were thus under strong short-run pressures to get out of debt as quickly as possible, or to avoid going into debt at all. Such pressures forced them to gamble on the short-run strategy of the two-year rotation, a strategy rejected by the better-connected pashas until well after World War I.

Even this sharp difference in the interest rates confronting the two classes does not fully capture the true cost for peasants of foregoing higher likely revenues now in order to maintain soil quality. Indebtedness for peasants meant not only high interest rates but, more seriously, the threat of land loss through foreclosure. Such a loss would have threatened the peasants' subsistence goal. As we shall see, real wages in agriculture were probably declining, and a landless peasant would have depended entirely on the uncertainties of the labor market for his food, whether hired as a tamaliyya or a tarahil laborer. A "food first" rule, therefore, would place avoidance of land loss before all other goals. Insofar as failure to repay debts would result in land loss, indebted peasants had an important incentive to switch to the more "present cash generating" two-year rotation.

However, the problem is more complicated. Banks did foreclose promptly and confiscate land for failure to repay debts. But this bolstered peasant reluctance to take loans from the banks.[161] Further, many banks simply refused to make small loans, rather than charge high interest rates to cover the high transactions costs of such loans. Until 1899 the minimum loan of the country's principal bank, the Credit Foncier, was L.E. 300; in that year it was reduced to L.E. 100, still well beyond the reach or needs of most small peasants.[162] The government did establish an Agricultural Credit Bank in 1902 precisely to deal with the problem of peasant credit; from 1902-08 it lent over L.E. 10,000,000 in small sums. However, in response to a rise in arrears, the government began to restrict the operations of this bank. It was further restricted by the promulgation of the "Five Feddan Law" in 1912, according to which no landowner with five feddans or less could have his land expropriated for failure to repay debts. This, of course, effectively closed the Bank.

TABLE 3.15
PEASANT DEBT, 1913

| Province | Percentage of 0-5 Feddan Holders in Debt | Percentage of 0-5 Feddan Holdings' Land Area Owned by Debtors | Percentage of Recorded Debt to Non-Banks (Moneylenders) |
|---|---|---|---|
| Asyut | 25.48 | 32.17 | 88.70 |
| Aswan | 16.60 | 24.69 | 70.33 |
| Behera* | 43.13 | 53.19 | 65.62 |
| Beni Suaif | 26.89 | 35.00 | 82.09 |
| Sharqiyya* | 44.92 | 54.95 | 68.64 |
| Daqahliyya* | 46.77 | 61.39 | 72.22 |
| Fayyum | 28.71 | 41.99 | 85.49 |
| Gharbiyya* | 42.79 | 56.49 | 77.72 |
| Girga | 34.29 | 40.77 | 90.64 |
| Giza | 26.14 | 31.92 | 93.12 |
| Qalyubiyya* | 39.32 | 47.17 | 76.55 |
| Qena | 32.37 | 38.43 | 69.76 |
| Minufiyya* | 41.80 | 55.48 | 72.47 |
| Minya | 26.82 | 29.87 | 84.03 |
| Total: | 37.13 | 46.80 | 75.81 |

Source:
 Egypt, Ministry of Finance, Department of Statistics, Annuaire Statistique, 1914, 509.

* Delta Provinces

---

In contrast to the banks, the primarily Greek moneylenders seem to have taken a larger share of the harvest rather than simply fore-close immediately.[163] Many peasants found it easy to become indebted to these moneylenders and difficult to get out of debt.[164] In addition to the usual reasons for this, such as local monopoly power, peasant illiteracy, and the like, Egyptian moneylenders had another advantage: as Greeks, they were classified as Europeans, which meant that any litigation against them would be judged in the Mixed Courts by Europeans. This made it even more difficult, costly, and, hence, unlikely, that a peasant could take any legal action against fraudulent practices by the moneylenders.

The structure of the credit market outlined above thus provides an additional clue to small-peasant choice of the land-depleting two-year rotation. Since the Greeks often did not take over the land, they were much less concerned with its long-run fertility than an owner would have been. Since it was not their asset, they were not terribly concerned with its depreciation. Peasants who were indebted to moneylenders would have been concerned to try to get as much revenue as possible now in order to pay off their debts, since there was always the fear that the moneylenders would foreclose eventually. The moneylenders encouraged this, as well as any other device to increase their immediate cash flow (i.e., sale of working capital, such as animals).[165] Consequently, those peasants who were in debt would have had a strong incentive to plant cotton as often as possible; that is, to switch to the two-year rotation. And those small holders who were not in debt had almost as much to fear from the prospect of falling into debts that cost 25-50% in interest.

Indebtedness partly explains peasant choice of technique for two reasons: (1) peasants greatly feared land loss because it would jeopardize their subsistence, and (2) moneylenders preferred more cotton now, rather than land seizure, as a means of debt repayment. It should be noted that moneylenders' preferences were also the product of history and rural social relations. It was dangerous for the moneylenders to foreclose immediately. We have seen that brigandage and arson were rampant in the 1890s and 1900s. The Greeks had been singled out and massacred by the mobs during the riots in Alexandria and in a number of Delta towns in 1882. This occurred on a larger scale during the revolution of 1919. Russell noted that Greek nazirs on country estates "have all been either shot or frightened away."[166] The very strong peasant desire for land and the high level of rural violence, plus the ability of moneylenders to retire to Alexandria with their funds, help to explain their preferences.

The plausibility of this two-part explanation for the peasant choice of technique is enhanced when we look at another aspect of the choice within the lexicographic framework. The two-year rotation increased peasants' risks. In the first place, of course, the two-year rotation risked the depreciation of the land's productivity. This, along with the reduced protein supply, should be considered the principal risk of the rotation shift. Further, the shift from a three- to a two-year rotation would mean that a larger percentage of a peasant's income was derived from the market, thus subjecting him to the risks of price fluctuations. The trend of cotton prices was upward; however, if one takes the standard deviation of prices as a percent of the mean as a risk indicator, then cotton was the second riskiest crop (see Table 3.16). Of course, this may have been below some "acceptance level"; we have no way of determining this. But the price data do suggest that the switch to the two-year rotation would have increased the risks for the peasants, since a larger percentage of their land was being planted in a relatively risky crop.

In summary, peasants chose a crop rotation which reduced their protein supply, lowered the productivity of their land, and increased the area allocated to a relatively risky crop. They did this although they faced, apparently, roughly the same price of labor as did the pashas, who retained the three-year rotation. The principal difference between the two groups, apart from their access to information, was in their position in the credit market. Peasants were in

TABLE 3.16
RISK INDICATOR: STANDARD DEVIATION OF PRICES
AS PERCENT OF THE MEAN, SIX AGRICULTURAL PRODUCTS,
1895-1913

| Crop | Standard Deviation as Percent of Mean |
|------|---------------------------------------|
| Cotton | 28.353% |
| Wheat | 14.831 |
| Beans | 21.226 |
| Maize | 20.566 |
| Barley | 36.230 |
| Rice | 0.839 |

Source:
s/p: Calculated from Egypt, Ministry of Finance, Department of
Statistics, Annuaire Statistique, 1914, 387.

debt to moneylenders and to banks. They chose the two-year rotation
to avoid losing their land, either immediately in the case of banks,
or more gradually in the case of moneylenders.

There is considerably less evidence on rich peasants than on
small ones for this period. Two forces seem to have led rich
peasants to choose the two-year rotation: (1) information, and (2)
debt. Relative input prices should have been roughly the same for
rich peasants and for pashas. Rich peasants lived in the countryside, and consequently had some of the same problems as did small
peasants in getting sound agronomical information. However, their
family members would often live in the cities,[167] and they themselves were close to government information channels, if not indeed
part of the network itself. On the other hand, they would have
needed working capital to exploit their estates. 'Amr noted that
rich peasants depended on outside financing for their agricultural
operations.[168] The mortgage debt of all farmers rose from seven to
sixty million pounds sterling from 1883 to 1914.[169] Most of this
borrowing, however, was by pashas; in 1894, more than 70% of mortgage
loans were secured on properties of over fifty feddans.[170] The
remainder must have been by owners of medium properties, since small
peasants did not usually use banks. We have seen that the ceiling on
Credit Foncier loans was lowered to L.E. 100 in 1899, which would
have brought such loans within reach of holders of between twenty and
fifty feddans. The rich peasants would also have been unaffected by
the "Five Feddan Law." A further indication of indebtedness among
middle proprietors is that one of the demands of the Nationalists in
1919 was the abolition of debt for all peasants owning less than
thirty feddans.[171] Finally the urban holders of medium properties

who rented their land for cash and then did not closely supervise their tenants probably contributed to the switch to the two-year rotation.

The scanty evidence prohibits any firm conclusions here. We simply note that many medium landowners switched to the two-year system, that some proprietors holding less than fifty feddans had taken out mortgage loans, that those owning less than thirty feddans appear to have been in debt, and that as country residents rich peasants may have lacked high-quality agricultural advice. This same country residence makes it unlikely that the rich peasants were not specifying the crop rotation to be used on their lands, but this may well not have been the case for urban middle proprietors for whom supervision costs were higher. Such cash-rental without close supervision presumably also contributed to the spread of the two-year rotation.

## THE DISTRIBUTIONAL CONSEQUENCES OF TECHNICAL CHANGE

We now turn to the distributional consequences of the technical changes and the resulting short-term growth in agricultural production. In order to determine how the various classes fared during this period of growth, it will be necessary to make some simplifying assumptions and to construct a model that will allow the use of all of the scraps of data that we have. The focus here is on small

peasants and pashas. Define $y_i = p \dfrac{Q_i^c}{T_i^c} \dfrac{T_i^c}{T_i^*} \dfrac{T_i^*}{N_i}$

where $y_i$ = per capita gross cotton revenue;

$Q_i^c$ = quantity of cotton;

$T_i^c$ = land area planted in cotton;

$T_i^*$ = total cultivated area;

$N_i$ = population;

$p$ = price of cotton (assumed to be the same for both peasants and pashas);

and $i = p$, $f$, where $p$ = "pashas", $f$ = "fellahin" or peasants. $Q^c/T^c$ is cotton yields, $T^*/T$ is the percentage of land in cotton, a measure of the rotation system, and $T/N$ is average land ownership per capita, each for the respective class.

We wish to examine what happened in this period to the ratio of the gross cotton revenues of the pashas ($Y_p$) to the gross cotton revenues of the peasants ($Y_d$), a proxy measure for the inequality of cotton incomes for the two classes. Since the revenues for each class are identical to the products shown in the expression above, it follows that the change in the inequality of cotton revenues is proxied by the movements in the terms on the right, dividing their values for pashas by their values for peasants. Thus the change in inequality of cotton revenues is equal to the percentage change in yields for the pashas plus the percentage change in the percent of land planted in cotton plus the percentage change in per capita land holdings, all for the pashas, minus the percentage change in yields for the peasants, the percentage change in the percent of land planted in cotton, and the percentage change in land held per person, all for the peasants. Now let us see what scraps of information we can plug into this model.

First, let us look at yields. We would expect that the British technical changes would have raised the peasants' yields more than those of the pashas.[172] This is because the pashas had monopolized the best lands for cotton cultivation along the banks of the canals during the pre-British period.[173] The peasants' yields under the balli method suffered from a lack of water.[174] Since one of the principal effects of the Delta Barrage was to increase water supply, the peasants presumably benefitted more from this technical change than did the pashas. Yields under the balli method used by the peasants were reportedly between 2 and 3 cantars per feddan in 1877.[175] There are a variety of estimates for yields on the pashas' lands. McCoan gives between 4 and 6, with occasionally 8 on the best land. On the other hand, on Isma'il's "carefully managed" estate, 5.75 are reported for 1871.[176] By 1913, with the national average at 4.4, Schanz reports yields of 4 cantars per feddan for the small peasants and 5 for "plantations," i.e., large estates.[177] On balance, it would appear that peasant yields rose between 33 and 100%, while the pashas' yields advanced little, if at all.

We have figures on per capita land ownership in Annuaire Statistique, the government's official published statistics. The average size of holdings in the 0–5 feddan class fell by 33 percent, from 1.5 in the 1890s to 1.0 feddans in 1913. The pasha class, which I am defining as those holding 50 feddans or more, showed little change. Any tendency to fragmentation through inheritance seems to have been offset by purchases of State Lands which were auctioned off during this period. In 1900 land per pasha was 188 feddans; in 1913, 192 feddans. If the 1900 figure is taken as representative of the earlier period, then there was a two percent rise in land per pasha.

There are no figures on the percentage change in the percent of land planted in cotton. If we assume that the pashas switched from a three- to a two-year rotation, then the percentage change is approximately 50 percent. If we assume that the pashas retained the three-year system, then there is no change in this figure for them. As noted above, contemporary evidence suggests that the latter is the more plausible assumption.

If we combine these various crude estimates, we can get a rough indication of the change in the distribution of revenue between the two classes. The most favorable estimate for the peasants would be to assume that their yields rose 100%; then we have 0 + 0 + 2 − 100 − 50 + 33 = −115, or a 115% improvement in the relative position of the peasants. If we assume that the peasants' yields rose by 33%, then we have 0 + 0 + 2 − 33 − 50 + 33 = −48, still a gain in the peasants' favor. Notice that this calculation is not very sensitive to

assumptions on pasha crop rotation. Taking the least favorable estimate for the peasants, we find that they gained by some 48%. Note
that this would be cancelled only if all of the pashas had switched
to the two-year system. Even assuming that all pashas who rented out
their lands allowed them to be placed under the two-year system (and
we have evidence to the contrary), such pashas would have to have
greatly outnumbered the more closely supervised 'ezbah estates to
arrive at such a result. We have seen that this is also an implausible assumption. It seems fair to conclude that the peasants' revenues rose more quickly than those of the pashas from 1890-1914.

This picture of changes in the class distribution of cotton
revenue requires some qualification. It seems likely that the gap
between pashas and peasants narrowed sharply from 1890 to 1900, but
widened thereafter. However, by 1914 the gains in equality of the
earlier period (1890s) had not yet been entirely reversed: hence the
(apparent) long-run gains in equality. The plausibility of this view
can best be seen if we scrutinize the changes in each of the three
variables of the model more closely. First, roughly two-thirds of
the peasants' shift in rotation (favorable to equality in the above
model) occurred before 1900. Second, peasant yields almost certainly
rose and fell with average cotton yields, whereas the pashas' yields
were probably more nearly stable. We have already seen that the
dramatic rise in the water supply was the main source of the gains in
peasants' yields, that the additional water favored them more than
the pashas, and that the main source of the increased water supply
was the British-built Delta Barrage. Since this was completed in
1891, it seems highly likely that peasant yields rose more rapidly
than those of the pashas before, say, 1900 when the adverse ecological effects of intensification appeared.

After that point, peasant yields probably fell more rapidly than
those of the pashas. This is because those who chose the two-year
rotation caused their own soil to deteriorate more rapidly than the
soil of their neighbors who retained the three-year system. The
clayey soil structure of much of the Delta inhibited the lateral flow
of water, which implied that the water table would have been higher
on lands using the two-year rotation than other lands.[178] Furthermore, peasants using the two-year rotation forfeited the desirable
physical, chemical, and biological features of a longer fallow,
whereas the pashas did not. Insect attacks were an externality,
affecting both pashas and peasants, but it seems highly likely that a
more rapid fall in peasant yields lies behind the observed fall in
average yields from the turn of the century. Third, the numbers on
the changes in land per person (the only numbers in the preceding
discussion which indicated widening inequality) are from the 1900-
1914 period. No earlier figures are available. But there is nothing
in either the contemporary sources or the secondary literature which
would indicate that peasant lands fragmented more rapidly from 1890
to 1900 than they did from 1900 to 1914. Nor does it appear likely
that land per pasha grew more rapidly from 1890 to 1900 than from
1900 to 1914. The available evidence, then, supports the contention
that there was a decline in the inequality of cotton revenues among
landowners from 1890 to 1900, and then a reversal of this trend from

1900 to 1914.[179]

What evidence we have suggests that the position of the landless class deteriorated relatively and possibly absolutely during the cotton boom. Many peasants were in this class; by 1907, at least 36% of the rural population was landless, compared with approximately 33% in 1880.[180] Information on wage rates for this period is, not surprisingly, highly fragmentary, questionable, and contradictory. I have assembled these numbers in Table 3.17. There is little to contradict the hypothesis that money wages were either constant or rose only slightly during the era. If this was so, real wages probably fell. For landless agricultural workers, the most relevant deflator is certainly the price of the principal foodstuff, maize. The trend of this price was upwards from 1895 to 1909, showing a downturn in that year (see Table 3.17). The table also shows the "real wage," or the ratio of wage rates to maize prices for the different years and estimates of wages that are available. Real wages appear to have fallen, or at best, to have remained more or less constant, although there are considerable variations from year to year. This general picture is confirmed by the published statistics on the average wages of unskilled urban construction workers. Although rural wages were reported to be some 60-100% below those in the cities,[181] we might expect the trend of wages for both urban and rural unskilled labor to be roughly the same.

The relative stagnation of wages may seem surprising, given that the combination of the extension of the cultivated area and the rotation shift would have raised the demand for labor. However, to the extent that the rotation shift was adopted by the small fellahin landowners, this technical change would not have raised the market demand for labor, since, as we have seen, the additional labor required by the new process would have been supplied by the peasant family. However, rich peasants who adopted the two-year system would have hired labor in some form. On the other hand, the supply of labor in the Delta was increasing. The population grew from about seven million in 1882 to over eleven million in 1907. As we have seen, the land-to-population ratio for small holders was decreasing in this period; consequently, the number of "landholders" at the bottom who had to supplement their income from the land by income from wages must have been increasing. Further, large numbers of Upper Egyptians migrated to the Delta on a seasonal basis. As argued above, it is highly likely that these tarahil workers put downward pressure on all wages in the Delta. These conditions meant that the supply of wage labor was increasing at least as rapidly as the demand, leading to a stable or declining money and real wage rate.

This brings us to the question of regional disparities which these technical changes and growth may have engendered. Landholders in Upper Egypt still using the basin system cannot have benefitted from the growth of cotton production. The crop value of their land was lower than that of land in the Delta: In 1906, the average gross value of Delta crops was L.E. 15.5 per feddan and L.E. 7.7 per feddan in Upper Egypt.[182] Since the revenue per unit of land was lower than in the Delta, and since the population-to-land ratio was higher,

TABLE 3.17
WAGES, RENT, MAIZE PRICES, REAL WAGES,
AND RENTAL-WAGE RATIOS,
1870-1914

|        | (1)Rent (LE/fdn/ year) | (2)Wage Rates (LE/day) | (3)Rent/ Wage = (1)/(2) | (4)Price of Maize (LE/ardeb) | Real Wage = (2)/(4) (ardeb/day) |
|--------|------------------------|------------------------|-------------------------|------------------------------|---------------------------------|
| 1870s  |                        | .02-.03                |                         |                              |                                 |
| 1877   |                        | .03                    |                         |                              |                                 |
| 1888   | 1.40                   | .04                    |                         |                              |                                 |
| 1889   |                        |                        |                         |                              |                                 |
| 1890   | 1.05;1.5               |                        | 26.25-42.9              |                              |                                 |
| 1891   |                        | .035                   |                         |                              |                                 |
| 1892   |                        |                        |                         |                              |                                 |
| 1893   |                        |                        |                         |                              |                                 |
| 1894   |                        |                        |                         |                              |                                 |
| 1895   | 3.6                    |                        |                         | .463                         |                                 |
| 1896   |                        |                        |                         | .524                         |                                 |
| 1897   |                        |                        |                         | .561                         | .03-.05                         |
| 1898   |                        |                        |                         | .708                         |                                 |
| 1899   |                        | .02-.03                |                         | .651                         |                                 |
| 1900   |                        | .03-.045               | 80.0-113.3              | .712                         | .042-.0632                      |
| 1901   | 4.0                    | .03; .03-.05           |                         | .705                         | .046-.07                        |
| 1902   | 4.0                    |                        |                         | .623                         |                                 |
| 1903   | 6.0-6.7                |                        |                         | .642                         |                                 |
| 1904   |                        |                        |                         | .645                         |                                 |
| 1905   |                        |                        |                         | .753                         |                                 |
| 1906   | 8.0-10.0               | .025                   | 320-480                 | .877                         | .0289                           |
| 1907   | 8.0-11.0               | .025                   | 320-440                 | .800                         | .031                            |
| 1908   |                        |                        |                         | .890                         |                                 |
| 1909   |                        | .03                    |                         | 1.038                        | .03                             |
| 1910   | 10.0                   | .035-.045              | 250-286                 | .744                         | .047-.0538                      |
| 1911   |                        |                        |                         | .800                         |                                 |
| 1912   | 8.0-10.0               | .047                   | 178-400                 | .580                         | .081                            |
| 1913   | 12.0-18.0              | .03-.045               | 267-720                 | 1.014                        | .0247-.0444                     |
| 1914   |                        | .025-.03               |                         |                              |                                 |

Sources:
    Rents: 1888: Gali, 131; 1890: Nahas, 104; Schanz, 44-45; 1895:
Hamed el-Sayyid Azmi, "A Study of Agricultural Revenue in
Egypt," L'Egypte Contemporaine, 1934, 714; 1901: Nahas, 143;
1902: Owen, 242; 1903: Report...on the Finances..., 1903, 71-
73; 1906: Despatch from the Earl of Cromer Respecting the Water
Supply of Egypt, 1907, 4; 1907: Reports on the Administration of
the Irrigation Service in Egypt for the Year 1907, 23-24; 1910:
Raoul de Chamberet, Enquête sur la Condition du Fellah Egyptien,

(1909), 33; 1912: Owen, 242 and Schanz, 44—45.

<u>Wages</u>: 1870s: Owen, 266; 1877: McCoan, 178; 1891: R. Wallace, "Opening Address on Egyptian Agriculture," University of Edinburgh, Agriculture Dept. Pamphlet, (1891), 20; 1889: Willcocks, (for irrigation work), 256; 1899: Owen, 266; 1901: Nahas, 133; <u>Ministry</u> <u>of</u> <u>Justice</u>, <u>Report</u>...<u>1901</u>, 84; 1906: Strakosch, <u>Erwachende</u> <u>Agrarlaender</u>: <u>Nationallandwirtscharft</u> <u>in</u> <u>Aegypten</u>... (1910), 71; 1912: Abdel Wahhab Pasha; Schanz, 45; 1914: Owen, 273—4, A Lambert, "Les salariés dans l'entreprise agricole égyptienne," <u>L'Egypte</u> <u>Contemporaine</u>, March, 1943.

Price of Maize: <u>Annuaire</u> <u>Statistique</u>, 1914, 387.

---

### TABLE 3.18
### NOMINAL AND REAL WAGES
### FOR UNSKILLED URBAN CONSTRUCTION WORKERS,
### ALL EGYPT AVERAGES,
### 1903, 1908, 1913

| Year | Wage(1) (LE/ day) | Wage(2) (LE/ day) | Pmaize (LE/ ardeb) | W(1)/Pm (ardeb/ day) | W(2)/Pm |
|------|---------|---------|--------|---------|---------|
| 1903 | .045 | .045 | .642 | .07 | .07 |
| 1908 | .05 | .05 | .890 | .06 | .07 |
| 1913 | .055 | .05 | 1.014 | .05 | .05 |

Wage (1): "manoevres"
Wage (2): "terrassement"

(The calculation is not sensitive to the choice of year for maize prices.
Using average prices for 1900—1904 (.64 E/ardeb),
1905—1909 (.88), and 1910—1913 (.85) as deflators gives
real wages for the three years of .07, .06, .06, respectively,
for both types of labor.)

Source:
<u>Annuaire</u> <u>Statistique</u>, 1914. Wages: 376—379; Maize Prices, 387.

---

the revenue per person from land was unambiguously lower in this region. The Sa'idis did get the agricultural wages in the Delta, but since these showed no increase during the boom in cotton exports, we may conclude that, on balance, the Sa'idis did not share in the increase in income generated by the process of technical change and the concomitant growth in cotton production.

Finally, however the monetary gains of technical changes may have been spread among the population, these same technical changes also spread endemic disease. Malaria increased due to the increased amount of water standing in the canals. Bilharzia (Schistosoma Haematobium and schistosoma mansoni) spread throughout the agricultural population due to the extension of perennial irrigation.

> It is clear that the increase in bilharziasis has been produced by the substitution of basin irrigation by perennial irrigation. The mollusks, which were unable to withstand conditions created by the system of innundation, have found in the present system an ideal breeding place.[183]

and

> The effect of introducing perennial supply for irrigation was to increase the incidence of bilharziasis (in an area of Egypt) from five to forty-five percent of the population, and in some areas even to seventy-five percent and at the same time to cause an increase of malaria of equally devastating proportions.[184]

Estimates vary, but perhaps fifty to seventy-five percent of the rural population came to suffer from the disease, which reduced labor productivity some twenty-five to fifty percent.[185] The cure is long, protracted and expensive. Some forty percent of the population was found to have hookworm (anklyostoma), whose spread may also be attributed to the conditions created by perennial irrigation.

> Komombo Estate in Upper Egypt, a desert in 1900, now extends to 40,000 acres of irrigated land and supports some 40,000 people. A recent medical survey indicated that some 84 percent of these people are infested with Bilharzia and 24 percent with anklyostoma.[186]

These diseases plague the fellahin even today.

CONCLUSION AND SUMMARY

In summary, we have seen how the British administration consolidated private property rights in land and labor, and then have examined the systems regulating their combination. I have argued that the 'ezbah system was the dominant mode of exploitation on the large estates, coupled with some cash renting to intermediaries or directly to small peasants. Whatever the form of the labor contract, most landlords carefully supervised their workers by using agents. This fact, their vastly greater resources, and their superior information enabled most of them to retain the three-year rotation. We have also seen why the small peasants chose the two-year rotation: they had been forced near to the margin of subsistence, and, like many of their landless brethren employed on the 'izab, had been forced into debt. Middle proprietors were largely of two sorts: rich peasants, who exploited their lands with the 'ezbah system or with sharecropping, and members of the urban middle class, who largely relied on cash renting. Debt, lack of information, and, in the second case, the form of land tenure, led many of these landowners to switch to the two-year system. The bi-modal land tenure system may be seen as partly responsible for the rotation shift and the resulting agro-ecological problems.

We have also seen how these technical changes affected income distribution. The general picture at which I have arrived here is substantially in line with Owen's conclusions.[187] The sharpest break is between those owning and those not owning land. Those who had been dispossessed by the social changes and upheavals of the pre-British era and those who lost their land to the banks were the losers. Since earlier social changes had left many peasants near the border of subsistence, and since such events had made possible the rise of the moneylenders in the village, they contributed to the peasant adoption of the two-year rotation. Therefore, they had consequences for the distribution of income among landowners, and for the future productivity of the land itself. The general trend among landowners was sharply toward equality in the 1890s, then slow reversal from about 1900 to 1914. Even for the 1890s, many of the beneficiaries were moneylenders, not small peasants. Insofar as rich peasants avoided debt, they probably increased their incomes substantially in the 1890's. However, the next decade must have been harder, since their yields probably declined along with those of other users of the two-year rotation. The usual result of capitalist growth in agriculture—polarization—appears to have occurred in Egypt at the turn of the century, with the important exception of the relative improvement of the position of those small peasants unencumbered by debt in the 1890s. These gains, however, were partially reversed in the following decade.

One can see similarities between the Egyptian case and the present day "Green Revolution" in the Third World. In both cases, technical change in agriculture led to considerable increases in short-run output. At the same time, in both cases, the introduction of the new technology which made the growth possible engendered disparities between regions and between classes. The regional disparities arise from the same reasons in both cases; the location-specific character of the new technology. The class disparities are also similar. The differences in access to information and to credit are prominent themes in the literature on the "Green Revolution." As in Egypt, these differences are the product of the highly unequal distribution of resources among the population. The somewhat more favorable outcome for the small landowners in Egypt as opposed to modern India or Pakistan can be explained by two facts: (1) British irrigation changes helped the small peasants more than the pashas, who already monopolized the best cotton lands by 1882, and (2) the peasants, not the pashas, chose the new crop rotation. This is, essentially, the reverse of the Green Revolution story, where landlords do the innovating and irrigation changes favor the wealthy. Yet even here there are some similarities: one might say that the absence of drainage changes hit the small landowners harder than the pashas, a sort of "dual" of the Green Revolution irrigation change. Finally, just as the initial distributional gains of landless laborers in the Punjab were later swamped by further technical change, i.e., mechanization, technical change in Egypt from 1920 to 1940 favored the pashas at the expense of the fellahin, continuing the trend from 1900 to 1914. Meanwhile, agricultural output per person remained lower than at the turn of the century. The nature of these technical changes and the reasons for their adoption are the subject of the next chapter.

## Footnotes

[1]    R.S. Tignor, (1966), Chs. 1 and 2.

[2]    Ibid., Ch. 2.

[3]    Ibid.

[4]    Baer, (1962), 11-12; 'Amr, (1958), 87-88.

[5]    Ibid.

[6]    Karl Marx, Capital, Vol. 1, chap. XXVIII. (London, 1867). The legislation, although severe, was not nearly as vicious as the laws against the English poor of the sixteenth and seventeenth centuries.

[7]    Egypt, Minister of Justice, Report for the year 1901, 113.

[8]    Ibid., (1899).

[9]    See E.J. Hobsbawm, Primitive Rebels, (N.Y. 1965).

[10]   Jacques Berque, Egypt: Imperialism and Revolution, Jean Stewart, trans., (London, 1972), 131; Sir Thomas Russell, Egyptian Service, 1902-1946, (London, 1949), 79ff.

[11]   Ministry of Justice, Report, 1908, 8; Russell, (1949).

[12]   Ministry of Justice, Report, 1908, 8; Report, 1904, 16; cf. E.J. Hobsbawm and George Rude, Captain Swing, (London, 1968), on arson as the last resort of the oppressed in rural England in 1830.

[13]   Russell, (1949), 33.

[14]   Baer, (1962), 32. However, "the new classification was only completed in 1907 and introduced in 1912." Ibid.

[15]   Baer, "The Village Shaykh...." in (1969).

[16]   Cromer, (1907).

[17]   See below, Table 3.17.

[18]   Annuaire du Ministère de la Justice pour l'année 1908, 214-215.

[19]   Ministry of Justice, Report, 1901, 49.

[20]   Gali, (1889), 137ff; Saleh, (1922), 72-73; Lozach and Hug, (1930), 39-40.

[21]   Report by His Majesty's Agent and Consul General on the Finances, Administration, and Condition of Egypt and the Soudan, 1903. Inclosure No. 1, "Statements Showing the Condition of the Fellahin," 72.

[22]   M. Schanz, Cotton in Egypt and the Anglo-Egyptian Sudan, (Manchester, 1913), 45; William Balls, Egypt of the Egyptians, (London, 1915), 183-184.

[23]    Ministry of Justice, Report, 1900, 48.

[24]    Baer, (1962), 224-225.

[25]    See Ibid., 44.

[26]    Ibid., 22.

[27]    Tignor, (1966), 243-244.

[28]    Ibid.

[29]    Owen, (1969), 239-240; Baer, (1962), 224-225.

[30]    See Owen, (1969), 273-275 on the management of the  Manzalawi
        estate  in Gharbiyya; Lecarpentier, L'Egypte Moderne, (Paris,
        1914), 35-26.

[31]    Schanz, (1913), 45-46; Balls, (1915), 183-184.

[32]    Owen, (1969), 244.

[33]    Gali, 136-137.

[34]    Nahas, (1901), 141; Willcocks, (1913), 256.

[35]    Agricultural Census of Egypt, 1929, 86; Agricultural  Census
        of Egypt, 1939, 12-13.

[36]    Ibid.

[37]    The Population Census of Egypt, 1917, lx.

[38]    Ibid.

[39]    Ibid., 364.

[40]    Muhammad el-Zalaki, An Analysis of the Organization of  Egyp-
        tian  Agriculture  and Its Influence on National Economic and
        Social Institutions. Unpubl. Ph.D. thesis, University of
        California, Berkeley, 1941, 35; Agricultural Census of Egypt,
        1929, 62-63.

[41]    Bent Hansen and Girgis Marzouk, Development and Economic Pol-
        icy in the U.A.R. (Egypt), (Amsterdam, 1965), 61.

[42]    The percentage of renters may have been  even  smaller.   The
        numbers  calculated above would indicate a ratio of "cultiva-
        tors on own land" to "relatives working on  family  land"  of
        1.05.   In  1929  and  1939,  the  same ratio is .79 and .78,
        respectively.  It  seems  unlikely  that  this  ratio  should
        change  so  much from 1907 to 1929-1939.  Calculated from the
        Agricultural  Census  of  Egypt,  1929,  62-63;  Agricultural
        Census of Egypt, 1939, 222-223.

[43]    Annuaire Statistique, 1914, 320-321.

[44]    Report...1903, "Inclosure...," 71.

[45]    Gali, (1889), 136-137; Willcocks, (1913), 256.

[46]    The Report...1903, "Inclosure..." gives the cash  portion  of
        the wages of such a worker as 25 P.T. per month, or roughly 1
        P.T. per day, since it reports that the  workers  worked  for

the owner 25 days per month. Ordinary day laborers received about 3 P.T. per day at this time. See below, Table 3.17. The reasons for this wage difference are discussed below.

[47]    Ibid., 71.

[48]    Nahas, (1901), 141; Husein Ali el-Rifai, La Question Agraire en Egypte, (Paris, 1919), 124–125.

[49]    Lecarpentier, (1914); el-Rifai, (1919), 124–125; Nahas, (1901), 137; Lozach and Hug, (1930), 191.

[50]    Nahas, 137.

[51]    'Amr, (1958), 129. But see below for a discussion of the capitalist features of this system.

[52]    Nahas, (1901), 145; A. Lambert, "Les Salariés dans l'entreprise agricole égyptienne", L'Egypte Contemporaine, 33, (1943); Lefter Mboria, La population de l'Egypte, (Cairo, 1938), 149; el-Rifai, (1919), 127.

[53]    L'Egypte Indépendante, (Paris, 1937), 294; Saleh, (1922), 70–72.

[54]    Nahas, (1901), 145.

[55]    Annuaire Statistique, 1914.

[56]    Nahas, (1901), 145–147.

[57]    Owen, 267.

[58]    Husni Husayn, "'Ummal At-tarahil fi-l-ard al-jadidah" (Tarahil Laborers on New Land), At-ta'lia, Jan., 1971, 20.

[59]    Lambert, (1943); L'Egypte Indépéndante, (1937), 294; Gabriel Saab, The Egyptian Land Reform, (1967), 147; Mahmoud Abdel-Fadil, (1975), 46–48.

[60]    Nahas, op. cit., 169.

[61]    Samir, Saffa, "Exploitation économique et agricole d'un domaine rural égyptien," L'Egypte Contemporaine, 1949, 40, 410–411; A. Lambert, (1943); 'Amr, (1958), 124; Husayn, (19719, 21.

[62]    Nahas, (1901), 146.

[63]    Saffa (1948), 429; Nahas, (1901), 136.

[64]    Nahas, (1901), 136; H. Wells, "L'enseignement agricole," L'Egypte Contemporaine, 2, (1911), 358; el-Rifai, (1919), 127.

[65]    Lambert, "Divers Modes de faire-valoir des terres en Egypte", L'Egypte Indépendante, (1938), 288, 290.

[66]    Lambert, (1943).

[67]    Nahas, (1901), 141.

[68]    Schanz, (1913), 65.

[69]     Saffa, (1948), 408. Schanz, 65 ff., asserted that a boy
         could pick from thirty to fifty pounds of cotton per day in
         1914. At five cantars of 99 pounds each, roughly 495 pounds
         had to be picked. However, he notes that only about one-half
         of this was gathered in the first picking, since the bolls
         did not all ripen simultaneously. This implies a minimum
         demand for five children per feddan. The actual demand was
         almost certainly higher, because of the risk of storms,
         insect attacks, etc., in which case all of the picking had to
         be done at once.

[70]     Lambert, (1943).

[71]     See below, on growing seasons and the crop rotation practiced
         on large estates.

[72]     L'Egypte Indépendante, (1938), 293; Nahas, (1901), 141.

[73]     Anouar Abdel-Malek, Egypt: Military Society, Charles Law
         Markmann, trans., (N.Y., 1968), 401.

[74]     "Social articulation is obtained when the modern sector is
         oriented to the production of wage-goods; and hence, the
         capacity to consume of the economic system develops through
         rising real wages in relation to rising labor productivity.
         Under social disarticulation...realization is obtained either
         through the external sector...or through partial consumption
         of the return to capital and rents. The relationship between
         realization and rising wages is lost. The rate of capital
         accumulation is maximized by minimizing wages." Alain de Jan-
         vry and Carlos Garramon, "The Dynamics of Rural Poverty in
         Latin America", Journal of Peasant Studies, 4, 3 (April,
         1977), 206-207.

[75]     Karl Marx, (1867), Vol I, esp. I, VI, and X, Sec. 2; Vol.
         III, Parts I and II. See also Michio Morishima, Marx's
         Economics: A Dual Theory of Value and Growth, (cambridge,
         1973), chap. VII, esp. 85-86.

[76]     Lambert, (1943).

[77]     See 'Amr, (1958), 28, where he contrasts "capitalist labor
         monopoly" with "feudal land monopoly": the former is for the
         purpose of exploitation, of use (ghard al-istighlal), the
         latter for skimming off the produce (ghard tasarruf).

[78]     Gali, (1889), 134.

[79]     Ibid.

[80]     Report...1903, "Inclosure...," 71-72.

[81]     L'Egypte Indépendante, (1938), 290.

[82]     See above, fn. 30.

[83]     Willcocks, (1913), 256; Lecarpentier, (191), 35; Gali,
         (1889), 138; Saleh, (1922), 72.

[84]    Willcocks, (1913), 256-257.

[85]    M. Th. F. Van Vloten, "La motoculture et les avantages des
        grandes exploitations agricoles," L'Egypte Contemporaine, 1,
        (1910), 652.  G. Martin and I.G. Levy, "Le marché égyptien et
        l'utilité de la publication des mercuriales", L'Egypte Con-
        temporaine, 1, (1910), 448.

[86]    Lecarpentier, (1914), 35

[87]    Lambert, (1938); L'Egypte Indépendante, (1938), 289.

[88]    Gali, (1889), 138; Saleh, (1922), 72.

[89]    Lozach and Hug, (1930), 39-40.

[90]    Mokbel Guimei, Le crédit agricole et l'Egypte, (Paris, 1931),
        151-152.

[91]    Lozach and Hug, (1930), 111; Ayrout, (1962), 18.

[92]    Guimei, (1931), 151.

[93]    See Nahas, (1901), 141, on economies of scale in 'ezbah
        supervision.

[94]    See Report...1903, "Inclosure..." which reports that medium
        proprietors exploiting small parcels of land (6, 7, and 11
        acres) "associated themselves with a partner," who received
        in the first case 1/5 of the cotton and wheat and 1/4 of the
        maize, 72-73.

[95]    Saleh, (1922), 73.

[96]    Gali, (1889), 138.

[97]    Ibid.; Saleh, (1922), 72.

[98]    Saleh, (1922), 72-73.

[99]    J.B. Piot-bey, "Coup d'oeil sur l'état actuel  du bétail  en
        Egypte," L'Egypte Contemporaine, 2, (1911), 201. Lambert,
        (1938); See above on cattle poisoning. V. Stuart,(1883), 29,
        however, observed such a system in use in 1883.

[100]   Lambert, (1938).

[101]   Report...1903, "Inclosure...," 73 (Case #5).

[102]   Crouchley, (1938), 48. Muhammad 'Ali extended perennial
        irrigation by deepening the canals, rather than by raising
        the water-level.

[103]   Owen, (1969), 187.

[104]   Annuaire Statistique, 1914, map.

[105]   See below on overwatering and Table 3.9. on water inputs
        under perennial irrigation. Owen provides evidence that the
        peasants abandoned the balli method by the late 1890's,
        (1969), 257.

[106]   J. Zannis, Le Crédit agricole en Egypte, (Paris, 1937), 18.

[107]   J. Barois, Irrigation in Egypt, (Washington, D.C., 1889), 89;
        Gali, (1889), 270; J.I. Craig, "Notes on the Cotton Statis-
        tics of Egypt," L'Egypte Contemporaine, 2 (1911), 166-198.

[108]   J. Anhoury, "L'économie agricole de l'Egypte," L'Egypte Con-
        temporaine, 32 (1941), 514; Saffa, (1948), 326; Owen (1969),
        248.

[109]   Willcocks, op. cit., passim; V.M. Mosseri, "La fertilite de
        l'Egypte," L'Egypte Contemporaine, 17 (1926), 93-124; E.R.J.
        Owen, "Agricultural Production in Historical Perspective," in
        P. Vatikiotis, ed., Egypt Since the Revolution, (N.Y., 1968),
        56-57.

[110]   Crouchley, (1938), 157-160.

[111]   Ibid., 157.

[112]   Willcocks, (1913), 124.

[113]   Fallow lasted from March to November under the three-year
        crop rotation.  The two-year rotation shortened the fallow
        period to one month, May/June.  See V.M. Mosseri, (1926),
        116-124; Samir Saffa, (1948), 326; Ministry of Agriculture,
        L'Egypte Agricole (Cairo, 1936), 21.

[114]   The above description is based upon V.M. Mosseri, (1926) and
        Mosseri, "Revue sommaire des récents travaux sur le maintien
        et l'amélioration de la qualité des cotons égyptiens,"
        L'Egypte Contemporaine, 17, (1926), 399.

[115]   Egypt, Ministry of Agriculture, La Culture du Coton en
        Egypte, (Cairo, 1950), 95. See also Crouchley, (1938), 156
        and Willcocks, (1913), 38.

[116]   Owen, (1968), 52.

[117]   See Willcocks, (1913), 261-262; Schanz, (1913), 67; and the
        Agricultural Journal of Egypt, 1911, 31. Schanz attributed
        the problem with late planting to "injurious autumn fogs and
        rains"; but later it was found that attacks of the cotton
        boll worm were the real problem with late planting.  Sieg-
        fried Strakosch, Erwachende Agraerlaender: National-
        landwirtschaft in Aegypten und im Sudan unter englischen Ein-
        flusse, (Berlin, 1910), 93, however, asserts that some
        peasants practiced intercalary planting of birsim with maize.
        This, presumably, would have helped to get more cuts of bir-
        sim in before the cotton planting date.  Such a practice
        would not have avoided the problems of more plants for the
        insects to feed on and additional water, both favorable to
        the proliferation of the pests, however.

[118]   Schanz, (1913), 63.

[119]   Owen, (1969), 193-194.

[120]   See Elizabeth Whitcombe, Agrarian Conditions in Northern
        India, Vol. 1, (Berkeley, 1972), especially 8, 75-78, 81,
        93-94.

[121]   M.A. Hollings, ed., The Life of Colonel Scott-Moncrieff,
        (London, 1917), 175.

[122]   Robert Wallace, "Opening Address on Egyptian Agriculture,"
        University of Edinburgh Agriculture Department pamphlet,
        1891, 9.

[123]   G.P. Foaden and F. Fletcher, Textbook of Egyptian Agricul-
        ture, (Cairo, 1908), 156.

[124]   Report...1910, 30-31.

[125]   Ibid., 19-20.

[126]   Ibid., 31.

[127]   Ibid., 27.

[128]   See ibid., passim. "Though the necessity of guarding against
        damage by increased irrigation was fully realized by Sir
        Colin Scott Moncrieff, it was impossible, under the severe
        financial pressure of these early days, to obtain the funds
        required for improved drainage...The truth is that, from the
        first, improved drainage never kept pace with irrigation."
        John A. Todd "The Agricultural Drainage of the Egyptian
        Delta", Appendix I of the Official Report of the Visit of the
        Delegation of the International Federation of Master Cotton
        Spinners' and Manufacturers' Associations to Egypt, October-
        November, 1912.

[129]   Owen, (1968). It should be remembered that upon completion
        of the Aswan dam there was a fairly quick "switch," in which
        canals which had formerly served as drains no longer could do
        so.   It is not clear that the British really recognized this
        aspect of the problem until Kitchener's 1912 drainage plan.

[130]   Tignor, (1966), 215-217; The actual threat which the Mahdi
        posed to Egypt was much exaggerated. See P.M. Holt, The Mah-
        dist State in the Sudan, 1881-1898, (London, 1970), 182ff.

[131]   Ronald Robinson and John Gallagher with Alice Denny, Africa
        and the Victorians, (London, 1961), chaps. 8, 9.

[132]   See Tignor, (1966), 215; Owen, (1969), 315.

[133]   Justin C. Ross, introduction to Willcocks, (1913), xiv-xv;
        Arno Schmidt, Cotton Growing in Egypt, (Manchester, 1912),
        28-29.

[134]   Tignor, (1966), 240.

[135]   Earl of Cromer (1908), 466-467. Emphasis added.

[136]   Muhammad Saleh, (1922), 69-75; Schanz, (1913), 28-29; S. Mer-
        ton, "Distribution of Cotton Seed," Agricultural Journal of
        Egypt (1918), 35-48; Francois Charles-Roux, La production du
        coton en Egypte (Paris, 1908), 164; Strakosch, (1910), 93.

[137]   Schanz, (1913), 31.

[138]   Ibid.

[139]   Arno Schmidt, (1912), 41.

[140]   Soil science began in Russia in the 1870s: Glinka published his The Great Soil Groups in German in 1914.

[141]   See the circular reprinted in Schmidt, (1912), 57 and Report...1910, 26-27.

[142]   The following discussion focuses on the small peasant proprietor's choice of the two-year crop rotation. We will return to the rich peasants below.

[143]   Bent Hansen, "Employment and Wages in Rural Egypt," American Economic Review, 59, (June, 1969), 298-314.

[144]   Ayrout, (1962), 49; Schanz, (1913), 64-67.

[145]   See above, fn. 53.

[146]   Robert Mabro, "Employment and Wages in Dual Agriculture," Oxford Economic Papers, 23 (Nov., 1971), 401-417.

[147]   For other studies of peasant and small farmer behavior using a "safety-first" model, see Jim Roumasset, Rice and Risk: Decision Making Among Low Income Farmers, (Amsterdam, 1976); Richard Day and Inerjit Singh, "A Microeconomic Chronicle of the Green Revolution," Economic Development and Cultural Change, 23, 4, (July, 1975), 661-686; G. Wright and H. Kunreuther, "Cotton, Corn and Risk in the Nineteenth Century," Journal of Economic History, 35, 3 (Sept., 1975), 5.

[148]   P.M. Tottenham, The Irrigation Service, Its Organization and Administration, (Cairo, 1927), 19.

[149]   Schanz, (1913), 63.

[150]   Owen, (1969), 250-251; Annuaire Statistique, 1914, 448-450.

[151]   See Table 3.12. The State Domain figures do not seem to be very representative of yields for the whole country. Indeed, they are some 20-25% below the yields published in 1911-1913 in Monthly Agricultural Statistics: 4 ardebs vs. 4.5-5.5 ardebs. However, the point for my purposes here is the change over time. There is evidence that the State Domain yields were consistently below the national average as early as the 1880s. See Owen, (1969), 250-251. Therefore, the earlier national yields were probably also higher than those of the State Domains. It seems fair to conclude that bean production fell along with the area planted in beans.

[152]   F.M. Lappe, Diet for a Small Planet, (N.Y., 1971), 78, 90.

[153]   Annuaire Statistique, 1914, 30.

[154]   The number of cattle and buffaloes fell by some 500,000, or roughly by 30%, from 1903 to 1914, due to disease and cattle poisoning. Recall that cattle plague was endemic in Egypt until a vaccine was found in 1913. There was an especially acute outbreak in 1903-04, J.B. Piot-Bey, op. cit., L.

Jullien, "Chronique agricole de l'année 1913," L'Egypte Contemporaine, 4 (1914), 583-594.

[155]   See Wright and Kunreuther, (1975), for a discussion of peasant risk-aversion and the choice of cotton vs. food crops.

[156]   Roughly three feddans. Baer, (1962), 76.

[157]   M.A. Rifaat, The Monetary System of Egypt, (London, 1935) 39ff.

[158]   Abdel-Latif, La Loi des Cinq Feddans, (Cairo, 1913); L. Polier, "Notes a propos de la loi des cinq feddans," L'Egypte Contemporaine, 4 (1913).

[159]   J.F. Nahas, (1901), 115; Mokbel Guimei, Le crédit agricole et l'Egypte, (Paris, 1913), 160.

[160]   Abdel-Latif, (1913); Strakosch, (1910), 64.

[161]   Abdel-Latif, (1913); L. Polier, (1913).

[162]   Nahas, (1901), 114.

[163]   Abdel-Latif, (1913).

[164]   Schanz, (1913).46.

[165]   R.A. Harari, "Banking and Financial Business in Egypt," L'Egypte Contemporaine, 27 (1936), 147.

[166]   Russell, (1949), 33.

[167]   L'Egypte Indépendante, (1938), 26.

[168]   'Amr, (1958), 120.

[169]   Issawi, (1961), 9.

[170]   Owen, (1969), 271.

[171]   'Amr, (1958), 129.

[172]   This argument applies to the 1890-1900 period. See below for qualification and discussion of the behavior of yields for the two classes during the first decade of the twentieth century.

[173]   Owen, (1969), 72-76.

[174]   Ibid.

[175]   J. McCoan, Egypt as It Is, (London, 1877), 184.

[176]   Owen, (1969), 129.

[177]   Schanz, (1913), 78-79.

[178]   Emile Catzeflies, "Le drainage des terres humides et salées du Delta égyptien," L'Egypte Contemporaine, 6 (1915-1916), 307-342.

[179]   Note that even from the 1890s the trend is toward equality in cotton revenues. For indebted peasants' lands, the 1890s

witnessed a reduction in the inequality between moneylenders and pashas. Further, since prices rose along with rents, the capital gains of the pashas were presumably important and significant: Since land per peasant was declining, but land per pasha stable or rising, the pashas would have done better here than would have the peasants.

[180]  Owen, (1969), 148. _Annuaire Statistique_, 1914, 34. The numbers here are for "ouvriers et domestiques de ferme." This assumes that _all_ of those listed as "cultivateurs et fermiers" owned land, which is unlikely. The actual proportion of landless in the rural population was probably considerably larger than 36%.

[181]  Ministry of Justice, _Report for the Year 1901_, 84.

[182]  Owen, (1969), 267.

[183]  R.T. Leiper, "La Production Experimentale des Bilharzioses Egyptiennes," _Bulletin de l'Institut Egytien_, (Cairo, 1917), quoted in Rivlin, (1961), 248.

[184]  E.B. Worthington, _Middle East Science_, (London, 1946), 73.

[185]  Mostafa Nagi, _Labor Force and Employment in Egypt_, (N.Y., 1971), 50.

[186]  Worthington, (1946), 153.

[187]  Owen, (1969), 268.

# 4 | The Nature of Technical Change in Egyptian Agriculture, 1920–1940

## INTRODUCTION

In this chapter I will examine the technical changes which occurred in Egyptian agriculture between the two World Wars. Recall that the switch to perennial irrigation and the adoption of the two-year crop rotation led to soil deterioration and the proliferation of pests. Both of these primarily affected cotton, or, at least, it was their effects on this crop which received the most attention from contemporary observers.[1] One can conceive of various responses to such problems. One would be a sort of "reswitching," in which the two-year rotation would be abandoned in favor of the older three-year system. Since after the land reform in the 1950s the government imposed the three-year rotation in the consolidated land reform areas (whereas before the reform the two-year rotation had been the rule),[2] we should at least entertain the possibility that such "reswitching" occurred in the interwar period. However, as we shall see, the available evidence does not support the notion that such a reversal of rotation systems occurred. Indeed, the evidence indicates the opposite: <u>further</u> intensification and <u>more</u> widespread adoption of the two-year system.

What technical changes <u>were</u> adopted, then? There were three principle ones: (1) improved drainage; (2) earlier sowing and closer spacing; (3) increased use of artificial fertilizer. In addition, some agricultural machinery, especially threshing and segregating machines, were imported. I shall argue that the first two changes were adopted as a response to the problems of soil deterioration and insect pests, respectively. Fertilizer use, on the other hand, presents a somewhat more complicated case. On the one hand, we shall see that there were price incentives for its adoption along the lines of Hayami and Ruttan's notion of "induced technical change." This would appear unrelated to previous production problems. Public policy played a role here: the Sidqi government placed a tariff on wheat imports, thereby bolstering the internal price. This, in turn, encouraged landowners to purchase fertilizer for use on their wheat fields. The spread of threshers and segregators in the 1930s seems to have had a similar motive. On the other hand, adoption of fertilizer does seem to be related to the previous ecological problems in two ways: first, use of nitrogen fertilizer led to earlier ripening of cotton, which in turn helped to avoid pest attacks. Second, the

111

marginal physical product of fertilizer used on cotton varied directly with the quality of the soil and therefore with the quality of drainage. There are reasons, then, for regarding the above three technical changes as a "package" whose adoption was related to previous, produced production problems discussed in Chapter 3.

The following chapter has two purposes: first, to demonstrate the above assertions, and second, to determine the pattern of resource use which the adoption of these technical changes implied. It will appear, in general, that they were land saving and capital and labor using. I will give attention to any evidence, such as the existence of economies of scale or of differential access to resources, which might provide clues to the distributional consequences of these technical changes.

THE SPREAD OF THE TWO-YEAR CROP ROTATION

Let us first examine farmers' choices of crop rotation. The available evidence suggests that those already using the two-year rotation (small and rich peasants) retained the system, and that more farmers adopted it. Let us examine first the quantitative and then the "qualitative" evidence. It will appear that there was no large-scale "reswitching"; rather, there was further intensification of land use.

It would be desirable to extend Craig's measure of the extent of the two-year system (see Table 3.2) into the interwar period in order to demonstrate this. Unfortunately, this is impossible. Therefore, I will use the other two proxies for the spread of the two-year rotation which were used in the previous chapter: the ratio of cropped to cultivated land, and the ratio of the area planted in cotton and to the cultivated area. A comparison of Tables 4.1 and 4.2 with Tables 3.6 and 3.7 provides evidence that the two-year rotation was retained and extended in the interwar period. However, the advance of the two-year system seems to have proceeded at a slower rate than for the pre-World-War I period, if 1917-1918 is used as a base: the measure of cropped area to cultivated area grows at an average annual rate of .53% for the pre-1914 period and at .355% for the interwar period. The picture is changed if 1918-1919 is chosen as the base year, in which case the average annual rate of change in the measure is 6.8%.

As a glance at Table 4.1 will show, however, this average conceals fluctuations in the measure. These fluctuations are even more pronounced if the cotton area/cultivated area measure is considered. These fluctuations may indicate a shift out of the two-year system and then back to it; there is some testimony in L'Egypte Contemporaine to support such an interpretation. The government attempted to limit cotton acreage in the 1920s, trying to raise the price by restricting supply (a policy which failed, incidentally), with unrestricted crops occurring only in 1919, 1920, 1924, 1925, and 1930.[3] Contemporary observers seemed to feel that one benefit of the government's limiting any one cultivator's cotton acreage to one-third of his total area would be to induce a return to the three-year system.[4] However, these laws were evaded through ruses such as including formerly uncultivated land in one's "cultivated area."[5]

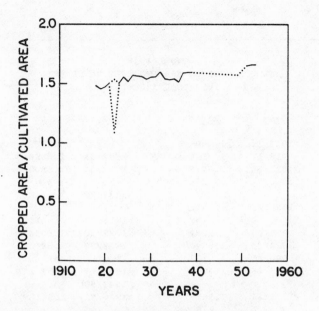

Fig. 4.1  Ratio of Cropped Area to Cultivated
Area  1917/18 - 1938/39  (Source: Table 4.1)

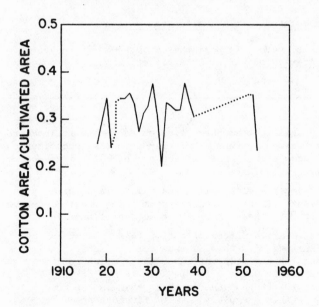

Fig. 4.2  Ratio of Cotton Area to Cultivated Area
1917/18 - 1952/53  (Source: Table 4.2)

TABLE 4.1
RATIO OF CROPPED AREA TO CULTIVATED AREA
1917-1918/1952-1953

| Year | Ratio | Year | Ratio |
|------|-------|------|-------|
| 1917/18 | 1.4806 | 1931/32 | 1.5989 |
| 1918/19 | 1.4519 | 1932/33 | 1.5384 |
| 1919/20 | 1.4715 | 1933/34 | 1.5300 |
| 1920/21 | 1.5054 | 1934/35 | 1.5401 |
| | | 1935/36 | 1.5110 |
| 1921/22 | 1.0842 | | |
| | (1.5362)* | 1936/37 | 1.5911 |
| 1922/23 | 1.5042 | 1937/38 | 1.5953 |
| 1923/24 | 1.5543 | 1938/39 | 1.5963 |
| 1924/25 | 1.5153 | | |
| 1925/26 | 1.5704 | 1948/49 | 1.5712 |
| | | 1949/50 | n.a. |
| 1926/27 | 1.5622 | 1950/51 | 1.6516 |
| 1927/28 | 1.5566 | | |
| 1928/29 | 1.5354 | 1951/52 | 1.6610 |
| 1929/30 | 1.5561 | 1952/53 | 1.6605 |
| 1930/31 | 1.5581 | | |

Source: Calculated from Annuaire Statistique

* Unreliable; number in parentheses obtained using editions of Annuaire Statistique after 1924/25, other number from using previous
issues.

It is possible that these figures do not reflect the switch
between rotation systems, but rather, the substitution of crops
within the same system. Hansen and Marzouk, for instance, give these
substitution possibilities:

Cotton competes with millet and maize...and with rice. In
the event of the rotation system being changed, cotton also
competes with wheat, barley, beans, and clover.[6]

The average relative prices of cotton to substitute crops fell over
the interwar period. This is true for the substitute crops within a
rotation and for the relevant substitutes if a rotation switch was
being considered. There is not much in the output price data which
would contradict the hypothesis that changes of crops within the
rotation system explain the fluctuations in the cotton
area/cultivated area ratio (see Table 4.3).

TABLE 4.2
RATIO OF COTTON AREA TO CULTIVATED AREA
1917/1918–1952/1953

| Year | Ratio | Year | Ratio |
|------|-------|------|-------|
| 1917/18 | 0.2491 | 1931/32 | 0.2002 |
| 1918/19 | 0.2970 | 1932/33 | 0.3351 |
| 1919/20 | 0.3445 | 1933/34 | 0.3282 |
| 1920/21 | 0.2413 | 1934/35 | 0.3192 |
|  |  | 1935/36 | 0.3200 |
| 1921/22 | 0.2743 |  |  |
|  | (0.3372)* | 1936/37 | 0.3766 |
| 1922/23 | 0.3443 | 1937/38 | 0.3358 |
| 1923/24 | 0.3443 | 1938/39 | 0.3044 |
| 1924/25 | 0.3550 |  |  |
| 1925/26 | 0.3316 | 1950/51 | 0.3527 |
|  |  | 1951/52 | 0.3513 |
| 1926/27 | 0.2735 | 1952/53 | 0.2347 |
| 1927/28 | 0.3130 |  |  |
| 1928/29 | 0.3279 |  |  |
| 1929/30 | 0.3753 |  |  |
| 1930/31 | 0.3068 |  |  |

Source: Calculated from Annuaire Statistique.

*   Unreliable; number in parentheses obtained using editions of Annuaire Statistique after 1924/25, other number from using previous issues.

The various proxies available indicate that there was further intensification of agriculture in the interwar period, largely in the form of a shift to the two-year rotation. Given the weaknesses of all of the quantitative measures, it would be wise to consider the available qualitative evidence here. First of all, as mentioned above, the government restrictions on the cotton area were interpreted as laws imposing the three-year system.[7] Discussions of the evasions of these restrictions mention that the large cultivators were the most conspicuous violators of the law, providing evidence that some large cultivators were adopting the two-year system. There is testimony that the small peasants continued to use the two-year rotation.[8] Jean Anhoury, writing in 1941, asserts that the two-year system was "the rule," with only a minority of the large cultivators retaining the three-year system. A publication of the Ministry of Agriculture in 1936 asserted that the two-year rotation has been "generally adopted."[9] Some landowners, apparently, did keep the old system: if we compare the cropped area/cultivated area ratio for the

period of the Korean War with the same measure from just before World
War II, there is a jump in the number from the latter to the former
(see Table 4.1). There is ample testimony that large cultivators
switched to the two-year rotation during the Korean War years due to
the high price of cotton.[10] The appropriate inference is that some
pashas retained the three-year system until the early 1950s.

There were also regional variations; less cotton was planted in
regions close to Cairo, such as Minufiyya, Qalyubiyya, Giza, and
Fayyum. Here increased fruit and vegetable cultivation for urban
markets tended to displace cotton and led to a retention of the
three-year rotation.[11] For the rest of the country, available evidence suggests that the peasants retained the two-year system, that
other farmers made the switch, and that most, but not all, of the
large landowners were among the new users of the two-year rotation.

Given the adverse soil effects of the two-year rotation, one may
ask why so many pashas adopted this system. Let us look first at the
ratio of rents to wages. If this is the relevant choice parameter
for technical choice between a more (two-year) and less (three-year)
labor using/land saving technique, we would expect to find that the
rental/wage ratio was rising, reflecting the increasing scarcity of
land relative to labor, as the pashas, (or anyone making crop decisions, for that matter) adopted the two-year system. That is, a rise
in the price of the scarce factor would induce some landowners to
adopt the technique which economized on the use of that factor.[12]
Table 4.4 presents the scattered data available on wages and rents.
As was true for the pre-World War I period, there is little in these
figures to support the hypothesis that changes in input prices
explain the pashas' decisions. The ratio falls, or, allowing for
data problems at least, does not rise, while the pashas shifted into
the more labor using/land saving technique. As a comparison of
Tables 4.4, 4.1, and 3.17 will show, the rental/wage ratio for the
interwar period is, in general, well below the same ratio for 1910-
1914, yet the ratio of cropped area to cultivated area (a proxy for
the rotation in use) is higher.

TABLE 4.3
RELATIVE PRICES, COTTON AND SUBSTITUTE CROPS,
1913–1937

| Year | $P_c/P_{maize}$ | $P_c/P_{wheat}$ | $P_c/P_{bean}$ | $P_c/P_{barley}$ | $P_c/P_{millet}$ | $P_c/P_{rice}$ |
|------|------|------|------|------|------|------|
| 1913 | 3.022 | 2.3492 | 2.4594 | 3.4478 | 3.2833 | 1.1890 |
| 1914 | 2.7614 | 1.8989 | 1.862 | 3.1466 | 3.0006 | 0.9604 |
| 1915 | 3.1852 | 2.1955 | 2.8972 | 3.2341 | 3.2929 | 1.2219 |
| 1916 | 4.0953 | 3.4931 | 3.6040 | 5.9141 | 4.5772 | 1.5123 |
| 1917 | 3.4503 | 2.1723 | 3.0089 | 4.7293 | 3.3904 | 1.3820 |
| 1918 | 2.9438 | 2.0792 | 2.6370 | 3.7782 | 3.1297 | 1.2993 |
| 1919 | 5.0182 | 4.1806 | 3.7793 | 6.0887 | 4.2929 | 1.6973 |
| 1920 | 2.9097 | 1.8297 | 1.4639 | 2.8917 | 4.9261 | 1.1583 |
| 1921 | 5.3632 | 2.9934 | 3.7956 | 6.2602 | 5.3187 | 1.3184 |
| 1922 | 4.5910 | 3.1669 | 2.6110 | 4.6990 | 4.3464 | 1.1023 |
| 1923 | 5.5532 | 4.7051 | 3.2333 | 7.3482 | 5.6838 | 1.8796 |
| 1924 | 4.0651 | 3.4977 | 2.8316 | 5.6163 | 3.8396 | 1.6607 |
| 1925 | 4.1271 | 2.4518 | 2.2960 | 4.4759 | 3.8876 | 1.2281 |
| 1926 | 4.2600 | 2.1959 | 1.4176 | 3.7064 | 4.1434 | 1.0934 |
| 1927 | 4.6338 | 3.7230 | 2.7541 | 5.8564 | 4.3806 | 1.6469 |
| 1928 | 3.8439 | 2.7755 | 1.7767 | 4.7834 | 3.6836 | 1.3819 |
| 1929 | 3.7245 | 2.4910 | 2.3435 | 4.9105 | 4.7073 | 1.1715 |
| 1930 | 2.2048 | 1.6225 | 1.1440 | 2.9903 | 2.3521 | 0.7935 |
| 1931 | 2.2591 | 1.3260 | 1.1009 | 1.9884 | 2.3644 | 0.5667 |
| 1932 | 3.6152 | 1.9941 | 2.5099 | 3.8075 | 3.8773 | 1.1164 |
| 1933 | 2.1858 | 2.0338 | 2.5949 | 4.9055 | 2.4282 | 1.1490 |
| 1934 | 2.4986 | 1.7674 | 1.7093 | 2.9384 | 2.9161 | 1.1271 |
| 1935 | 3.7100 | 1.9284 | 1.7845 | 3.2176 | 4.3337 | 1.2873 |
| 1936 | 3.4240 | 2.4987 | 2.2475 | 5.6849 | 3.5008 | 1.2788 |
| 1937 | 2.0065 | 1.7019 | 1.6165 | 2.9193 | 2.2161 | 0.8345 |

Source:
Calculated from Egypt, Ministry of Finance, Dept. of Statistics, Annuaire Statistique.

$P_c$ = price of cotton

It is, of course, quite possible that both the rental and the wage figures contain biases. It might be well to use the price of cotton, the main cash crop, as an indicator of the value of land. Then the relative scarcity of land and labor is given by the ratio of cotton prices to wage rates. These are presented in Table 4.4. As can be seen, much the same picture emerges: pashas retained the three-year system during the period of rising land scarcity, and adopted the two-year rotation during the period of declining relative land scarcity. Changes in the relative scarcity of land and labor, however measured, do not seem to have motivated the pashas' crop rotation decisions.

Another possible reason why the pashas might have chosen the two-year rotation would be a change in output prices. If the price of an output produced relatively intensely in a process or technique rises, then at the margin that process will be chosen and we will observe an increase in its use. In the case of a switch from a three-year to a two-year crop rotation, the relevant price is that of cotton, since the two-year system produced more cotton per year than the three-year system. However, the price of cotton does not rise markedly in comparison with any (not to mention all) of the relevant substitute crops (see Table 4.3). If the pashas were switching to the two-year system solely on the basis of short-run profit maximization, we would expect to find either a sharp rise in the rental/wage ratio or in the relative price of cotton. But we observe neither.

There is a further problem facing any purely "relative price" explanation of the shift. Such an explanation could not tell us why these pashas selecting the two-year system were willing to get higher profits now at the expense of lower yields, and, hence, lower profits later. In addition, it does not appear that pashas were constrained by debt and/or lack of information on the effects of such a rotation shift, as were the small peasant innovators of the pre-1914 era.

Nor do tenurial arrangements seem to be the explanation. First and most important, all of the sources emphasize the close degree of supervision of large estates in the interwar period. Rotation, irrigation, and drainage questions were all decided by the proprietor, regardless of the type of land/labor contract used.[13] Second, there is no evidence whatsoever that large estates were renting out more of their land in the interwar period than they were before World War I. The 'ezbah system described in the last chapter continued to dominate large estates; in 1939, 70-76% of estates between 50 and 500 feddans, and 91% of estates larger than 500 feddans, were exploited directly.[14] Such a breakdown by size of farms is not available for 1929; in that year 78% of the land area was cultivated directly. Pashas had been using and continued to use, direct exploitation under the 'ezbah system. It would appear that tenurial arrangements cannot explain the pashas' choice of the two-year crop rotation.

I would like to suggest that the pashas chose the two-year rotation at this time because some of the production problems associated with perennial irrigation and the two-year system were ameliorated during this period. The loosening of these constraints raised the possible long-run output and revenue of the two-year system. This

TABLE 4.4
RENTAL RATES, WAGE RATES,
AND RATIO OF RENTS TO WAGES,
AND COTTON PRICES TO WAGES,
1870s–1938

| Year | Rent (L.E./Feddan/yr) | Wage Rates (L.E./day) | Rent/Wage | $P_c$/wage[j] |
|------|------|------|------|------|
| 1870s | | .02–.03 | | |
| 1877 | | .03 | | |
| 1889 | | .04 | | 66.35 |
| 1890 | 1.05;1.5 | | 26.25–42.9 | |
| 1891 | | .035 | | 65.89 |
| 1895 | 3.6 | | | |
| 1900 | | .035–.045 | 88.9–133.3 | 59.21–88.97 |
| 1901 | 4.0 | 0.3 | | 71.35 |
| 1902 | 4.0 | | | |
| 1906 | | .025 | | 136.2 |
| 1907 | | .025 | | 152.88 |
| 1910 | 10.0 | .035–04 | 250–286 | 112.05–128.06 |
| 1912 | 8.0–10.0 | | 178–400 | 80.2–121.97 |
| 1913 | 12.0–18.0 | .03–.045 | 267–720 | 126.8–152.16 |
| 1914a (early) | | 0.25–.03 | | |
| 1920s | 12.8–32[b] | .07–.08[i] | 160–457 | 85.59 |
| 1922 | | .04–.05[d] | | 122.84–153.55 |
| 1927 | 7.09[e] | .05[e] | 141.8 | 118.72 |
| 1928 | 7.0–11.0[c] | .045[i] | 155.5–244.4 | 115.02 |
| 1928–29 | 11.0 | | | |
| 1930–31 | 5.3[g]:7.1[f] | .04[g] | 132.5:177–5 | 50.4 |
| 1931–32 | 4.53[g] | .03[g] | 151 | 81.87 |
| 1932 | 3–5[c] | | 100–166 | |
| 1932–33 | 5.1[c] | .025[i] | 204 | 91.12 |
| 1937 | 5–80[f] | | | |
| 1937–38 | 6.5[f.h] | .03[h] | 216.6 | 71.2 |

Sources:

a. See Table 3.12

b. Based on Crouchley's statement that postwar rents were "60–100% above 1914 levels," Economic Development of Egypt.

c. A. Lambert, "Divers Modes de Faire-Valoir les Terres en Egypte," "L'Egypte Contemporaine, 1938, 196.

d. Muhammad Saleh, La Petite Propriété en Egypte,, 1922 71.

e. Minost, "L'action contre la Crise," "L'Egypte Contemporaine, 190, 573.

f. M.R. Ghonemy, Resource Use and Income in Egyptian Agriculture, unpub. Ph.D. thesis, N. Carolina State College, 1953.

g. Hamed el-Sayyid Azmi, "A Study of Agricultural Rev- enue in Egypt...,"L'Egypte Contemporaine, 1934, 714.

h. Mr. Anis, "The National Income of Egypt," L'Egypte Contem- poraine, 1950, 753, 759.

i. A. Lambert, "Les Salaires dans l'entreprise Agricole Egypti- enne," L'Egypte Contemporaine, 1943, p.229.

j. Annuaire Statistique

---

made its adoption more attractive at the margin.[15] The next task, then, is to examine the measures which were taken to counteract the problems associated with more intensive land use. As before, I will be especially concerned to see if the adoption of these measures created any additional dilemmas for the small cultivators. We will want to see if the new techniques, viewed here as improvements on and complements to the two-year rotation system and perennial irrigation, in their turn favored the use of labor and/or capital. This will provide a link with the technical changes discussed in Chapter 3 and a background for a consideration of the social consequences of the adoption of these new techniques in Chapter 5.

TECHNICAL INTERRELATEDNESS AND RESOURCE USE PATTERNS

There were three principal technical changes during this period: improved drainage, new planting techniques, and the application of chemical fertilizers. All of these were complementary to use of the the two-year rotation; they were changes which attempted to cope with the production problems created by the adoption of that cropping pat- tern and of perennial irrigation. Let us consider these innovations in more detail.

Recall that the switch to perennial irrigation and the two-year rotation stimulated the rise in the water table, the salination of the soil, and the adverse physical and chemical effects upon the soil of the shortening of the fallow (see Chapter 3). These problems, in turn, were due to inadequate drainage. Sa'ad Zaghlul stressed the importance of raising the productivity of the land, and of extending drainage and irrigation in particular, in his opening address to the newly formed Egyptian parliament in 1924.[16]

The Egyptian government tackled the drainage problem in the interwar period. First pumping stations were constructed:

The Delta was divided into a number of zones, corresponding with its natural drainage areas between the lines of higher land which mark the ancient water-courses. After the war, work was undertaken on the drainage of these areas by the construction of big drainage canals in which the water was to be maintained at a suitably low level by pumps designed to lift the water from the terminal ends of the canals and pump it into the sea or northern lakes. In this way, in 1930 no less than ten big pumps were in operation along the northern fringe of cultivation in the Delta. Since that date the electrification of these pumping stations has been undertak-

en. There are now (1938) seventeen drainage stations...
providing drainage to over one million feddans in the north
of the Delta.[17]

A large scale program of drain construction was undertaken (see Table
4.5). By 1933-1934 some L.E. 13.5 million had been spent for this
purpose. From 1930 to 1938 the area of drained lands in the Delta
rose from 980,000 feddans to 2.2 million feddans, at a cost of L.E.
9.5 million.[18] At the same time, further irrigation works were
built: canals, the Nag Hamadi dam in 1928, and the second heightening
of the Aswan dam in 1937.

TABLE 4.5
PUBLIC DRAINS, 1922-1939 (STOCK)

| Year | Kms. | Year | Kms. |
|------|------|------|------|
| 1922 | 6,523 | 1931 | 7,726 |
| 1923 | 6,756 | 1932 | 7,709 |
| 1924 | 6,786 | 1933 | 8,040 |
| 1925 | 6,678 | 1934 | 8,200 |
| 1926 | 7,030 | 1935 | 8,563 |
| 1927 | 7,088 | 1936 | 8,972 |
| 1928 | 7,369 | 1937 | 9,168 |
| 1929 | 7,453 | 1938 | 9,708 |
| 1930 | 7,528 | 1939 | 10,246 |

Source: Annuaire Statistique.

122

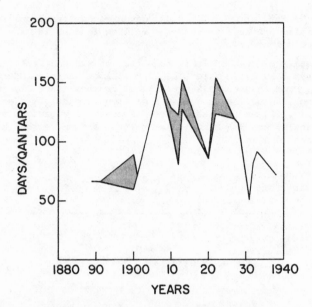

Fig. 4.3  Relative Price of Cotton and Labor
1890-1938  (Source: Table 4.4)

A word about the nature of the public finance of these drains as well as the canals, might be in order here.  First, public drains and canals were those which served the lands of two or more villages.[19] There was no user fee for either drains or irrigation canals:

Direct support for the Irrigation Service is derived from the Land Tax and supplemented by revenue from the government owned railroads, customs collections, etc.  There is no direct water tax.[20]

Rather, payment of the land-tax was a "guarantee" of the fellah's obtaining his share of available irrigation water.[21] Regional allocation of drains was, apparently, decided on technical grounds, which seems reasonable enough given the interrelated nature of Delta drainage and the generally good natural drainage of the Sa'id. Finally, if a person's land were to be selected to have a public drain or canal transverse it, he was to be repaid the market value of the land.[22]

Some of the well-to-do landowners in the Northern Delta had installed pumps and built drainage canals on their own land.[23] Such works were complements to the government drains, since the former emptied into the latter. However, the private sector lagged behind the public in this crucial area.[24] The reasons for this lag are not difficult to see.  An adequate drainage network required (1) capital expenditure and (2) the use of as much as ten percent of the land area.  Due to the indivisible nature of a drain, the small cultivators would have found the percentage of their land so occupied near the maximum. In an area characterized by very small holdings, it was no doubt difficult to get agreement on whose parcel of land would be turned into a drain. The externalities problem may have been serious.[25] Given the indivisible nature of the input and, therefore, the economies of scale in its use, not to mention the capital costs of construction,it is not surprising to find that small holders lagged far behind large landlords in installing field drains.

Le voyageur remarque en quelques endroits de petites propriétés qui sont incultes à cause de la faute de drainage. Malheureusement, le petit propriétaire n'apprécie pas le drainage à sa juste valeur, et voudrait-il entreprendre un tel travail, il n'en aurait pas les moyens. Les grandes propriétaires seuls ont pu procéder à cet amenagement grâce à l'étendue de leurs domaines.[26]

A possible alternative system to field drains which would have circumvented the problem of taking up cultivated area was tile (or pipe) drainage. This, however, was prohibitively expensive, and was only discussed, in a general way, before World War II.[27] Experiments in 1956 showed that such drains were effective for removing water, required almost no upkeep, increased yields by approximately thirty percent, but cost about L.E. 28 per feddan to construct[28] --approximately equal to the yearly budget of a fellah family of four.[29]

The lack of secondary, private drainage slowed the "production" of high quality land. It does seem clear, however, that the government's projects contributed substantially to the removal of the problem of the rise in the water table, primarily by "unblocking" the lines of natural drainage. We may conclude that the problem of drainage was ameliorated but hardly solved. Furthermore, it would appear that drainage was better on the pashas' lands than it was in those areas where small peasant holdings predominated.

The second technical change during the period was a change in planting techniques. Cotton was planted earlier and the seeds were spaced more closely than before. Both of these changes promoted earlier ripening, thus helping to avoid attacks of the pink boll-worm. According to agricultural experts, the optimum spacing of the ridges (i.e., that which maximized yields per feddan) had changed from seventy-five centimeters in 1915 to sixty-five centimeters in 1935, and the space between the holes from 45 to 35 cm. for the same years.[30] Spacing may have become even closer in the latter half of the 1930s.[31] There is only qualitative evidence on the extent of the adoption of this technique; it appears to have been widespread.[32] Earlier planting seems to have been adopted in the early and mid-1930's. These measures were successful in inhibiting the attacks of the pink boll-worm.[33]

Poorer peasants probably could not plant their cotton earlier. Their practice in the two-year rotation was to take several cuts of birsim to guarantee food for their animals. Insofar as they adopted earlier planting, they would either have had 1) to take fewer

TABLE 4.6
PLANTING DATES FOR COTTON, 1930's

| % of cotton crop sown by: | End of Feb. | March 15 |
|---|---|---|
| 1930 | 7 | 31 |
| 1936 | 14 | 60 |

Source:
    Nassif, "L'Egypte, est-elle surpeuplée," L'Egypte Contemporaine,
    33 (1942), 729.

cuttings of birsim, or 2) not to plant birsim before cotton and change birsim's place in the rotation. The first possibility is unlikely: fewer cuttings would imply a fall in the food supply for the animals, unless alternative sources of food were expanded. As we shall see in the next chapter, the latter was not the case. Consequently, we would expect the number of animals to fall if the number of cuttings of birsim were reduced. However, the opposite occurred: the number of buffaloes rose from 645,537 in 1921 to 956,036 in 1937, while the number of cattle rose from 595,964 to 983,219 for the same years (for more detail, see Table 4.7). The quality of these animals and their food supply are discussed in the next chapter. A reduction in the number of birsim cuttings appears highly unlikely.

The evidence on birsim's place in the rotation is ambiguous. All descriptions of either the two- or three-year rotation from before 1914 which I have seen have birsim preceding cotton. Interwar evidence is scanty: a 1933 source has birsim before cotton in the two-year system.[34] On the other hand, Anhoury, writing in 1941, has cotton coming after maize, not birsim, in the two-year system.[35] This would be consistent with a tale of peasants adopting earlier planting during the course of the 1930s. Saffa's description of the two-year rotation in 1948 has birsim before cotton, but he is describing a large estate, which did not have the same fodder requirements, since such estates often substituted steam pumps for animal power in irrigation.[36] In the 1970s, birsim preceded roughly 80% of the cotton crop.

If birsim was not planted before cotton, there would have been an increased need for nitrogen fertilizer for the cotton. Mosseri estimated the nitrogen fixation of a hectare of birsim as the equivalent of between 250-500 kilograms of nitrate of soda.[37] Some "evidence in reverse" for this hypothesis is provided by Nagy, who says that if a cotton field was not manured, then birsim would have preceded it.[38] As Nagy implies, this may not have presented any very serious problem if it was met by the increased supply of natural fertilizers. We note, however, a certain complementarity between

TABLE 4.7
CATTLE AND BUFFALOS, 1919-1937

| Year | Buffaloes | | Cattle | |
|------|-----------|--------|---------|--------|
| | Delta | Sa'id | Delta | Sa'id |
| 1919 | 368,511 | 165,473 | 295,484 | 199,883 |
| 1920 | 393,009 | 186,041 | 325,618 | 230,703 |
| 1921 | 422,231 | 216,119 | 316,973 | 262,978 |
| 1922 | 404,664 | 205,946 | 312,611 | 258,483 |
| 1923 | 415,081 | 232,066 | 329,265 | 285,260 |
| 1924 | 449,401 | 265,777 | 359,867 | 306,255 |
| 1925 | 459,729 | 253,202 | 352,857 | 300,875 |
| 1926 | 492,135 | 265,327 | 393,520 | 314,189 |
| 1927 | 496,507 | 253,493 | 409,081 | 310,796 |
| 1928 | 510,129 | 271,573 | 445,431 | 329,686 |
| 1929 | 558,754 | 257,287 | 446,314 | 337,272 |
| 1930 | 535,072 | 254,774 | 458,984 | 300,771 |
| 1931 | 556,227 | 260,850 | 468,776 | 307,792 |
| 1932 | 591,609 | 285,796 | 523,254 | 368,709 |
| 1933 | 571,644 | 278,657 | 535,874 | 358,158 |
| 1934 | 606,107 | 277,503 | 547,963 | 359,145 |
| 1935 | 606,464 | 288,676 | 563,338 | 375,730 |
| 1936 | 630,267 | 296,920 | 592,901 | 386,417 |
| 1937 | 623,794 | 326,088 | 568,352 | 403,174 |

Source:
   Ministry of Agriculture, Monthly Agricultural Statistics, various issues.

earlier planting and fertilizer use for small holders. It will appear below that such a complementarity existed regardless of the size of holdings.

Changing the place of birsim in the rotation would also have meant that the birsim would have been cultivated on fields adjacent to the cotton fields. This would have given the leaf worm pest more plants to feed on as it developed, before moving on to the cotton plants.[39] The closer spacing of the cotton plants may also have contributed to the attacks of this pest by multiplying the number of plants per feddan. These attacks were quite severe, occurring in thirteen of the twenty years from 1920-1939, inclusive. The average loss appears to have been on the order of eighty pounds per acre or about 18% of the average yield per acre for 1920–1939 in the areas affected.[40] The attacks of this pest raised the demand for child labor, since the principle method of combatting the leaf-worm attacks was to have children pick the worms off the leaves of the plant. This method required anywhere from one to five children per feddan.[41]

The greater demand for child labor was only one of a number of complementarities involved in the new planting techniques. If small peasants changed the place of birsim in the rotation, they would have needed to use more nitrogen fertilizer. Such a demand was reinforced by the fact that nitrogen fertilizer induced more rapid growth in plants.[42] Anything that induced growth and earlier ripening was needed to combat the boll-worm attacks. Any increase in fertilizer use would have been accompanied by an increase in the demand for labor, for the two inputs were complementary. In addition, the closer spacing of the plants must have required more labor, since the techniques of making the ridges and implanting the seed remained unchanged. More irrigation labor was needed, since more frequent waterings, and especially earlier waterings, stimulated the growth and earlier ripening of cotton plants.[43] This, in turn, would have been complementary with increased drainage. Drainage was complementary with closer spacing in another way: the closer spacing of the plants inhibited the lateral growth of the roots of the cotton plant. This could be compensated for by the downward extension of the roots, provided that the water table was low enough.[44] Finally, the additional quantity of seed per feddan would have required more capital, as would any increased demand for fertilizer to hasten plant growth.

In summary, the new planting techniques were a response to the problem of pest attacks, a problem which was exacerbated by the extension of perennial cultivation and the adoption of the two-year crop rotation. The new techniques economized on land, and made intensive use of both child and adult labor, water, fertilizer, drainage, seed, and capital.

Another seed-related change was a shift in the cotton variety. The story here is largely that of the rise and fall of Sakellerides, named after its breeder. The yields of this variety were some 25% below that of the main other variety, Ashmouni.[45] It enjoyed popularity because the quality of its fiber fetched a high price. Its main popularity was during World War I and the early 1920s. Its quality, like that of Mit Afifi before it, began to deteriorate through uncontrolled interbreeding and was largely replaced by other varieties from the late 1920s on. The output consequences of this are examined in Chapter 5. So far as I can determine, adoption of the new variety did not change other input requirements.

The final major technical change of importance in this period was the adoption of artificial fertilizers. Total imports rose dramatically in the interwar period (see Table 4.8). This was primarily a demand for nitrogen fertilizer: between eighty and ninety percent of imports were of this type. Since Egyptian soils are rich in potash, such fertilizers were rarely necessary. Phosphates were also used, primarily for birsim.

There was some question as to which crops benefitted from the application of fertilizers. Primarily, the question concerns the utility of nitrogen fertilizer for cotton. Nitrogen fertilizer clearly raised the yield of cereals such as wheat, maize, and barley (see Table 4.9). Birsim did not require nitrogen fertilizer, but did benefit from the application of phosphates.

The case of cotton was less obvious. Cartwright and Dudgeon[46] took an agnostic stand on its effects. D.S. Gracie, a government agronomist, and his colleagues argued on the basis of field experiments that the limiting factor for cotton yields was the mechanical nature of the soil: permeability to water and air, "properties which are in the main a reflection of the amount of deterioration the soil has undergone.[47] Within the limits imposed by this and other soil features, nitrogen fertilizer "can and does cause important increases in yield."[48] Original yield level, determined by the quality of soil, and responsiveness to fertilizer were found to be positively and significantly correlated. The maximum percentage increase from nitrogen fertilizer was constant; therefore, the marginal physical productivity of fertilizer was higher in areas with a higher "pre-fertilizer" yield level.

---

TABLE 4.8
FERTILIZER IMPORTS, 1913 - 1938

| Year | Metric Tons of Fertilizer | Year | Metric Tons of Fertilizer |
|------|---------------------------|------|---------------------------|
| 1913 | 71,654 | 1926 | 243,073 |
| 1914 | 72,610 | 1927 | 225,421 |
| 1915 | 61,243 | 1928 | 275,370 |
| 1916 | 25,432 | 1929 | 327,863 |
| 1917 | 36,940 | 1930 | 317,722 |
| 1918 | 3,071 | 1931 | 261,696 |
| 1919 | 57,718 | 1932 | 234,557 |
| 1920 | 120,246 | 1933 | 295,672 |
| 1921 | 43,747 | 1934 | 422,399 |
| 1922 | 118,207 | 1935 | 561,615 |
| 1923 | 101,755 | 1936 | 572,438 |
| 1924 | 179,087 | 1937 | 641,838 |
| 1925 | 258,306 | 1938 | 513,799 |

Source:
  Egypt, Ministry of Agriculture, Monthly Agricultural Statistics, Various Issues.

---

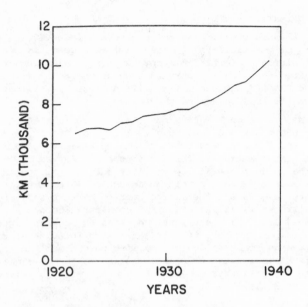

Fig. 4.4 Drainage Canals, 1922-1938
(Source: Table 4.4)

Consequently, there were regional variations in the use of fertilizer for cotton, although its use was widespread.

Cotton has always been the most heavily manured crop; in the early 1920's it was receiving practically all the fertilizer imported; even in 1937...it did not get less than 340,000 tons of a total of 566,000 tons and it may have received considerably more.[49]

The heaviest use of fertilizer was in Upper Egypt; the cotton crop there was "by far the most heavily manured crop in Egypt," receiving "fully half of the nitrogen fertilizer imported into Egypt."[50] Its use seems to have been especially heavy in certain areas; "growers in Minya and Asyut provinces used to speak of giving six hundred kilos per feddan in the 1920's."[51] Within the Delta, use of fertilizer for cotton increased from north to south.[52] In 1935-1940 Delta cotton was receiving at least one hundred kilos of fertilizer per feddan on the average, an average which, as they point out, conceals differences within the region.

One of the reasons for the regional differences was purely geographical: temperature differences. Gracie et al. suggest that the difference between their results and those of experimenters before World War I, who generally found little relation between cotton yields and the application of nitrogen fertilizer, may have been due to the rise in mean temperatures from 1904 to 1940.[53] Given their insistence on the importance of the soil's mechanical condition and

TABLE 4.9
RESPONSIVENESS OF CEREALS AND BIRSIM
TO FERTILIZER, 1937

A.  Cereals
    (yield per feddan, ardebs)

| Crop | Kgs. of Nitrogen fertilizer | | | |
|---|---|---|---|---|
| | 0 | 100 | 200 | 300 |
| Wheat | 4.9 | 6.8 | 7.8 | 8.3 |
| Barley | 7.5 | 10.6 | 12.6 | 13.3 |
| Maize | | | | |
| after birsim | 8.8 | 10.7 | 12.0 | 13.0 |
| before birsim | 6.6 | 9.1 | 10.9 | 12.0 |

B.  Birsim (tons) and Kgs. of Superphosphate

| 0 | 100 | 200 |
|---|---|---|
| 28.1 | 31.5 | 32.4 |

Source:
    State Domain experiments of 1937; report in Nassif, "L'Egypte,
    est-elle sur-peuplée?" L'Egypte Cotemporaine, 1942, 376.

hence, of the pre-fertilizer yield level as a factor determining the productivity of fertilizer and given the importance of drainage for maintaining the soil's condition, it is clear that differences in drainage quality also contributed to regional variations in fertilizer application. The lands of Upper Egypt, newly converted to perennial irrigation and having relatively good natural drainage, did not suffer from drainage problems to the extent of those in the Delta; this quite probably contributed to the higher productivity of fertilizer in the former area.

Gracie and Khalil's policy recommendations underscored this link between fertilizer and drainage:

The main, if not the only, practical steps which can be taken (assuming average seasons and the permanence of the pink boll-worm) to increase the possibilities of nitrogenous manuring of cotton in Egypt must obviously be directed to raising the general yield level. This can only be accomplished by the prevention and remedying of soil deterioration by the general adoption of intensive drainage.[54]

The early British failure to provide sufficient drainage had long-lasting consequences.

Nitrogen fertilizer was not a perfect substitute for animal manure. Nassif[55] asserts that twenty cubic meters of animal manure was approximately equal to one hundred kilos of nitrate "en general." On the other hand, on Saffa's estate, animal manure and nitrates were used together on the maize crop. The Khedival Agricultural Society was urging its members to employ both sorts of fertilizer, animal to "improve the land and raise its fertility" and artificial to stimulate plant growth. This seems quite plausible, and is consistent with Gracie's views. The application of animal manure would tend to increase the porosity of the soil to water and air, which, as we have seen, were key elements in determining the level of yields. Although the degree of substitutability or complementarity no doubt varied from locality to locality, on balance it would appear that animal manure and nitrogen fertilizers were complements.

There is some support for this notion in the statistics on livestock. The number of buffaloes and cattle nearly doubled from 1919-1937 (see Table 4.7), while the cropped area rose by only some ten percent from 1920-1937. Some of the animal dung was used as fuel by the fellahin, but the growth rate of animals greatly exceeded that of the population, which rose from 12,751,000 in 1917 to 15,900,000 in 1937. It seems highly likely that an increase in the application of animal manure per unit of crop area occurred. It appears that farmers were increasing their inputs of both animal and artificial manures, rather than substituting one for the other.

We have already seen that nitrogen fertilizer use was complementary with labor-using technical changes of closer spacing and earlier planting. This complementarity was reinforced by the fact that the use of fertilizer was itself a labor-using activity. Animal fertilizer had to be collected, usually by children (see, for instance, the description of work tasks in Ammar)[56] It was normally applied by adults.[57] For instance, the application of 25-30 cubic meters of animal manure on a feddan before the planting of maize required the labor of two men. The application of artificial manures was also labor-using; here too, children were used.[58]

In summary, the following pattern of technical change and resource use in the interwar period emerges:[59] new techniques of closer spacing and earlier planting were complementary with the use of labor, water, and seed, as well as fertilizer. Fertilizer, in turn, was a labor-using activity. The increased demand for seed and fertilizer induced an increased demand for credit. Both the new planting techniques and the use of fertilizer for cotton were complementary with drainage. The large number of complementarities involved reminds one of the "package" of new inputs required for implementing the technical changes in under- developed agriculture in our own time, the "Green Revolution."

TECHNICAL CHANGE AS A RESPONSE TO PAST ACTIONS

Two of the technical changes of the interwar period, earlier planting and the extension of drainage, were clearly related to the events of the pre-World War I era. The extension of perennial irrigation and of the two-year crop rotation system before the War undermined soil quality and contributed to the spread of insect pests. Drainage and earlier planting were direct responses to these problems. The motivation behind the spread of artificial fertilizers is more complex. On the one hand, its adoption was both directly and indirectly related to the production problems which were produced by the technical changes of the pre-1914 period: directly, because the use of nitrogen fertilizers accelerated the growth of cotton plants and consequently helped to reduce losses to pests; indirectly, because the productivity of fertilizer for cotton depended upon the quality of the soil (and, therefore, on the quality of the drainage).

On the other hand, a case can also be made that the adoption of fertilizer was induced by changes in input prices. Hayami and Ruttan[60] explain the diffusion of artificial fertilizer use in both Japan and the United States in this way. Their procedure is to compare the price of fertilizer with the price of the substitute, land. As Williamson[61] has pointed out, this is incorrect in terms of capital theory, since fertilizer use is a flow and land is a stock. The relevant flow index for land services, in this view, would be rents. Using rent as the shadow price of land avoids the problems of changes in the capital markets causing changes in land prices. As we shall see in the next chapter, changes in the credit structure in Egypt were considerable during the interwar period, so we would do well to pay attention to this point. The numbers for rent that we have are, of course, highly suspect; the prices of fertilizer are average prices. However, if we construct a series of relative price of land services and fertilizers, we find some support for the "induced technical change" view for the 1930s (see Table 4.10). For any sequence of years, the direction of change of relative price index and fertilizer imports is the same.

However, it is not at all obvious that the very large difference in fertilizer imports at the end of the 1920s and at the end of the 1930s can be accounted for by differences in the relative prices. Given the very severe data problems, it would be wise to look at the relative price of cotton to fertilizer. Recall that cotton received at least 60% of the fertilizer imported from abroad. If the adoption of fertilizer was being "induced" by relative price shifts, we would expect to see a marked upward trend in the price of cotton relative to the price of fertilizer. However, we observe nothing of the sort (see Table 4.10)

On the other hand, fertilizer was also used for wheat. Here the evidence supports the "inducement" hypothesis. Government policy helped here, for the Sidqi government placed a tariff on wheat in 1930 to discourage imports and to bolster prices. This seems to have contributed considerably to the increase in fertilizer use in the 1930s (see Table 4.10 and 4.9). In view of the evidence that fertilizer was being heavily used on cotton lands, however, such price changes cannot be the whole story. We may conclude that although the evidence lends some support for the "inducement" hypothesis, it is shaky enough to warrant serious consideration of the connection

TABLE 4.10
RENTS, COTTON, WHEAT, AND FERTILIZER PRICES,
AND THEIR RATIOS, 1920-1938

| Year | Rent | $P_{cotton}$[a] | $P_{fertilizer}$[b] | $R/P_F$ [c,d] | $P_F/P_C$ | $P_F/$[e] $P_{wheat}$ |
|------|------|------|------|------|------|------|
| 1920 | 12.8-32.0 | 6.900 | 25.35 | 0.5-1.26 | 3.67 | 7.741 |
| 1921 | | 6.858 | 15.28 | | 2.22 | 7.680 |
| 1922 | | 6.142 | 12.83 | | 2.08 | 7.617 |
| 1923 | | 7.959 | 11.96 | | 1.50 | 8.140 |
| 1924 | | 7.897 | 12.00 | | 1.52 | 6.118 |
| 1925 | | 6.093 | 11.11 | | 1.82 | 5.148 |
| 1926 | | 4.306 | 10.68 | | 2.48 | 6.272 |
| 1927 | 7.09 | 5.936 | 10.04 | 0.706 | 1.69 | 7.250 |
| 1928 | 7.0-11.0 | 5.176 | 9.58 | 0.73-1.15 | 1.85 | 5.915 |
| 1929 | 11.0 | 4.072 | 9.05 | 1.22 | 2.22 | 6.373 |
| 1930 | | 2.410 | 8.71 | | 3.61 | 6.748 |
| 1931 | 5.3; 7.1 | 2.016 | 7.76 | 0.068-0.91 | 3.84 | 5.879 |
| 1932 | 4.53-3-5 | 2.456 | 8.12 | 0.56 | 3.31 | 7.589 |
| 1933 | 5.1 | 2.278 | 6.66 | 0.77 | 2.92 | 5.943 |
| 1934 | | 2.650 | 5.75 | | 2.16 | 3.836 |
| 1935 | | 2.726 | 5.00 | | 1.83 | 3.504 |
| 1936 | | 2.870 | 5.15 | | 1.79 | 4.483 |
| 1937 | 5-8.0 | 2.154 | 5.81 | 0.86-1.37 | 2.69 | 4.590 |
| 1938 | 6.5 | 2.136 | 6.25 | 1.04 | 2.92 | |

Sources:
a.  Rents. LE/feddan/year.  See Table 4.4
b.  LE/cantar. Annuaire Statistique.
c.  Average price of nitrogen fertilizer, LE/m. ton, calculated from Egypt, Ministry of Finance, Egyptian Customs Administration, Annual Statement of Foreign Trade.
d.  Rental figures are for early 1920s.  See Table 4.4
e.  Calculated from Annuaire Statistique and "c".

between fertilizer and drainage.  Perhaps the most reasonable conclusion would be that price changes induced short-term, year-to-year fluctuations in fertilizer imports, whereas the improvements in drainage and the adoption of earlier planting determined the long-term, secular trend of increasing fertilizer use for cotton.  (See Appendix 4A.)

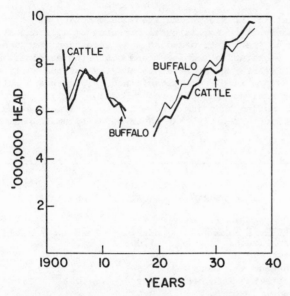

Fig. 4.5 Cattle and Buffaloes 1903-1937
(Source: Table 4.7)

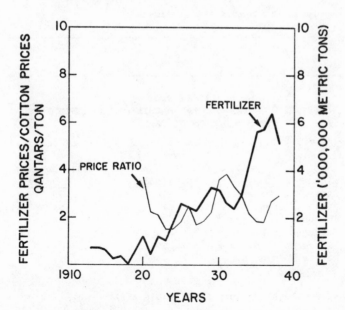

Fig. 4.6 Fertilizer Imports and Ratio of Fertilizer
and Cotton Prices, Interwar Period
(Source: Tables 4.6 and 4.8)

The wheat tariff also stimulated the adoption of some agricultural machinery. The Agricultural Censuses of 1929 and 1939 provide data on ploughs (both native and steam), threshing machines (native and other), segregating machines, and winnowing machines. Note that mechanical threshers and segregators show by far the most rapid rate of growth. This is plausibly explained by government policy and by the fact that native threshers produced a low quality of grain. It is also worth noting that the kinds of farm equipment owned by pashas increased much more rapidly than those owned by small peasants.

TABLE 4.11
AGRICULTURAL MACHINERY, 1929 and 1939

| Machine or Tool | Number (1929) | Number (1939) | % Change 1929-39 |
|---|---|---|---|
| Steam Ploughs | 1,008 | 1,795 | 78 |
| Native Ploughs | 564,144 | 603,903 | 7 |
| Threshing Machines | 569 | 2,123 | 273 |
| Segregating Machines | 746 | 2,083 | 179 |
| Winnowing Machines | 2,373 | 2,494 | 5 |
| Native Threshers (Nuraq) | 302,023 | 301,705 | 0 |

Source: Agricultural Census, 1939, 110-111.

TABLE 4.12
OWNERSHIP OF AGRICULTURAL MACHINERY AND IMPLEMENTS,
PASHAS AND SMALL PEASANTS, 1939

| Machine or Tool | % of total owned by Pashas (50+ feddans) | % of total owned by Small Peasants (0-5 feddans) |
|---|---|---|
| Steam Ploughs | 84 | 3 |
| Native Ploughs | 15 | 53 |
| Threshing Machines | 87 | 3 |
| Segregating Machines | 72 | 9 |
| Winnowing Machines | 69 | 7 |
| Native Threshers (Nuraq) | 15 | 43 |

Source: Agricultural Census, 1939, Table XXIX.

SUMMARY AND CONCLUSION

In conclusion, the interwar period saw a further intensification of land use. Viewing land as a produced input, it appears that some of the technical change, the extension of drainage, was an attempt to undo the effects of British hydraulic policy and peasant crop-rotation choice on land production. This was essential not only for its impact on yields directly, but also because the main cash crop, cotton, was more responsive to fertilizer on well-drained land than on land which had undergone deterioration from inadequate drainage. Closer spacing and earlier planting were also attempts to remedy problems caused by earlier technical changes. Insofar as perennial irrigation, the increased use of water, and the increased number of plants per feddan per season characteristic of the two-year rotation fostered the growth of pests, attempts to hasten the maturity of plants and to avoid these attacks were responses to earlier, produced production problems. The same may be said for the increased use of nitrogen fertilizer to hasten plant growth. The nature of technical change in the interwar period supports Owen's contention that "much of the investment in the agricultural sector between the two wars was necessary to repair damage already done to soil fertility."[62] It also supports the "disequilibrium" view of technical change: as agriculture intensified, certain disequilibria were created. Subsequent technical change may be viewed as, in large part, a set of interrelated, mutually reinforcing responses to these disequilibria. There was an important feedback of past actions on future technical innovations. The consequences of these technical changes for output and distribution are the subject of the next chapter.

APPENDIX 4A: REGRESSION ANALYSIS OF FERTILIZER DEMAND, 1920-1938.

Some support for the above arguments on the determinants of nitrogen fertilizer use may be obtained through a regression analysis of the demand for fertilizer. Two models, a "naive" or instantaneous adjustment model, and a simple dynamic adjustment model based on the work of Griliches,[63] are used.

The "Naive" model is specified as follows:

$$\ln FERT = a_1 + b_1 \ln(P_f/P_{crop}) + e_1 \quad (1)$$

where FERT = nitrogen fertilizer consumption, using imports = consumption, which was the case during this period; $P_f$ = price of fertilizer, measured as average prices; $P_{crop}$ = national crop prices. Both cotton and wheat prices are used in this and the dynamic model.

When the price of cotton is used, the coefficient "b" is zero and has the wrong sign; the $R^2$ is trivially low. However, when the price of wheat is used, the coefficient (-2.13) has the correct sign, is highly significant (1% level), and $R^2$ = .64.

A second naive model may be obtained by adding drainage (as measured in Table 4.5) as an independent variable:

$$\ln FERT = a_2 + b_2 \ln (P_f/P_{cotton}) + c_2 \ln DRAINS + e_2 \quad (2)$$

The coefficient for relative prices is still insignificant, while the coeffient for drainage is positive, and significant at the 1% level. Since the variables are measured in logarithms, the coefficient "$c_2$" gives the elasticity of demand for fertilizer in response to improved drainage. The high value of this elasticity (4.75) is consistent with the agronomic evidence and testimony reported above in the text.

Such a specification implies that farmers adjust instantaneously to price changes; it is, in fact, a long-run equilibrium model. Following Griliches [64] we develop a simple dynamic adjustment model as follows:

$\underline{True}$ long run demand is given by:

$$\ln FERT^*_t = c_3 + d \ln (P_f/P_{crop}) + e_3. \quad (3)$$

where Fert$^*_t$ = desired fertilizer consumption in long run equilibrium, other variables as before. Then the adjustment of actual use toward desired used is given by equation (4).

$$\ln FERT_t - \ln FERT_{t-1} = r(\ln FERT^*_t - \ln FERT_{t-1}), \quad (4)$$

where Fert$_t$ = actual fertilizer consumption at time (year) t, FERT$^*$ = desired fertilizer consumption in long run equilibrium, and r = the adjustment coefficient ($0 \leq r \leq 1$). Substituting (3) into (4), we obtain:

$$\ln FERT_t = rc_3 + dr \ln (P_f/P_{crop}) + (1-r)\ln FERT_{t-1} + re_t$$
(5)

Because the variables are expressed as logarithms, we may again interpret the coefficients as elasticities; "dr" gives the short-run elasticity, and "d" gives the long-run elasticity of fertilizer consumption with respect to the relative prices. (Use of the Durban-Watson statistic indicated that autocorrelation of the residuals was not a problem.)

The following results were obtained:

| Crop | Elasticity of N Fertilizer Consumption | | Adjustment Coefficient |
|------|-------------|----------|------------|
|      | Short run | Long run |            |
| Cotton | Insignificant | −1.879 | .141 |
| Wheat | −1.165 | −2.463 | .473 |

Several observations and $\underline{caveats}$ may be made here. First, the results are broadly consistent with those of the "naive model". Second, the dynamic model as specified here remains itself a fairly naive one; however, experiments with slightly different specifications (e.g. fertilizer use per feddan) did not yield significantly different results. Third, the elasticity of fertilizer consumption with respect to the price of wheat is $\underline{very}$ high, exceeded only by the elasticity of fertilizer use with respect to the price of rice in

Taiwan from 1950 to 1966.[65] This and the high adjustment coefficient (which implies that nearly one-half of the entire adjustment
took place in one year), is consistent with the general position
taken throughout this book and, indeed, by most students of Egyptian
agriculture, that Egyptian farmers are quick to respond to incentives.

Finally, it must be emphasized that the data base is quite weak,
using aggregate, macro-level data, while we have seen above that
there were very important regional and micro-level differences in
fertilizer use. Nevertheless, the results are broadly consistent
with the argument that although fertilizer use was induced by changes
in the price of wheat (itself the object of government policy), cotton prices were much less influential because of the strong impact of
drainage upon the marginal product of fertilizer for that crop.

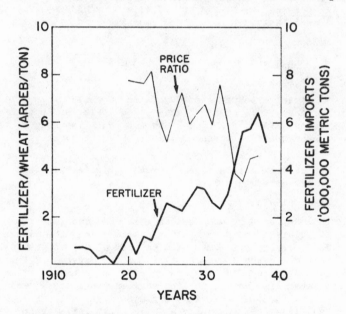

Fig. 4.7 Fertilizer Imports and Ratio of Fertilizer
and Wheat Prices, Interwar Period
(Source: Annuaire Statistique)

Footnotes

[1]     However, G. Manville Fenn in The Khedive's Country (London, 1904), 154, noted that the two-year rotation harmed the cereal crops as well due to the "lateness of sowing" which the two-year rotation required.

[2]     Gabriel Saab, (1967), 33-34.

[3]     H.K. Selim, Twenty Years of Agricultural Development in Egypt, 1919-1939, (Cairo, 1940), 73. Compare these years with Table 4.2.

[4]     S. Avigdor, "L'Egypte Agricole," L'Egypte Contemporaine, 21 (1930); E. Minost, "L'Action contre la Crise," L'Egypte Contemporaine, 22 (1931); L. Jullien, "Chrônique Agricole de l'année 1927," L'Egypte Contemporaine, 18 (1927).

[5]     Minost, (1931), 421.

[6]     B. Hansen and G. Marzouk, (1965), 55.

[7]     see note 4.

[8]     V. Israel, "Le problème du blé en Egypte, L'Egypte Contemporaine, 20, (1929), 251; Géographie du bassin du Nil, (Paris, 1933), F.A.L. des ecoles chrétiennes; M. Cassoria, "Chrônique de l'annee 1922", L'Egypte Contemporaine, 14 (1923).

[9]     Ministry of Agriculture L'Egypte Agricole, 21. The same view is presented in H.K. Selim, (1940), 115.

[10]    Saab, (1967), chap. 2.

[11]    In 1920, some 28,000 feddans were planted in fruit trees. Stimulated in part by tariff protection, the figure rose to 33,000 feddans in 1930, and to over 66,000 in 1937, concentrated in the Southern Delta and the Fayyum. Selim, (1940), 150-151. In 1937, the percentage of the cultivated area planted in cotton for the various provinces was as follows: Behera, 42%, Gharbiyya, 45%, Daqahliyya, 45%, Sharqiyya, 38%, Menufiyya, 30%, Qalyubiyya, 27%, Giza, 26%, Fayyum, 33%, Beni suef, 42%, Minya, 47%, Asyut, 43%, Gira, 32%, Qena, 6%, Aswan, 7%. Ibid., 114.

[12]    This is the "inducement" hypothesis. See Yujiro Hayami and Vernon Ruttan, Agricultural Development, An International Perspective, (Baltimore, 1971).

[13]    Lambert, (1938); L'Egypte Indépendante, 289-290.

[14]    Calculated from Agricultural Census, 1939, 12-13.

[15]    Of course the relevant question is not whether the removal of constraints on production raised revenue alone, but whether it raised revenue by more than it raised costs. There are two points worth noting in this context: (1) part of the cost

of drainage was borne by the government, and (2) there seem to have been economies of scale in its installation and use. Both of these points are discussed in the text below.

[16]     'Amr, (1958), 130.

[17]     Crouchley, (1938), 219–220.

[18]     Selim, (1940), 56. The quality of this drainage is question-
         able, however, as discussed below.

[19]     Egypt, La Loi sur les digues et Canaux. Décret de 1905.

[20]     Platt and Hefny, Egypt: A Compendium, (N.Y., 1958), 18.

[21]     William L. Balls, (1915), 157.

[22]     Egypt, La Législation en Matière Immobilière en Egypte.

[23]     L. Jullien, (1927).

[24]     Crouchley, (1938); Selim, (1940) 14, 55–56.

[25]     There is, of course, the possibility of peasant cooperation
         to solve the problem within a given village. Although there
         is some testimony about peasants' exchanging labor services,
         when it came to issues involving land, they seem to have been
         highly individualistic and noncooperating. "The Egyptian
         village is not a community in the social sense, not an organ-
         ism, but a mass." Ayrout, (1962), 113.

[26]     M. Saleh, (1922), 21–22.

[27]     Crouchley, (1938), 245.

[28]     Abdel-Mawia Basheer, Approches to the Economic Development of
         Agriculture in Egypt, unpub. Ph.D. dissertation, University
         of Wisconsin, 1961.

[29]     William Cleland, The Population Problem in Egypt; (Lancaster,
         Pa., 1936); Saffa, (1948), 396. See below, Chapter 6, on the
         current drive to install tile-drainage.

[30]     D.S. Gracie, Ministry of Agriculture, Technical and Scien-
         tific Service Bulletin #152, (Cairo, 1935).

[31]     William Balls, Yields of a Crop, (London, 1953), 62.

[32]     See, for instance, ibid., 11.

[33]     Ibid.

[34]     F.A.L. des écoles chrétiennes, (1933).

[35]     J. Anhoury, "Les grandes Lignes de l'économie Agricole de
         L'Egypte," L'Egypte Contemporaine, 32 (1941).

[36]     Saffa, (1948).

[37]     V.M. Mosseri, "La fertilité de l'Egypte," (1926) 113.

[38]     I.E. Nagy, Die Landwirtschaft im heutigen Aegypten und irhe
         Entwicklungsmoeglichkeiten, (Vienna, 1936), 67.

[39]    M. Schanz, (1913), 68. This was a problem in any case, how-
        ever, because of neighbors' fields.

[40]    William Balls, (1953), Figure 51.

[41]    Egypt, Ministry of Agriculture, La Culture du Coton en
        Egypte. (1950), 95. This remains a common practice to this
        day.

[42]    Mahmoud Tawfik Hefnawi, "Manuring of Vegetables," Ministry of
        Agriculture, Horticultural Section Leaflet #15 (Cairo, 1920);
        Balls, Gracie, and Khalil, Ministry of Agriculture, Technical
        and Scientific Service Bulletin #249, (Cairo, 1936), 24.

[43]    Gracie and Khalil, (1935), 7.

[44]    Balls, (1953), 69.

[45]    E. Dudgeon, "Cotton Cultivation in Egypt," L'Egypte Contem-
        poraine, 14 (1923), 529.

[46]    Cartwright and Dudgeon, Ministry of Agriculture, Miscellane-
        ous Publications, (Cairo, 1922).

[47]    D.S. Gracie, et. al. (1935), 32.

[48]    Ibid.

[49]    Balls, Gracie, and Khalil (1936), 31.

[50]    Ibid, 24, 31.

[51]    Ibid.

[52]    Ibid., 19-20.

[53]    Ibid., 31.

[54]    Gracie, and Khalil (1935), 32.

[55]    Nassif, "L'Egypte, est-elle surpeuplée?" L'Egypte Contem-
        poraine, 33 (1942).

[56]    H. Ammar, Growing up in an Egyptian Village. (London, 1954).

[57]    Schanz, (1913), 37.

[58]    Saffa, (1948), 37.

[59]    The spread of threshing machines is discussed below.

[60]    Hayami and Ruttan, (1971).

[61]    Jeffrey G. Williamson, Review of Hayami and Ruttan, in Jour-
        nal of Economic History, 33 (June, 1973).

[62]    Owen, (1969).

[63]    Zvi Griliches, "The Demand for Fertilizer: An Economic
        Interpretation of Technical Change", Journal of Farm Econom-
        ics, 40 (August, 1958), and "Distributed Lags, Disaggregation
        and Regional Demand Functions for Fertilizer" Journal of Fam
        Economics 41 (February, 1959).

[64]    Ibid. The specification of the Griliches model given here follows that of C. Peter Timmer, "Fertilizer and Food Policy in LDCs", Food Policy, 1 (February, 1976).

[65]    Timmer (1976).

# 5 | Stagnation and Inequality, 1920–1940

## INTRODUCTION

In this chapter we shall examine the impact of social structure and technical change on the growth of agricultural production and the distribution of land and income between the World Wars. This period is not characterized by any of the sweeping social changes or rapid growth of output that preceded World War I. The main picture is one of stagnation. Indeed, the term "agricultural involution" could be well applied to Egypt during the interwar decades, for, as we shall see, per capita agricultural production declined from the pre-war level, although there was some reversal of this trend towards the end of the period. Nevertheless, the pre-World War I levels of per capita production were not regained by that time. The first part of this chapter will review the evidence on agricultural production per person between the wars, and discuss the causes for that decline.

The second theme is how this (reduced) per capita output was shared among the population. I have found nothing that would contradict Hansen's[1] findings of a decline in rent's share and the constancy of labor's. However, the "factor shares" picture is not my main concern, although I will offer a few observations on its relationship with the technical changes described in the last chapter. Rather, I will try to examine (1) the distribution of cotton revenue between small holders and pashas and (2) the fate of the rich peasants and the landless agricultural workers. The interwar era lacks the dramatic social changes which characterized the pre-British period, but the number of renters and especially sharecroppers declined in the 1920s. There were also some attempts to reform the credit structure. Both of these changes will receive attention in the discussion of income distribution. In general, it appears that the wealthy proprietors and rich peasants probably did better than the small holders, and that the situation of the landless class was desperate indeed.

## THE STYLIZED FACTS OF EGYPTIAN AGRICULTURE, 1913–1940

The "stylized facts" of interwar Egyptian agriculture seem clear: output per worker in agriculture fell from prewar levels. The extent and timing of the fall differ depending, of course, on what sort of index number is used. O'Brien's[2] index weights production by the average gross value of the crops from 1913–1962.

(Table 5.1) Hansen and Marzouk[3] reproduce the index of Muhammmad el-Imam, which is a chained ideal Fisher index. The main discrepancies between these two measures are not very important, since they agree on the basic picture: precipitous fall in World War I, stagnation thereafter. Imam's index, however, shows a greater increase in the 1930s than does O'Brien's.

The dislocations of World War I appear to have been the cause of the rapid fall in output at the time. Animals, men, and crops were requisitioned for the British war effort at irregular and unpredictable times. Indeed, here the British returned to the old abuses which they had so piously decried in the 1880s and after. As in Isma'il's day, the provincial Mudir was told to furnish a certain number of men to serve in the army (now, the "Egyptian Labor and Camel Transport Corps"). The village 'umdah then decided which local villagers would be drafted to fulfill the government's quota.[4] This practice began in the summer of 1915 and continued until the end of the war. The recruits labored in Egypt, in Syria, and in France: they also served in the ill-fated Gallipoli campaign. At the time of the latter, the British also began to requisition crops. At first contracts for clover and hay were placed with Greek intermediaries, who, apparently, made handsome profits.[5] Soon, however, the British

---

TABLE 5.1
INDEX NUMBERS OF OUTPUT PER WORKER
IN AGRICULTURE
1895/1899-1960/1962

| Years | Index |
|-------|-------|
| 1895-99 | 100 |
| 1900-04 | 102 |
| 1905-09 | 98 |
| 1910-14 | 96 |
| 1915-19 | 76 |
| 1920-24 | 74 |
| 1925-29 | 79 |
| 1930-34 | 73 |
| 1935-39 | 74 |
| 1940-44 | 64 |
| 1945-49 | 69 |
| 1950-54 | 73 |
| 1955-59 | 88 |
| 1960-62 | 95 |

Source:
Patrick O'Brien, "The Long Term Growth or Agricultural Production in Egypt," in P.M. Holt, ed., Political and Social Change in Modern Egypt, 191.

---

either resorted to forced levies via the village 'umdahs or seized crops such as wheat, maize, barley, and birsim directly themselves. They also seized animals for transport. Small peasants were hit especially hard by the war. Not only were their crops, their animals, and they themselves requisitioned by the British army; also, the sudden fall of the price of cotton in 1914 led to forced sales of jewelry, cattle, and movable property to pay their taxes.[6] Credit was also restricted. Probably most important of all for agricultural productivity, the extension of the drainage system came to a complete halt.[7] Further damage to the irrigation system occurred during the revolt of March–April 1919, when peasants blocked irrigation canals and flooded lands.[8]

For all of the disastrous impact of World War I and the revolt of 1919 on agricultural productivity, it is worth noting that O'Brien's index begins to decline before World War I and does not show any substantial improvement until the Nasser era. For both indexes, output per worker in agricultural from 1920–1940 is well below the pre–1914 era. Whatever the fluctuations, the overall picture for the period is one of stagnation.

SOURCES OF GROWTH OF AGRICULTURAL LABOR PRODUCTIVITY

Hayami and Ruttan provide an analytical framework for studying changes in agricultural labor productivity.[9] Using cross-sectional data, they demonstrate that differences in labor productivity in agriculture can be imputed to three main sources: (1) differences in "resource endowment," (2) differences in the use of "industrial inputs," and (3) differences in "human capital." They emphasize that the "resource endowment," land, is produced, not given by nature:

Land (measured by hectares of agricultural land) used for agricultural production cannot be regarded as a mere gift of nature. It represents the result of previous investment in land clearing, reclamation, drainage, fencing, and other developmental measures.[10]

It should be clear by now that this was (and is) the case in Egypt. The failure to "produce" enough high quality land, the failure to provide adequate drainage, undermined agricultural productivity. Such a failure was especially critical because of the increasingly unfavorable man/land ratio. Hayami and Ruttan point out that machinery is, in general, used only in the context of a favorable land/labor ratio; lacking this in Egypt, machinery was not used, except for irrigation and wheat-threshing on large estates.[11] Fertilizer, the other main industrial input in agricultural production, primarily raises output per unit of land. If this grows more rapidly than land per worker declines, output per worker rises. But Egyptian cotton needed high quality land if the crop was to respond well to fertilizer. In the Egyptian context, some kinds of land and fertilizer were not substitutes. It seems likely that a good deal of the rise in output per worker near the end of the interwar period can be attributed to increased inputs of drainage as well as fertilizer, leading to a rise in yields (see Table 5.2). For cotton, an additional source of yield changes in the interwar period was the replacement of the Sakellerides variety by higher-yielding types.[12]

The final source of productivity growth, "human capital" or technical and, especially, elementary education, was neglected in the interwar period in Egypt. Article 19 of the Constitution of 1923 made elementary education compulsory, but the law was never enforced.[13] There can be little question that rural social relations played an important role here. In addition to the usual problems of a fellah family's not being able to spare the labor of its children, schools above the fourth grade were located in towns and

TABLE 5.2
YIELDS* PER FEDDAN OF THE PRINCIPAL CROPS
1920-1939

| Year | Maize | Wheat | Beans | Barley | Millet | Rice | Cotton |
|------|-------|-------|-------|--------|--------|------|--------|
| 1920 | 6.86 | 5.02 | 4.46 | 5.78 | 8.42 | 4.47 | 3.30 |
| 1921 | 6.07 | 4.78 | 4.46 | 5.71 | 8.07 | 3.80 | 3.37 |
| 1922 | 6.45 | 4.55 | 4.44 | 5.68 | 7.97 | 2.88 | 3.73 |
| 1923 | 6.78 | 4.98 | 4.43 | 5.64 | 8.05 | 4.26 | 3.81 |
| 1924 | 6.78 | 4.55 | 3.95 | 5.44 | 8.23 | 4.22 | 4.07 |
| 1925 | 7.00 | 4.94 | 4.45 | 5.72 | 8.45 | 4.33 | 4.14 |
| 1926 | 7.01 | 4.57 | 3.87 | 5.71 | 8.77 | 1.356** | 4.29 |
| 1927 | 6.97 | 5.05 | 4.65 | 6.00 | 8.33 | 1.383 | 4.01 |
| 1928 | 6.92 | 4.42 | 3.80 | 5.55 | 9.02 | 1.362 | 4.64 |
| 1929 | 6.82 | 5.28 | 4.56 | 5.95 | 9.17 | 1.411 | 4.63 |
| 1930 | 6.94 | 4.92 | 4.31 | 5.74 | 8.58 | 1.395 | 3.97 |
| 1931 | 6.71 | 5.26 | 4.17 | 5.97 | 8.95 | 1.185 | 3.78 |
| 1932 | 7.01 | 5.62 | 4.87 | 6.21 | 8.85 | 1.249 | 4.53 |
| 1933 | 6.68 | 5.28 | 4.47 | 5.95 | 8.23 | 1.354 | 4.75 |
| 1934 | 7.14 | 4.87 | 4.09 | 5.98 | 9.21 | 1.401 | 4.36 |
| 1935 | 7.66 | 5.56 | 4.55 | 7.01 | 9.87 | 1.603 | 5.11 |
| 1936 | 7.49 | 5.88 | 4.89 | 7.22 | 8.75 | 1.574 | 5.32 |
| 1937 | 7.56 | 6.01 | 4.94 | 7.35 | 9.08 | 1.513 | 5.57 |
| 1938 | 7.49 | 5.88 | 4.78 | 7.35 | 8.62 | 1.629 | 4.67 |
| 1939 | n.a. | 6.15 | n.a. | 7.55 | 9.26 | 1.738 | n.a. |

Source:
   Annuaire Statistique, Monthly Agricultural Statistics.

Units:
   Cantars (1=99.049 lbs.) for cotton; ardebs (1=5.444 bushels) for all others, except rice, 1926-39, which is given in daribas (1=934.5 kilograms) for those years.

**Year in which rice numbers switch from
   ardebs to daribas. "

cities. There were no school buses or other transportation facilities from the villages to these towns. Further, these schools
required western dress (a complete suit!) and shoes.[14] Beyond the
obviously prohibitive cost of providing such clothing, there is evidence that landlords did not take kindly to such signs of upward
mobility. Ghonemy reports cases in which a landlord was "infuriated,
insulted and outraged" to find one of his workers wearing shoes,
feeling that such behavior meant that the worker would no longer be a
good farm laborer.[15]

Although the contempt of the wealthy for the poor is a feature
of most societies, it seems especially pronounced in those producing
export crops on large estates, be they plantations or 'izab.[16]
'Ezbah workers were often in debt to their landlords, were supervised
with the kirbaj, and lived in fear of losing their jobs to sink into
the even worse condition of the daily-wage laborers. They consequently affected "an attitude of reverential awe, servility and mistrust" toward their overlords, occasionally negating both this defer-
ence and their dependent status with acts of violent resistance.[17]
The pashas believed that peasants could "be driven only with the
lash," and acted accordingly.[18] Such social relations were about as
likely to lead to mass education and literacy in interwar Egypt as
they were in the United States' South. Education of the fellahin
served no useful function for the ruling class in a labor-intensive
agriculture producing cotton for export. Nor could the widespread
incidence of bilharzia, anklyostoma, and pellagra and related protein
deficiencies have helped the educational prospects of the fellahin.
The combination of few resources, an oppressive class structure, malnutrition and disease blocked any significant contribution of education to increased agricultural labor productivity.

TOTAL FACTOR PRODUCTIVITY AND THE GROWTH OF LIVESTOCK

The picture of total factor productivity is even more discouraging than that of output per worker. The data problems make it virtually impossible to construct a reliable measure, but some indications
of the basic trend are available. Since the increase in all of these
inputs would be in the denominator of a total factor productivity
index, and since the numerator would be increased only by the inclusion of animal output (meat and milk), it seems highly likely that a
total factor productivity story would be even more discouraging than
the output per worker tale.

Given the enormous rise in the number of farm animals, one might
ask whether these animals should be conceived as an input or as an
output. Given the circular nature of agricultural production, the
question is artificial. Buffaloes and cattle served multiple functions for small holders, who owned some 50% of the animals in 1939.
They were the primary power source for land preparation and irrigation with saqiyas; they provided milk and cheese for home consumption, and were a store of value for emergencies. The rapid increases
in their number is consistent with the evidence of further intensification of production presented in Chapter 4. Broadly speaking, total
factor productivity would have fallen more rapidly than output per
worker if non-labor inputs grew more rapidly than labor. For some

inputs, such as "high quality land", there are no data. However, there is some information which would support the contention of a more rapid fall in total factor productivity than in output per worker. (see Table 5.3)

Feeding these animals was, however, increasingly difficult. Buffaloes and oxen grazed on birsim for at least half of the year, and then were fed a diet of straw, bean stalks, and cut birsim. [19] The figures for the area planted in birsim are presented in Table 5.4. From 1919–1920/1936–1937 the area allocated to this crop increased at an annual compound rate of approximately 1.5%, much less than the rate of increase of buffaloes (3.2%/yr.) or cattle (3.7%/yr).[20] There are no statistics on birsim yields. Since the crop responded well both to phosphates and to animal manure, there may have been a "virtuous circle" of more animals grazing on birsim, more manure being dropped on the fields, giving higher yields of birsim, hence more fodder, etc.[21] Some support for this notion may be found in a comparison of the birsim requirements for a buffalo given by Foaden and Fletcher in 1908 (3/4 feddan) with that of Ayrout in 1938 (1/2 feddan).[22] Buffaloes also "ate the coarse grasses and water rushes growing abundantly along the banks of the irrigation canals."[23] Since the number of canals increased markedly during the period, this food source presumably also rose.

However, the evidence on the other sources of animal fodder, necessary for the summer months, contradicts any picture of fodder supply keeping up with animal numbers. Bean output and area fluctuate, but, if anything, the figures indicate a fall in bean production (see Table 5.5). The maize area also fluctuated, but was generally lower in the 1930s than in the 1920s: the average area for 1920–

---

TABLE 5.3
COMPOUND GROWTH RATES OF INPUTS
IN AGRICULTURAL PRODUCTION,
INTERWAR PERIOD

| Input | Years | Rate of Growth (%/yr) |
|-------|-------|------------------------|
| Agricultural Labor Force | 1917–37 | 1.9 |
| Rural Population | 1917–37 | 1.1 |
| Drains | 1917–37 | 2.66 |
| Buffaloes | 1917–37 | 3.2 |
| Cattle | 1919–37 | 3.7 |
| Fertilizer Imports | 1920–38 | 8.1 |
| Fertilizer Imports | 1917–37 | 14.28 |

Source:
Calculated from Annuaire Statistique. See Tables 4.5; 4.7; 4.8.

---

TABLE 5.4
BIRSIM AREA,
1919/20-1936/37

| Year | Area (feddans) |
|------|------|
| 1919/20 | 1,356,180 |
| 1920/21 | 1,458,191 |
| 1921/22 | 1,396,975 |
| 1922/23 | 1,379,037 |
| 1923/24 | 1,387,563 |
| 1924/25 | 1,345,826 |
| 1925/26 | 1,437,199 |
| 1926/27 | 1,490,048 |
| 1927/28 | 1,509,466 |
| 1928/29 | 1,475,148 |
| 1929/30 | 1,463,982 |
| 1930/31 | 1,518,706 |
| 1931/32 | 1,639,780 |
| 1932/33 | 1,579,271 |
| 1933/34 | 1,437,262 |
| 1934/35 | 1,510,398 |
| 1935/36 | 1,741,398 |
| 1936/37 | 1,762,477 |

Source: Annuaire Statistique.

1925, 1925-1930 and 1932-1938 was 1,907,075; 1,989,144; and 1,610,044 feddans, respectively.[24] However, (1) it is not clear how relevant this is since the animals were fed the stalks, not the seed, and I have found no information on the relationship between the stalks and seed yield; (2) if it is assumed, for want of evidence to the contrary, that seed yield is a good proxy for stalk yield, we find that average yearly output was lower in the 1930s than in the 1920s (see Table 5.6) Barley, which was almost entirely an animal food, presents a similar story (see Table 5.7). Only wheat, a source of straw, shows a rise in production—up some 62.6% from 1919 to 1939.[25] The same problems of interpreting the maize figures apply here as well, however.

It appears then that, (1) the animals got most of their fodder from birsim, canal rushes, and wheat straw (tibn) and (2) they were poorly nourished.

...Fodder is...in most cases scanty and poor. The result of these conditions is...that all animals are in general allotted green fodder, such as cattle, buffaloes, sheep and goats have degenerated markedly and need to be cross-bred with other breeds. The prominent bones of the cattle reveal the deterioration of this once strong breed.[26]

An important reason for the growth in the number of animals was the spread of disease control. A vaccine for cattle plague was invented in 1912,[27] and in 1936 the Ministry of Agriculture was claiming that cattle plague had been eliminated from Egypt.[28] The same source also echoed Nagy's complaint that the livestock were poorly fed. This is not surprising, since, as we shall see, the

TABLE 5.5
AVERAGE YEARLY PRODUCTION OF BEANS
1919/1924-1932/1938

| Years | Bean Production (ardebs) |
|-------|--------------------------|
| 1919-24 | 2,042,689 |
| 1925-28 | 1,902,512 |
| 1932-38 | 2,044,957 |

Source: Annuaire Statistique.

TABLE 5.6
AVERAGE YEARLY PRODUCTION OF MAIZE,
1920/1925-1932/1938

| Years | Maize Production (ardebs) |
|-------|---------------------------|
| 1920-25 | 12,581,147 |
| 1926-30 | 13,795,963 |
| 1932-38 | 11,720,075 |

Source: Annuaire Statistique.

TABLE 5.7
AVERAGE YEARLY CONSUMPTION OF BARLEY
1919/1924-1931/1937

| Years | Barley Consumption (tons) |
|-------|---------------------------|
| 1919-24 | 219,720 |
| 1925-28 | 214,442 |
| 1931-37 | 200,976 |

Source:
Annuaire Statistique. Consumption = Production + Imports - Seed - Exports.

owners of these animals, frequently small peasants, were none too well fed themselves in the interwar period.

TECHNICAL CHANGE AND DISTRIBUTION: FACTOR SHARES

This brings us to the question of distribution in the interwar period. Unfortunately, the evidence is even scantier than for the British period. Nevertheless, I will try to sketch the overall picture. First, I will review the "factor shares" story, then discuss the question of the relative position of pashas and small landholders. Finally, I will present evidence on wages and on the number and conditions of the landless agricultural workers.

The factor shares picture is discussed thoroughly by Bent Hansen.[29] Over the whole period from 1897-1961, he finds a fall in rent's share; labor's share, on the other hand, appears to have been more or less constant. He attributes this fall in rents to increased fertilizer use and to an expansion of the cropped area. These two combine, in his view, to provide "substitutes for farm land."

I have no quarrel with his basic findings. Nor do I question the fact that fertilizer did, in the case of numerous important crops, provide a "substitute for farm land." But, apparently, this was not the case for cotton. A slightly more accurate description of events would be that (1) fertilizer, a substitute for land for certain purposes, came to be used on a large scale, and, (2) at the same time, drainage works were rapidly expanded, producing more high quality land. Thus, not only was the demand for land (conceived of as all of those characteristics of a piece of ground except for the quantity of nitrogen and phosphate) shifting in because of a fall in the price of a "substitute"; the supply was shifting out as more land was produced by the provision of more intensive drainage. These two processes, plus the adoption of the more land-intensive rotation, reinforced one another, reducing the relative scarcity of land.

This picture is consistent with the differences between the 1920s and 1930s. In the earlier decade the share of rents is upward, starting from the outbreak of World War I. The trend continues until the end of the decade and the onset of the Depression.[30] It is worth recalling that fertilizer imports grew rapidly from 1913 to 1929 (see Table 4.8), at the same time that rent's share rises. Drainage, on the other hand, expanded much more slowly in that decade (see Table 4.5). It is during and after the 1930's that the fall in rent's share appears; this is the same decade in which public drainage works were expanding most rapidly. Of course, fertilizer use was also expanding, but the use of fertilizer and drainage were complementary for the principal cash crop, cotton. The "factor shares" picture seems consistent with a story which emphasizes the importance of drainage.

TECHNICAL CHANGE AND DISTRIBUTION: SMALL PEASANTS AND PASHAS

I will return to the factor shares story in the discussion of the position of the landless agricultural workers. But first let us examine the distribution of cotton revenue between the pashas and the small landholding peasants.

As in Chapter 3, we shall concentrate on the changes in cotton revenue per person for the two classes. Such changes result from the combined effect of changes in the percentage of land planted in cotton (or, the rotation system used), changes in cotton yields, and changes in land area per person. The first issue, that of rotation employed, was discussed in Chapter 4 (see 4.2) Small peasants retained the two-year system—there was probably no change in the percentage of their land planted in cotton. On the other hand, some pashas were switching to the more cotton-intensive rotation. Here we find change in favor of the pashas in terms of the <u>direction</u> of change in cotton revenue.

The story of land per landowner for the two groups is slightly more complex. Figures on this are presented in Table 5.8. Two points emerge. First, average peasant landholdings steadily declined. Second, land per pasha fell during World War I, and was stable thereafter. Although legally fragmentation affected everyone equally, in actual practice it hit the small peasants harder. This is due at least in part to pasha land purchases of newly reclaimed land in the Northern Delta in the late 1920s. Thus, although the Gini index for landownership indicates stability or some slight trend towards equality in landholding (see Table 5.9), the pashas held their own in per capita terms.

Furthermore, for large numbers of peasants, discussed below in the section on agricultural wage workers, such fragmentation fundamentally changed their life and work. If a peasant held less than three feddans, he could not support his family, and had to enter either the market for land, trying to rent a parcel, or for labor, hiring himself out. In either case, he entered into relations of dependency and lost a degree of decision-making power over his work. Thus, not only did fragmentation as reflected in statistical measures not affect the pashas and the small peasants equally, but there was also a qualitative difference in how fragmentation affected them. In terms of the revenue model, the pashas seem to have done better than the peasants in changes in their land per person.

It is not clear what happened to the yields of the two groups; no numbers are available. However, an argument can be made that here, too, the pashas did better than the small peasants. First, recall the importance of well-drained land for cotton yields. Two points are in order here. First, since the small peasants adopted the water-using two-year crop rotation in the pre-World War I period when drainage was most inadequate, their lands presumably had deteriorated more than those of the pashas, who had, in general, retained the three-year system. Second, the existence of indivisibilities and hence economies of scale in drainage meant that the peasants did not invest in drainage to the same extent as did the pashas. Third, the cost of field drainage installation would have been prohibitive for a fellah, amounting to roughly one-third of the net revenue of a five feddan farm. [31] Consequently, it seem highly likely that their lands suffered more than those of the pashas from drainage problems. Assuming that the peasants behaved rationally, and given the interrelatedness of drainage and fertilizer

TABLE 5.8
AVERAGE LAND HOLDING PER PERSON, SMALL PEASANTS AND PASHAS,
1914—1939

(T*/N) (feddans/landholder)

| Year | 0-5 feddans | 50 feddans |
|------|-------------|------------|
| 1914 | 1.01 | 193.63 |
| 1915 | .99 | 192.00 |
| 1916 | .98 | 191.63 |
| 1917 | .97 | 188.88 |
| 1918 | .95 | 183.12 |
| 1919 | .93 | 179.70 |
| 1920 | .90 | 167.37 |
| 1921 | .91 | 171.26 |
| 1922 | .90 | 176.71 |
| 1923 | .89 | 176.76 |
| 1924 | .87 | 176.88 |
| 1925 | .87 | 170.46 |
| 1926 | .86 | 177.90 |
| 1927 | .86 | 176.28 |
| 1928 | .87 | 179.22 |
| 1929 | .85 | 181.11 |
| 1930 | .84 | 180.02 |
| 1931 | .84 | 181.22 |
| 1932 | .84 | 183.37 |
| 1933 | .83 | 181.39 |
| 1934 | .82 | 181.27 |
| 1935 | .82 | 181.29 |
| 1936 | .82 | 181.85 |
| 1937 | .82 | 178.73 |
| 1938 | .82 | 175.69 |
| 1939 | .82 | 178.00 |

Source: Calculated from Annuaire Statistique.

TABLE 5.9
GINI COEFFICIENT FOR LANDOWNERSHIP, 1900-1952
(excludes landless)

| Year | Coefficient |
|------|-------------|
| 1900 | .6668* |
| 1913 | .7726 |
| 1920 | .7744 |
| 1933 | .7608 |
| 1945 | .7594 |
| 1952 | .7580 |

*Calculated on a different basis: 0-1 and 0-5 feddan holders aggregated into one category.

Source: Calculated from Annuaire Statistique.

application, it seems likely that the pashas' yields improved more rapidly than those of the small peasants.

Such a presumption is reinforced when one considers the credit situation. During the 1920s the small peasants' only source of credit was local moneylenders. The Law of Five Feddans prevented the fellah from offering his land as collateral for loans. Since small peasants had to borrow to obtain fertilizer (or, for that matter, adopt any of the technical changes discussed in Chapter 4, all of which were capital-using), and since they would have had to pay the very high rates of interest (20-30% per year) which moneylenders charged, fertilizer cost them more than it did the pashas. Therefore, we would expect them to use less fertilizer than the pashas.

In 1931 the government established the Credit Agricole Egyptien. This bank made loans to small fellahin and to cooperatives. Initially, it charged the former 7% per year and the latter 5%. The rates were lowered to 6% and 4%, respectively, in 1933, and to 5% and 4% in 1939.[32] Annuaire Statistique presents detailed figures on these loans, telling us not only how much fertilizer was purchased with the loans, but also for which crops the fertilizer was to be used (see Table 5.10). It is interesting to compare the tonnage figure for cotton for 1937 with the statement of Balls, Gracie, and Khalil that, in that year, cotton "did not receive less than 340,000 tons" of fertilizer.[33] Thus the bank was financing about 19% of the fertilizer used on cotton. Now although there are serious data biases, Annuaire Statistique figures give 32% of the land area being held in plots of 0-5 feddans in that year. This indicates that many peasants were not borrowing from the Credit Agricole Egyptian. Most small peasants continued to rely on village moneylenders in the 1930s.[34] Such a view is reinforced by comparing the figures for the number of borrowers with the number of small holders in the 0-5 feddan class, (see Table 5.11) assuming that all borrowers were small fellahin. These figures are not to be trusted entirely, however, since the Annuaire Statistique numbers overstate the number of proprietors.[35]

This bias, however, is quite probably counteracted by a different bias in Table 5.11. In fact, numbers of borrowers from the bank were not small holders (as defined in this study), but rather were either (1) "middle" or rich peasants owning more than five feddans or (2) large landowners. The law establishing the Credit Agricole Egyptien explicitly empowered the bank to sell fertilizer and seed to all cultivators, regardless of the size of their holdings. Large landowners also borrowed through the mechanism of cooperatives whose borrowings amounted to 20-29% of the banks loans for fertilizer from 1936 on.[36] These cooperatives were dominated by large landlords, who used them to obtain fertilizer at the very low interest rates charged by the bank.[37] The evidence suggests, then, that the small peasants did not use as much chemical fertilizer as did the large landlords.

This is consistent with the fact that peasant lands had probably been more subject to deterioration than those of the pashas. Further, the peasants feared using the Credit Agricole Egyptien. Mention is made of the "complicated bank loan processes and restrictions."[38] Second, they feared losing their land, whereas moneylenders showed "une certaine indulgence" toward the repayment of loans. The fact that payments on loans from the bank were collected by tax collectors at the same time as the land tax no doubt contributed to peasant mistrust. The fellahin's fear of government institutions continued to play its role in the countryside.

Finally, since the peasants were probably not using as much fertilizer as the pashas, they would have had difficulty adopting earlier planting. Recall that if the number of cuts of birsim were not to be changed (highly unlikely, given the fodder shortage), then

TABLE 5.10

FERTILIZER TONNAGE PURCHASED FOR USE ON DIFFERENT CROPS WITH CREDIT FROM THE CREDIT AGRICOLE EGYPTIAN, 1933-1939

| Year | Wheat | Beans | Cotton | Rice | Maize |
|------|-------|-------|--------|------|-------|
| 1933 | 28,834 | 297 | 10,036 | 96 | 9,543 |
| 1934 | 42,841 | 1,509 | 35,318 | 952 | 19,163 |
| 1935 | 49,524 | 2,728 | 40,099 | 4,515 | 25,331 |
| 1936 | 37,482 | 1,425 | 54,653 | 5,449 | 22,331 |
| 1937 | 47,482 | 1,435 | 63,975 | 2,925 | 16,435 |
| 1938 | 44,895 | 1,413 | 60,070 | 5,425 | 19,014 |
| 1939 | 52,972 | 1,320 | 54,134 | 6,230 | 22,633 |

Source: Calculated from Annuaire Statistique.

TABLE 5.11
PEASANTS BORROWING FROM THE CREDIT AGRICOLE EGYPTIAN
AS PERCENTAGE OF THE TOTAL NUMBER OF PEASANTS
HOLDING LESS THAN FIVE FEDDANS,
1932-1939

| Year | Percentage |
|------|------------|
| 1932 | 8.9% |
| 1933 | 6.7 |
| 1934 | 12.31 |
| 1935 | 13.01 |
| 1936 | 12.05 |
| 1937 | 12.30 |
| 1938 | 12.25 |
| 1939 | 12.75 |

Source: Calculated from Annuaire Statistique.

adoption of earlier planting required changing birsim's place in the crop rotation. This, in turn, implied an increased need for nitrogen fertilizer. Since the peasants had relatively less nitrogen fertilizer than the pashas (we presume), the adoption of earlier planting would not have increased their yields to the same extent as earlier planting would have for the pashas. Of course, failure to adopt such planting would also have prevented their yields from advancing. Whichever alternative they chose, the peasants were likely to have had lower increases in yields than the pashas.[39]

It appears, then, that the pashas' gross revenue from cotton grew more rapidly than did that of the small peasants. In terms of the model employed in Chapter 3, the peasants kept using the two-year rotation (% change in $Tc/T^*>0$); some of the pashas were switching (% change in $Tc/T^*>0$). If 1914 is chosen as the base year, pasha $T^*/N$ fell by 8%, that of the peasants by 20%; if 1920 is used, pasha $T^*/N$ for the pashas rises slightly, but that of the peasants falls by 8%. Finally, there is no reason to suppose that peasants managed to offset these disadvantages by increasing their yields more rapidly than the pashas. Quite the contrary. It seems very likely that the pashas were getting at least as much fertilizer and were able to drain their lands more easily. Their lands had also undergone less deterioration than had those of the peasants. The pashas were better able to adopt the practice of earlier planting. It seems highly likely that the position of the small fellahin relative to the pashas was deteriorating.

What of the rich peasants? The evidence here is even scantier than for the other two landowning groups. We have already seen that middle proprietors adopted the two-year rotation before World War I, and that there is no evidence of any reswitching. Like small peasants, the percentage of their land planted in cotton presumably did not change. Likewise, land per "middle proprietor" was also quite stable. The average holding for 5-10 feddan holders was 6.93 feddans in 1914, 6.84 in 1939; for 10-20 feddan holders, the averages are 13.8 and 13.6, and for 20-50 feddan holders, the averages are 30.5 and 30.1 in 1914 and 1939 respectively.[40] This is more stable than either the pashas' or small peasants' average.

Although there is no direct information on rich peasants' yields, let us see whether any of the forces which prevented poor peasants from adopting the three major technical changes of the period also constrained rich peasants. First consider earlier planting. We have seen that the small peasants would not adopt this technique because they needed to take extra cuttings of birsim to feed their animals. Here rich peasants were better off; small peasants, with some 30% of the land, owned roughly 50% of the cows and buffaloes, while rich peasants, with 30% of the land, held only 34% of the work animals. Since the animal/land ratio was lower for rich peasants than for poor peasants, they had less reason for taking extra cuttings of birsim and thereby delaying the planting date of cotton. Further, they would have changed the place of birsim in the rotation more easily, since, as we shall see, they could purchase nitrogen fertilizer more easily.

Drainage presents a mixed picture. On the one hand, because middle proprietors had switched to the two-year rotation before the war, their lands had probably deteriorated as had those of small peasants. On the other hand, rich peasants' farms were larger and their resources greater, so the problems of fixed costs and indivisibilities must have been less acute. Further, their larger farm size reduced the costs and difficulties of arranging agreements on whose land would be traversed by drainage canals. And should such a problem have arisen, their considerable local political and social power, strengthened by the system of indirect Parliamentary elections, would have guaranteed a solution favorable to their interests.[41] Since their lands were probably better drained than those of small peasants, the marginal physical product of fertilizer on their cotton fields would have been higher.

It was also easier for them to get credit. In the 1920s the Agricultural Bank made loans to holders of 5-10 feddan farms, and the Credit Foncier and the Egyptian Land Bank made loans to those holding 10 feddans and more.[42] The expansion of credit by the Credit Agricole Egyptien in the 1930s must have helped them also. They were eligible for the loans[43] and their prominence in their localities ensured them favorable treatment by the cooperatives. If credit was more available to rich peasants, fertilizer would, in turn, be cheaper, and we would expect them to use more of it than did the poor fellahin. It seems fair to conclude that the rich peasants' yields grew faster than those of small peasants. It is likely, therefore, that the gap between rich and poor peasants widened in the interwar period.

The evidence suggests that these middle proprietors began to exploit their estates more directly between the wars: We have seen evidence that these proprietors relied heavily on share and cash rental systems before World War I. By 1939, however, some 73% of lands held in farms of 5 to 50 feddans were exploited directly.[44] Such a breakdown by farm size is not available from 1929, when 78% of the total cultivated area was exploited directly. But there is evidence that the change came in the 1920s. First, from 1929-1939, there was hardly any change in the cultivated area which was exploited directly, moving from 78% to 79%, respectively. Second, the census of 1917 lists 506,181 persons as "cultivators of land taken on lease." By 1929 that number had fallen to 234,687, although the population occupied in agriculture had increased from 2.8 million to 3.5 million. In 1937 the number had fallen again, to 210,384, while the agricultural work force grew to 4.28 million. The striking change is clearly in the 1920s, when nearly 80% of the total decline from 1917 to 1937 occurred. Sharecropping in particular seems to have declined. Although Saleh in 1922 described this system as the normal mode of exploitation for rich peasants, writers in the 1930s speak of the system as being "very rare"[45] or "infrequent."[46] At that time sharecropping seems to have been confined to poor or middling land where the rural population density was relatively low, such as in the Northern Delta or on Delta lands abutting the desert.[47] By 1939, when the agricultural census differentiated cash from share rents for the first time, 75% of leased lands were rented for cash. [48] Sharecropping (metayage) was practiced, then, on only about 5% of the cultivated area.

It would appear that the shift away from renting was toward some variety of wage-labor system, whether it was the 'ezbah system or simply a straight cash wage system. The 1917 census lists 414,162 persons as "agricultural laborers (wage earning)"; the 1927 census lists 1,435,214 as "labourers." Given the difficulties surrounding the 1917 census (see Chapter 3) with respect to the definition of "laborers" as well as the probability that agricultural laborers were often drafted into service in World War I, it is helpful to compare the data from 1927 with those of 1907. In 1907, 832,785 workers were listed as "ouvriers et domestiques de ferme." Thus, from 1907 to 1927, the number of "paid workers" rose some 80% while the population occupied in agriculture grew by only 46%. It would appear, then, that a shift out of renting, and especially out of sharecropping, into the 'ezbah system or some other form of direct exploitation occurred between 1907 and 1927, probably between 1917 and 1927.

Of course, this "shift" may be nothing more than a "statistical artifact." However, the changes are quite striking, so it is possible that such a change actually did occur. If so, the reasons for the change are obscure. The literature on agricultural land and labor contracts emphasizes that the returns to the contracting parties under different kinds of contracts equalizes over time.[49] There is nothing in the relative price data which would suggest that direct exploitation rather than some rental contract would have become more attractive in the interwar period (see above, Table 4.4, on the ratios of rents and cotton prices to wages).

Perhaps the credit expansion of the interwar period contributed to the demise of sharecropping. Credit for rich peasants expanded in the 1920s,[50] and in the 1930s rich peasants used the Credit Agricole Egyptien, which made advances at below the market rate of interest to cover the cultivation costs of cotton and wheat. Making such short-term loans was a principal activity of agricultural cooperatives, in which rich peasants and other landlords were prominent.[51] These advantages might have contributed to the decision not to switch <u>back</u> to sharecropping during the period of falling cotton prices of the early 1930s, but of course, only the credit expansion of the 1920s can have contributed to the actual shift out of renting from 1917 to 1927. The same may be said for government rent control policies: they may have encouraged farmers to stay with systems of direct exploitation. But since the laws controlling agricultural rents were passed in the early 1930s, they can have played no role in the shift of the 1920s. In any case, it was argued in parliament that since the laws primarily affected those holding long leases, they would affect only large "intermediary" renters, not the small peasant renters.[52] Finally, although the expansion of credit <u>may</u> have contributed to the demise of sharecropping, it is worth noting that sharecropping did not entirely disappear, but was confined to marginal (i.e. high-risk) land zones.

The reasons for the (apparent) change and then stability in the systems of land tenure from 1917 to 1937 remain something of a puzzle. We can only note the trend, wonder whether or not it is a statistical artifact, and observe that farmers did not respond to the fall in their revenues in the 1930s caused by the fall in cotton prices by switching back to sharecropping.[53] Instead they lowered wages (see Table 4.5) and, if urban residents, went back to their villages to direct agricultural operations personally.[54]

TECHNICAL CHANGE AND DISTRIBUTION: THE FATE OF THE LANDLESS CLASS

The picture of relative deterioration of the position of those on the bottom of rural society is unrelieved when one looks at the situation of the landless agricultural workers. We have seen that the share of wages remained constant, while the share of rents fell. The course of wage rates is shown in Table 5.12. In general, nominal wage rates were fairly steady in the 1920s and fell sharply in the Depression. Of course, much of the fall can be attributed to the collapse of the prices of agricultural commodities which occurred in the early 1930s (see Table 5.13). Given the very spotty nature of the data, we may conclude that the trend of this (deflated) wage was either roughly constant or slightly upward.

This is consistent with the information presented in Chapter 4 on the nature of technical changes in interwar Egyptian agriculture. Recall that drainage, earlier planting, and fertilizer use were all labor-using technical changes. Since the adoption of artificial fertilizer did not imply any reduction in the use of farmyard manure, we may conclude that the demand for labor increased over the whole period.[55] At the same time the population grew rapidly: Annuaire Statistique gives the population occupied in agriculture as rising from 2.82 million in 1917 to 3.5 million in 1927 to 4.28 million in

TABLE 5.12
NOMINAL WAGES, MAIZE PRICES AND REAL WAGES,
1900-1938

| Year | Wage Rate (L.E./day) | Price of Maize (L.E./ardeb) | $\frac{w}{Pm}$ (ardeb/day) | $\frac{w}{Pm}$ (kg/day) |
|------|------|------|------|------|
| 1900 | .03-.045 | .712 | .04-.06 | 3.88-8.68 |
| 1901 | .03 | | .04 | 5.96 |
| | | | | |
| 1906 | .025 | .887 | .03 | 4.05 |
| 1907 | .025 | .800 | .03 | 4.34 |
| | | | | |
| 1910 | .035-.04 | .744 | .047-.0538 | 6.58-7.53 |
| | | | | |
| 1912 | .047 | 1.09 | .043 | 6.02 |
| 1913 | .03-.045 | 1.014 | .0247-.0444 | 3.46-6.22 |
| 1914 | .025-.03 | .899 | .278-.0333 | 3.89-4.66 |
| | | | | |
| 1920 | .07-.08 | 2.0592 | .034-.039 | 4.76-5.46 |
| | | | | |
| 1922 | .04-.05 | 1.1619 | .034-.043 | 4.76-6.02 |
| | | | | |
| 1927 | .05 | 1.1127 | .045 | 6.3 |
| 1928 | .045 | 1.1695 | .038 | 5.32 |
| 1929 | .055 | .9498 | .058 | 8.12 |
| | | | | |
| 1931 | .04 | 0.7749 | .052 | 7.28 |
| 1932 | .03 | 0.5902 | .051 | 7.14 |
| 1933 | .025 | 1.0427 | .024 | 3.36 |
| 1934 | .027 | 1.0602 | .025 | 3.50 |
| | | | | |
| 1937 | | 1.0733 | | |
| 1938 | .03 | n.a. | .028 | 3.92 |

Source:
Wage Rates, see Table 3.12 and 4.4.  Maize Prices Annuaire Statistique.

1937.

It should be emphasized that a large percentage of these people
were seeking wage work.  First, as fragmentation proceeded, increasing numbers of peasants had to enter the labor market.  Second, land
losses  for debt and especially for failure to pay taxes continued in
the interwar period.  Until 1926, if a peasant accumulated more  than
L.E.  2  in tax arrears, he lost his land.  In that year the level of

TABLE 5.13
WAGE RATES, COTTON PRICES, AND THEIR RATIO,
1920-1938

| Year | Wages (L.E./day) | Price of Cotton (L.E./cantar) | $^W\!/Pc$ (cantars/day) |
|------|------------------|-------------------------------|-------------------------|
| 1920 | .07-.08 | 5.9916 | .012-.013 |
| 1922 | .04-.05 | 5.3343 | .0075-.0094 |
| 1927 | .05 | 5.1560 | .0097 |
| 1928 | .045 | 4.4954 | .010 |
| 1931 | .04 | 1.7506 | .23 |
| 1932 | .03 | 2.2337 | .014 |
| 1933 | .025 | 2.2791 | .011 |
| 1937 |  | 2.1536 |  |
| 1938 | .03 |  | .014 |

Source:
 Wages, see Table 4.4; Cotton Prices, Annuaire Statistique.

arrears for which land was to be seized was reduced to L.E. 1.[56]
From 1927 to 1937, some 44,000 fellahin lost their lands.[57] The
early years of the depression under the Sidqi government seem to have
been especially bad, when tax collectors revived the use of the kirbaj and forced the peasants to sell cattle, implements, and land to
pay the taxes.[58] There was resistance to such official
violence,[59] but it was never organized on any scale. Law Number 65
of 1942 legalized industrial trade unions but forbade the organization of agricultural workers.[60] By that time there were at least
one-and-a-half million landless peasants.[61] To this number should
be added the 1.75 million who held less than one feddan. Actually,
we should add all of those who held less than three feddans, the
minimum necessary for self-sufficiency. Such figures are not available, however. Given that the average holding in the 1-5 feddan
class was 2.10 in 1920 and also in 1930, and had fallen to 2.05 feddans by 1938, it seems reasonable to suppose that any upward bias in
the number of 0-1 feddan holders is more than offset by the exclusion
of a large number of fellahin holding between one and three feddans
who would also have been seeking wage work. Overall, more than 75%
of the rural population had too little land to live on. Many of
these must have entered the market for labor, increasing its supply.[62]

The situation of these landless and land-poor was grim indeed in
the interwar period. A relevant index would be the wage deflated by
the price of maize. As Table 5.12 shows, real wages fell sharply in
the Depression. In addition, per capita consumption of the principal

foodstuffs declined in the 1930s after a rise in the 1920s (see Table 5.14).[63] There is certainly no evidence to indicate that such deficiencies were made good by consuming other sorts of food. Even small peasants owning some land apparently often sold the butter, milk, and cheese produced by their livestock, as well as any eggs, rather than consume them themselves.[64] In any case, production was low: in 1929, 1.7 gallons of fresh milk for fresh consumption per person were produced; the figures for butter and cheese are 2.6 and 3.3 pounds per person, respectively.[65] Egg production was 4 1/2 dozen per person per year.[66] Rural consumption of eggs was about 50% that of the urban population, and the people in the countryside consumed from 2/3 to 3/4 less meat than city dwellers. Further, in the rural sector, more than 66% of the population consumed less than the rural average.[67]

Such a situation most certainly led to malnutrition. A study by the Cairo University Medical School in the 1930s estimated that peasant diets lacked 20% of necessary proteins.[68] Pellagra was widespread, affecting perhaps one-third of the Delta population,[69] which is not surprising given the fact that the bulk of the peasants' food was corn bread:

It is associated with a corn (maize) diet and is primarily due to a dietary deficiency of niacin. Although corn has more niacin than some other staple foods, it seems that this is not all utilized, most likely because it is in a bound form unavailable to the body. The human body can convert the amino acid tryptophan into niacin...the main protein in corn is sein which is very low in tryptophan content.[70]

There is a vegetable source of niacin—beans[71]—but, per capita consumption of this food fell sharply in the interwar period. These nutritional problems were aggravated by the synergistic effects which exist between protein and vitamin deficiencies and the parasitical diseases bilharzia (schistosomiasis) and anklyostoma. The lot of the fellahin in the 1930s was miserable indeed.

TECHNICAL CHANGE AND DISTRIBUTION: THE SA'ID

The one bright spot in the distributional picture is the improvement in the relative position of the Sa'id. A sharp contrast should be drawn here between cotton-growing lands and those still under the basin system. The former areas did well: they had begun to produce cotton more recently and had superior drainage.[72] They used much more fertilizer per acre of cotton than did Delta cultivators. Their yields were presumably correspondingly higher, although the available figures are not of much use since the varieties grown in the two regions were different. The area which could receive perennial irrigation expanded by some 200,000 feddans due to the construction of new canals. On the other hand, the spread of this irrigation system brought not only higher incomes, but also bilharzia, anklyosoma, and malaria. The basin lands continued to be left behind: the location-specific nature of technical change remained: no perennial irrigation, no cotton. On the whole, however, the region almost certainly improved its cotton revenue position relative to the

TABLE 5.14
PER CAPITA CONSUMPTION OF BEANS AND MAIZE,
1917–1938

| Year | Beans (pounds) | Maize (kilograms) |
|------|------|------|
| 1917 | 45.95 | 123.7 |
| 1918 | 48.15 | 122.3 |
| 1919 | 48.65 | 118.7 |
| | | |
| 1920 | 39.26 | 128.2 |
| 1921* | 43.38 | 120.8 |
| 1922 | 42.53 | 119.3 |
| 1923 | 52.57 | 118.0 |
| 1924 | 35.11 | 120.3 |
| 1925 | 40.76 | 136.8 |
| 1926 | 31.55 | 138.9 |
| 1927 | 39.92 | 131.0 |
| 1928 | 36.46 | 129.6 |
| 1929 | 44.29 | 118.5 |
| | | |
| 1930 | 34.34 | 117.4 |
| 1931 | 30.78 | 129.7 |
| 1932 | 47.52 | 124.9 |
| 1933 | 34.66 | 93.2 |
| 1934 | 31.29 | 98.2 |
| 1935 | 33.56 | 100.5 |
| 1936 | 34.10 | 97.6 |
| 1937 | 33.29 | 100.6 |
| 1938 | | 94.1 |

Source:

Annuaire Statistique. Consumption = Production – seed + imports – exports.

* unreliable numbers.

Delta.

SUMMARY

In conclusion, the record of the interwar years is a disappointing one. Despite increased drainage, more farm animals, improved water supply, and large-scale imports of chemical fertilizers, crop yields did not grow rapidly enough to compensate for the fall in land per person. Many of the technical changes were responses to problems which had been created earlier. The real problem in Egypt was the production of high quality land, and previous events had made this difficult and costly. The technical changes did manage to increase the demand for labor enough to keep pace with the growth of population. Despite all this, by 1940 output per worker in agriculture was still below the 1914 level. Although the situation appears to have been improving in the late 1930s, in the interwar era Egypt was experiencing a kind of agricultural involution.

This is a gloomy picture. It becomes more gloomy still when one turns to distribution. The only bright spot is the improvement in the relative position of the Sa'id. The share of labor remained constant, but the landless and landpoor were ill-nourished, disease-ridden, and subject to discrimination at the hands of the wealthy. It was extremely difficult, if not impossible, for them to obtain an education. Their numbers rose not only through natural increase, but also as small peasants lost their lands for failure to pay taxes, for debts, and through fragmentation. Among landowners, the gap between the largest and the smallest widened. Of considerable importance for later years, both technical change and the political system strengthened the position of the richer peasants.

The technical changes of this period either maintained the preexisting distribution (with respect to wage earners), or worsened it (between large and small farmers). The changes did little to affect the strength of moneylenders in the villages. Government credit policy probably weakened the moneylenders' position somewhat, but it hardly eliminated them. As the position of small peasants deteriorated, they became increasingly similar to the landless agricultural workers. By the postwar era, writers speak of the two classes in the countryside, the rich and the poor. The small peasants, the 'ezbah workers, and the tarahil laborers were mired in the "swamps of poverty, of ignorance, and of endemic disease"[73] as the burden of declining or stagnant agricultural output per capita weighed heavily on their shoulders.

Footnotes

[1]     Bent Hansen, "The Distributive Shares in Egyptian Agricul-
        ture, 1897-1961," International Economic Review, 9 (June,
        1968).

[2]     in Holt, ed. (1968).

[3]     Hansen and Marzouk, (1965).

[4]     M. Sabry, La révolution égyptienne, (Paris, 1919), 9.

[5]     Sir Valentine Chirol, The Egyptian Problem, (London, 1930),
        136.

[6]     'Abd al-Rahman Al-Rafi'i, Thawrah 1919 (The Revolution of
        1919), (Cairo, 1955), 54-55; Sabry, (1919), 15; 'Amr, (1958),
        129

[7]     Report by His Majesty's Agent...for the Years 1914-1919, 40.

[8]     Chirol, (1930), 186, quoting British Military Bulletins, p.
        186.

[9]     Hayami and Ruttan, "Agricultural Productivity Differences
        Among Countries," American Economic Review, 60, (Dec., 1970).

[10]    Ibid., 896-897.

[11]    We shall see in Chapter 6 how and why farm mechanization has
        advanced in the 1970s despite an even less favorable
        land/labor ratio.

[12]    It should be noted that Sakel did not begin to be used on a
        wide scale until after the pre-World War I decline in yields
        discussed in Chapter 3.

[13]    See, inter alia, Khalil Saleh Reda, The Underpriviledged of
        Egypt; Twelve Million Farmers. Unpub. M.S. thesis, the
        University of Wisconsin, 1942.

[14]    M.R. Ghonemy, Resource Use and Income in Egyptian Agricul-
        ture, unpublished Ph.D. dissertation, North Carolina State
        College, 1953.

[15]    Ibid., 132-135. See also Ayrout, (1962), chap. III, on the
        attitudes of large landowners to their workers and tenants.

[16]    Cf. Eugene Genovese, Roll, Jordan, Roll: The World the
        Slaves Made, (N.Y., 1974), for the attitudes of slave holders
        in the United States South.

[17]    See ibid., for a masterly analysis of the Master-Slave
        dialectic of contempt-submission suddenly alternating with
        fear-revolt in the historical context of the United States
        South. See esp. 561-566 on education.

[18]    Ayrout, (1962), 33.

[19]    For example, Saffa, Ayrout, Foaden and Fletcher, Nagy, Jul-
        lien, Anhoury, and others.

[20]    Converting to Livestock Units (buffaloes = 1 L.U., cattle =
        .8 L.U.), the area of birsim for livestock fell from 1.3 fed-
        dans per L.U. in 1920 to 1.02 feddans per L.U. in 1937.

[21]    Compare the reverse process of the decline in pastures, a
        fall in the number of animals, and a decline in yields in
        Western Europe in the thirteenth and early fourteenth centu-
        ries. See B.H. Slicher van Bath, The Agrarian History of
        Western Europe, A.D. 500-1850, (London, 1963), 89.

[22]    Foaden and Fletcher, A Textbook of Egyptian Agriculture, Vol.
        2, (Cairo, 1910), 784; Ayrout, (1962), 38-39.

[23]    Foaden and Fletcher, (1910), 784.

[24]    Calculated from Table 5.6 and 5.2.

[25]    Annuaire Statistique, 1940.

[26]    Nagy, (1936), 136.

[27]    Jullien, "Chrônique agricole de l'année 1913," L'Egypte Con-
        temporaine, 4, 1913, 583-594.

[28]    Ministry of Agriculture, L'Egypte Agricole, 1936, p. 83.

[29]    Hansen, (1968).

[30]    Ibid.

[31]    L'Egypte Indépendante, (1937), 283. The calculation of net
        revenue does not include amortization costs on implements and
        livestock.

[32]    M.H. Abbas, The Role of Banking in the Economic Development
        of Egypt, Ph.D. unpublished dissertation, University of
        Wisconsin, 1954, 109.

[33]    Balls, Gracie, and Khalil, Ministry of Agriculture Technical
        and Scientific Service Bulletin #249.

[34]    L'Egypte Indépendante, (1937), 284.

[35]    See Gabriel Baer, (1962), 73ff, for a discussion of this
        point.

[36]    Platt, "Land Reform in Egypt," A.I.D. Spring Review of Land
        Reform (Washington, D.C., 1970), 18.

[37]    Abbas, (1954), 105; Ayrout, (1962); Zannis, (1937).

[38]    Ghonemy, (1953), 64.

[39]    This dilemma may explain why many peasants did not adopt ear-
        lier planting and had lower yields as a result.

[40]    Annuaire Statistique, various issues.

[41]    Cf. Leonard Binder, In a Moment of Enthusiasm: Political
        Power and the Second Stratum in Egypt, (Chicago, 1978), Chap.
        6.

[42]    E. Minost, "Essai sur la richesse foncière de l'Egypte" L'Egypte Contemporaine, 21 (1930), 348-349.

[43]    At first only individuals holding forty feddans and under could receive loans from the Crédit Agricole Egyptien, but this was later raised to ninety, and finally to two hundred feddans in 1937. Abbas, (1943), 103-104.

[44]    Agricultural Census, 1939, 12-13.

[45]    L'Egypte Indépendante, (1937), 293.

[46]    Lambert, (1943).

[47]    Ibid.

[48]    Agricultural Census, 1939, 18.

[49]    See, e.g., Steven N.S. Cheung, The Theory of Share Tenancy, (Chicago, 1969).

[50]    E. Minost, (1930).

[51]    Ibrahim Rashid Bey, "The Cooperative Movement in Egypt," L'Egypte Contemporaine, 30 (1939), 488-489.

[52]    Laws of the early 1930s reduced rents and accorded delays in payments. Law #103 of 1931 reduced rents for the year 1929-1930 if rent had not been linked to the price of cotton. Law #110 of 1931 reduced rents by 30% and granted a delay of one year if the lease was longer than one year. These laws were extended in 1932 (Law 551) and 1934 (Law #121) A. Yallouz, "Chrônique Législative de l'année 1933-34," L'Egypte Contemporaine, 225 (1934), 768-770.

[53]    There is some evidence that those renting in cash demanded payment in kind (not necessarily in shares) in 1931. Lambert, (1943).

[54]    Zannis, (1937), 29.

[55]    Of course, there were fluctuations. A fair amount of unemployment must have been generated in 1931 when the price of cotton fell so drastically that the cotton area was markedly reduced. The use of winnowing and threshing machines would have also reduced the demand for labor. It is unlikely that such mechanization was sufficiently extensive to have had a significant aggregate impact on the demand for labor.

[56]    L.A. Hugh-Jones, "The Economic Condition of the Fellahin," L'Egypte Contemporaine, 20 (1929), 410-411.

[57]    Ayrout, (1962), 24.

[58]    'Amr, (1958), 132.

[59]    M.F. Elkaisy Pasha, "The Public Security in Egypt in 1927," L'Egypte Contemporaine, 19 (1928), 23-24; L'Egypte Indépendante, (1938), 290.

[60]    H. Husayn, (1971), 21.

[61]    Doreen Warriner, Land and Poverty in the Middle East, (London, 1948), 34.

[62]    Of course, they also entered the market for land, trying to rent small parcels. But we have seen that most lands were exploited directly. However labor was remunerated, the rise in the number of peasants who did not own enough land to live on contributed to the increase in the supply of labor to the pashas and rich peasants.

[63]    Bent Hansen and Michael Wattleworth have constructed a consumption index for all cereals and pulses. This index also shows decline in the 1930s. Hansen and Wattleworth, "Agricultural Output and Consumption of Basic Foods in Egypt, 1886/87-1967/68", International Journal of Middle Eastern Studies, 9,4 (November, 1978).

[64]    Anthony Galatoli, Egypt in Midpassage, (Cairo, 1950), 44; Jacques Berque, Histoire Sociale d' un Village Egyptien, (Paris, 1957), 32-33.

[65]    El Zalaki, (1943), 250ff.

[66]    Galal Amin, Food Supply in Relation to Economic Development, with Special Reference to Egypt, (London, 1966), 63-64.

[67]    Ibid.

[68]    Ibid.

[69]    Ibid., quoting W.H.O. and F.A.O. sources; M.B. Ghali, The Policy of Tomorrow, Isma'il el-Faruqi, trans., (Washington D.C., 1953), 42.

[70]    Michael C. Latham, et. al., Scope Manual on Nutrition, (Kalamazoo, Mich., 1970), 41.

[71]    Ibid., 72.

[72]    Gracie and Khalil (1935).

[73]    'Amr, (1958), 7.

# 6

## Egyptian Agriculture Transformed, 1940–1980

As in all colonial areas, World War II inaugurated a prolonged period of political, social and economic crisis in Egypt. The cut in the trade links and the need to feed the British army led to further, extensive government intervention in the agricultural economy, to widespread changes in cropping patterns, to severe problems of food supply, and to rampant inflation. The problem of Egyptian sovereignty was exacerbated by the ultimatum of 1942, when the British sent tanks to the Abdin Palace and dictated new terms to the King. The Wafd's collusion in the renewal of the British occupation delegitimized them in the eyes of many Egyptians, contributing to the rising stature of more radical groups, especially the Muslim Brethren. Many peasants moved to the cities, pulling up agricultural real wages, which had fallen from 1939 to 1942. Meanwhile, large landlords turned increasingly to renting out their estates for cash.

The end of the War and the cotton boom of the early 1950s did not resolve these crises, but, indeed, seem to have exacerbated them. The increase in cotton prices led to sharp rent increases, which led to further peasant unrest. Meanwhile, the debacle of the Palestine war deepened the political crisis. Neither the King, nor the political parties could solve or even confront the mounting internal and external crises of Egyptian society; that task fell to the Nasser regime. Let us have a quick look at the agricultural and social changes of the period 1940–1952. This will provide background for the main concern of this chapter, social and technical change under Nasser and Sadat.

The principal agricultural changes of the War years were 1) the drastic shift in cropping patterns (from cotton to cereals), 2) the equally drastic decline in fertilizer imports, 3) the fall, and then increase, in real wages, and 4) the (probable) increase in direct renting of large estates. The key to the first two changes was, of course, the fall of France and the Italian entry into the War which severed Egypt's trade links. The growth of Axis power in the Mediterranean and the severe shortage of Allied shipping capacity prevented the export of cotton and the import of fertilizer on anything like the pre-war scale. At the same time, there was a mounting supply problem in the country, as British troop levels rose from approximately 50,000 in both Egypt and the Sudan in 1939 to nearly 200,000 in 1942.[1]

The food crisis was the catalyst for change. As the data in Tables 6.1 and 6.2 show, cereal production fell from 1939 to 1941, largely as a result of the catastrophic decline in fertilizer imports. Government policy was initially passive, however: the decrees on land areas (described below) came only after the worst was over and Rommel's advance had been stopped at El-Alamein (July, 1942). Indeed, both the British and the Egyptian governments purchased the cotton crop in 1940 and 1941.[2] In the absence of area restrictions, this led to the rough stability in cotton's (and therefore also cereal crops') share of the cultivated area. (See Table 6.1) Since fertilizer imports, and, therefore, grain yields had fallen, while the area remained little different, cereal output fell. Farmers and middlemen hoarded grain; the newly constituted Middle East Supply Center received warnings in early 1942 that the "grain stocks in private hands were not expected to last until March" as the Afrika Korps approached Alexandria.[3]

The government's response to the crisis was two-fold. To deal with the immediate emergency, small amounts of grain held for the army were periodically released. Meanwhile, the government organized a large-scale system of compulsory deliveries: in effect, the government nationalized the grain trade.[4] This system was retained throughout the War, and was supplemented by a series of decrees which dictated a considerable increase in the amount of land devoted to cereals, and a corresponding reduction in the amount of cotton planted.[5] Finally, they sweetened the pill for smaller farmers by distributing fertilizers through the Egyptian Agricultural Credit Corporation.[6] Such large-scale government intervention into agricultural production was a new departure for twentieth-century Egyptian governments, and established precedents for the policies of the Nasser regime.

The area decrees brought about a sharp shift in cropping patterns. The traditional two- and three-year crop rotations were thoroughly disrupted as cotton plantings drastically declined and as cereals came to be continuously cropped in many cases. As a result of repeated cereal planting, overall cropping intensity did not decline. (The social consequences of the shift out of cotton will be examined below.) The birsim area was also maintained, and the number of farm animals increased after an initial drop in 1940.[7] But the increased supplies of farm manure, the larger traction power, and the maintenance of the clover crop could not compensate for the decline in the fertilizer imports and intensive cereal cultivation. Unsurprisingly, crop yields declined.

Some of these changes were reversed after the War. Fertilizer imports more than doubled in 1947 (see Table 6.2). The revival of trade brought the renewal of cotton cultivation, but as Table 6.1 shows, this occurred only slowly. The government buoyed up prices by entering the futures market in 1946; the cotton price doubled.[8] At the same time, Wartime area restrictions remained in place, but were widely evaded. This was especially so when the Korean War stimulated

TABLE 6.1
CROP AREAS, PRODUCTION, AND LAND YIELDS, 1939–1952

CROP

| YEAR | COTTON Area (000 feddans) | Pro-duction (000 qantars) | Yield (qantar/ feddan) | WHEAT Area (000 feddans) | Pro-duction (000 ardebs) | Yield (ardeb/ feddan) |
|---|---|---|---|---|---|---|
| 1939 | 1,625 | 8,692 | 5.35 | 1,446 | 8,892 | 6.15 |
| 1940 | 1,685 | 9,170 | 5.35 | 1,506 | 9,071 | 6.02 |
| 1941 | 1,644 | 8,065 | 4.91 | 1,502 | 7,492 | 4.99 |
| 1942 | 706 | 4,233 | 6.00 | 1,576 | 8,411 | 5.34 |
| 1943 | 713 | 3,569 | 5.01 | 1,917 | 8,611 | 4.49 |
| 1944 | 853 | 4,640 | 5.44 | 1,651 | 6,307 | 3.82 |
| 1945 | 982 | 5,221 | 5.31 | 1,647 | 7,881 | 4.79 |
| 1946 | 1,212 | 6,066 | 5.01 | 1,586 | 7,752 | 4.89 |
| 1947 | 1,254 | 6,370 | 5.08 | 1,630 | 6,962 | 4.27 |
| 1948 | 1,441 | 8,900 | 6.07 | 1,516 | 7,203 | 4,75 |
| 1949 | 1,692 | 8,704 | 5.14 | 1,417 | 7,782 | 5.49 |
| 1950 | 1,975 | 8,500 | 4.30 | 1,372 | 6,785 | 4.95 |
| 1951 | 1,980 | 8,750 | 4.08 | 1,500 | 8,060 | 5.39 |
| 1952 | 1,970 | 9,922 | 5.04 | 1,402 | 7,260 | 5.18 |

CROP

| YEAR | MAIZE Area (000 feddans) | Pro-duction (000 ardebs) | Yield (ardeb/ feddan) | RICE Area (000 feddans) | Pro-duction (000 dariba) | Yield (dariba/ feddan) |
|---|---|---|---|---|---|---|
| 1939 | 1,548 | 10,888 | 7.03 | 547 | 950 | 1.74 |
| 1940 | 1,540 | 10,925 | 7.09 | 509 | 706 | 1.39 |
| 1941 | 1,527 | 9,158 | 6.00 | 448 | 606 | 1.35 |
| 1942 | 1,983 | 10,369 | 5.23 | 673 | 995 | 1.48 |
| 1943 | 1,951 | 9,836 | 5.04 | 642 | 725 | 1.13 |
| 1944 | 1,890 | 11,135 | 5.89 | 620 | 862 | 1.39 |
| 1945 | 1,879 | 12,124 | 6.45 | 630 | 917 | 1.45 |
| 1946 | 1,653 | 10,157 | 6.15 | 632 | 993 | 1.55 |
| 1947 | 1,608 | 10,010 | 6.23 | 776 | 1,351 | 1.74 |
| 1948 | 1,551 | 10,066 | 6.49 | 786 | 1,384 | 1.76 |
| 1949 | 1,494 | 8,929 | 5.98 | 703 | 1,236 | 1.76 |
| 1950 | 1,451 | 9,329 | 6.43 | 700 | 1,314 | 1.88 |
| 1951 | 1,660 | 9,928 | 6.12 | 490 | 647 | 1.36 |
| 1952 | 1,704 | 10,571 | 6.30 | 374 | 536 | 1.48 |

TABLE 6.1 CONT.

CROP

| YEAR | MILLET Area (000 feddans) | Production (000 ardebs) | Yield (ardeb/ feddan) | BARLEY Area (000 feddans) | Production (000 ardebs) | Yield (ardeb/ feddan) |
|------|------|------|------|------|------|------|
| 1939 | 412 | 3,819 | 9.26 | 263 | 1,985 | 7.55 |
| 1940 | 374 | 3,241 | 8.67 | 268 | 2,009 | 7.50 |
| 1941 | 429 | 3,352 | 7.81 | 256 | 1,754 | 6.86 |
| 1942 | 824 | 6,886 | 8.36 | 321 | 2,306 | 7.17 |
| 1943 | 729 | 5,524 | 7.58 | 419 | 2,620 | 6.25 |
| 1944 | 728 | 5,454 | 7.49 | 331 | 1,890 | 5.71 |
| 1945 | 684 | 5,241 | 7.66 | 359 | 2,180 | 6.08 |
| 1946 | 551 | 3,752 | 6.81 | 245 | 1,480 | 6.05 |
| 1947 | 343 | 4,127 | 7.61 | 237 | 1,414 | 5.96 |
| 1948 | 525 | 3,989 | 7.60 | 220 | 1,387 | 6.30 |
| 1949 | 499 | 4,053 | 8.13 | 168 | 1,152 | 6.84 |
| 1950 | 393 | 3,040 | 7.73 | 117 | 758 | 6.49 |
| 1951 | 420 | 3,442 | 9.08 | 120 | 830 | 7.02 |
| 1952 | 433 | 3,413 | 9.03 | 137 | 981 | 7.19 |

CROP

| YEAR | BEANS Area (000 feddans) | Production (000 ardebs) | Yield (ardeb/ feddan) | SUGAR Area (000 feddans) | Production (000 qantars) | Yield (qantar/ feddan) |
|------|------|------|------|------|------|------|
| 1939 | 385 | 1,974 | 5.13 | 72 | 54,899 | 759 |
| 1940 | 394 | 2,005 | 5.09 | 76 | 55,952 | 733 |
| 1941 | 369 | 1,836 | 4.98 | 78 | 52,650 | 678 |
| 1942 | 359 | 1,789 | 4.98 | 88 | 55,638 | 633 |
| 1943 | 381 | 1,832 | 4.81 | 87 | 48,656 | 557 |
| 1944 | 425 | 2,080 | 4.90 | 96 | 57,832 | 605 |
| 1945 | 392 | 1,989 | 5.07 | 96 | 58,855 | 611 |
| 1946 | 381 | 1,935 | 5.08 | 92 | 56,323 | 611 |
| 1947 | 382 | 1,686 | 4.42 | 94 | 60,302 | 644 |
| 1948 | 398 | 1,852 | 4.66 | 91 | 55,592 | 609 |
| 1949 | 424 | 2,069 | 4.88 | 85 | 50,281 | 588 |
| 1950 | 356 | 1,277 | 3.59 | 81 | 56,290 | 691 |
| 1951 | 320 | 1,495 | 4.68 | 90 | 62,534 | 727 |
| 1952 | 355 | 1,610 | 4.53 | 92 | 75,561 | 785 |

TABLE 6.1 CONT.

| | CROP | |
|---|---|---|
| | BIRSIM | FRUIT |
| YEAR | Area | Area |
| | (000 feddans) | (000 feddans) |
| 1939 | 1,757 | 75 |
| 1940 | 1,832 | 69 |
| 1941 | 1,832 | 66 |
| 1942 | 2,079 | 71 |
| 1943 | 1,652 | 71 |
| 1944 | 1,965 | 79 |
| 1945 | 1,875 | 69 |
| 1946 | 1,967 | 73 |
| 1947 | 1,990 | 81 |
| 1948 | 1,977 | 81 |
| 1949 | 2,028 | 84 |
| 1950 | 2,180 | 87 |
| 1951 | 2,120 | 90 |
| 1952 | 2,202 | 94 |

Source: Annuaire Statistique, various issues.

a large-scale boom in commodities, including cotton. By 1952 the cotton area had more than surpassed the pre-War levels.

Two social changes during the period stand out: changes in real wages and the shift toward renting. Real wages may be seen in Table 6A1. (See Appendix 1 to this chapter) It is evident that the sharp decline in real wages from 1939 to 1942 was the result of cost-of-living increases: the shortage of cereals had severe consequences for the poor. It is also noticeable that real wages began to recover in the same year, and with the exception of 1946, rose fairly steadily until 1952. The demand for labor was reduced by the shift from cotton to cereals, and by the decline in fertilizer. But such demand-side effects were more than compensated for by the migration of fellahin to the cities, where they sought employment with the British forces. By the War's end, some 200,000 Egyptians were so employed; many others flocked to the cities hoping to find such work.[9] Such migrants did not return to the countryside after the end of the War, despite the fact that demobilization disemployed thousands: By 1950 the agricultural labor force was smaller than that in 1939 by over 500,000 individuals.[10] As we shall see, the importance, even dominance, of events outside of agriculture for agricultural real wages becomes a common theme in the whole post-War era.

The second major social change was the increase in farm tenancy. We have seen in Chapter 5 that only some 17% of the cultivated area was leased, either for cash or on shares, in 1939. In 1949, however, out of a total cultivated area of over 5,900,000 feddans, some

TABLE 6.2
FERTILIZER IMPORTS, 1938-1950

| Year | Fertilizer Imports |
|------|--------------------|
|      | (tons)             |
| 1938 | 513,799            |
| 1941 | 5,240              |
| 1943 | 158,629            |
| 1944 | 272,491            |
| 1945 | 260,125            |
| 1946 | 214,437            |
| 1947 | 457,755            |
| 1948 | 509,214            |
| 1949 | 621,148            |
| 1950 | 683,506            |

Source:
  Monthly Bulletin of Agricultural Economics Statistics and Legislation, #7. Egyptian Department of Agriculture, 1951, 11.

3,610,622 feddans (60.7% of the total) were rented. Nearly two-
thirds of all rentals were for cash (2,175,196 feddans or 36.5% of
the total cultivated area), the rest being leased on shares.[11] This
represents a major departure from previous decades.

There are several possible explanations for this dramatic
increase in farm tenancy. First, the alleged change may represent a
"statistical artifact". We have seen how the exploitation of large
estates before the War relied extensively on the hybrid land/labor
system which I have called the "'ezbah system". We have seen how
some crops (i.e. maize and clover) were, in fact "leased" to the
estates' permanent agricultural workers (tamaliyya). It may be that
such practices were now considered to be "leases" whereas formerly
they had not been. This seems doubtful, however, since the prevail-
ing system of rental after the War was for cash, and because contemporary accounts are filled with discussions of the problems of
peasant renters, especially those renting cotton lands for cash.[12]
We have seen that cotton was usually grown directly by the landlords
under the 'ezbah system. This fact plus the magnitude of the change
point toward a real shift in tenurial institutions.[13]

I would suggest two hypotheses to explain the rise of cash renting: 1) the changes in cropping patterns enforced by the government
in World War II undermined the economic rationale for the 'ezbah system, and 2) the mounting political crisis created considerable uncertainty among landowners about their property rights in land. Both
provided incentives for landlords to withdraw from direct production
into cash rentals. We have seen that the crop rotation of cotton,
wheat, maize, and birsim induced landlords or their representatives

to concentrate their scarce supervisory energies on the cash crops, especially cotton, leaving maize and birsim largely to the devices of the year-round workers, the tamaliyya. The need for year-round labor and the need to make concessions, in the form of subsistence plots, to the tamaliyya were also dependent on the growing of cotton, which requires fairly steady labor inputs throughout the eight-month growing season.[14] The disruption of cotton growing sharply reduced the need for year-round labor. The regulations of World War II thus eliminated one of the principal economic props of the direct-exploitation, 'ezbah system.

This tendency may have been reinforced by the rise in real wages. It may have been that farmers were caught between compulsory deliveries at prices fixed by the state and rising wages; they may have sought to offer their lands for rent under such conditions. The quite rapid diffusion of tractors may have had a similar cause, as farmers sought to replace increasingly scarce labor with machines.[15] Farmers may also have offered leases as a means of retaining laborers who would otherwise leave for the cities. Since the crops which were now virtually the only ones grown had formerly been leased out under the 'ezbah system, landowners may have switched to cash tenancy as a way of reducing the losses which the cessation of cotton growing imposed upon them.

It is interesting to note, however, that rental systems were retained after cotton growing revived in the post-war years. Once the old, two-year crop rotation returned, why didn't the larger farmers resume direct exploitation of their estates? Political factors were crucial here. Essentially, I would hypothesize a major shift in time-horizon on the part of large landowners due to perceived threats to their property rights. There were several elements to this perception. First, the continuing political crisis in the cities led to increasing use of the land question by various political factions. When the Wafd was attacked for mismanagement of the relief efforts for the Qena and Asyut malaria epidemic and famine, Nahas countered by stressing the unequal distribution of landownership in those provinces.[16] Land reform was becoming increasingly salient as a political issue, with land reform bills even being introduced in Parliament in 1945, 1948 and 1950.

Peasant resistance and class conflict contributed to landlords' reluctance to farm their estates directly. There was a sharp increase in rural crime at the end of the War.[17] More overt forms of political action appeared shortly thereafter. In 1946 over one thousand tenants and laborers attacked and destroyed the office of the Kom Ombo estate in Aswan. In 1947, in Shoha in Daqahliyya there were revolts against absentee landlords.[18] Such conflicts mounted in intensity between 1949 and 1952: At Kafr Negm, the 7,000 feddan estate of Prince Muhammad 'Ali, peasants complained of high rents, but to no avail. They then burned crops and destroyed agricultural machinery. The police searched and pillaged the peasants' homes, arresting five, one of whom was later murdered. At the Badrawi estate the landlords' men attacked peasants' houses, seizing possessions to pay for rent arrears. The peasants were fired upon when they came to the estate house to protest. They then stoned the house and burned the grain stores. Five hundred soldiers broke up the demonstration, with considerable violence. At Mit Fadalla peasants went on strike, refusing to harvest cotton, since the rents were above the level of cotton prices. Arrests, beatings, and peasant attacks on the estate office ensued. Again soldiers were called out to suppress the fellahin.[19]

All such incidents were widely reported and commented upon, which must have done much to increase landlords' uneasiness about their future. One response would have been to withdraw as much as possible from direct production, renting out their lands for cash.[20] Landlords typically rented not to poor peasants directly, but to intermediaries.[21] These were usually local rich peasants. Such farmers were already in a strong position as a result of their farming skills, their resources, and their political position under the electoral system. The withdrawal of the large landlords from any direct role in the countryside further consolidated their position.

But although the shift in tenure forms aggrandized them and protected the large landlords, it did not end peasant resistance. Indeed, it may have exacerbated it. The conjuncture of the shift to cash rents with the rise in cotton prices during the Korean War boom led to a cycle of increased rents, increased violence, increased absenteeism, and further rent hikes. Something more drastic than a shift in the mode of farm payment was required to solve the rural social crisis. The response to the multiple crises of the Egyptian political economy came with Free Officers' Coup of July 23, 1952.

## THE POLITICAL CONTEXT OF NASSER'S AGRICULTURAL POLICIES

The political situation in which the Free Officers came to power may now be summarized.[22] External and internal crises from 1940 to 1952 had de-legitimized the established order. The Wafd, whose basis of support in the countryside had been eroded by Sidqy in the 1930s, had been further weakened in the eyes of most Egyptians by its overtures to the urban rich. Further, it had forfeited the respect of almost all Egyptian nationalists when Nahas agreed to form a Cabinet under British guns in 1942. The Palace fared little better: it had been discredited by its inept and corrupt handling of the war with the Zionists in 1948. Nor had it proved able to cope with the mounting internal problems of the country. The internecine conflict between Palace and Wafd had produced mutual exhaustion and stalemate. At the same time, the Palace was not too weak to repress effectively the major political alternatives, the Muslim Brethren, and the (considerably weaker) Marxist Left. To a certain extent, the Free Officers were able to seize power because all other political groups had been so exhausted by their mutual struggle. The Officers, as a secret, conspiratorial group, had not had to endure such competition.

This political context is essential for understanding Nasser's agricultural policies. First, the regime originated in a military conspiracy: as military officers, they were accustomed to hierarchical methods of operation. Such an orientation was reinforced by the need to retain maximum flexibility to deal with the treacherous flux of regional politics. Second, their social origins were largely the well-to-do peasantry and their urban cousins. But the Officers were not the Party of such classes. Rather, they were "Bonapartists": a military clique which lacked any firm base of support. They had been able to seize power because of their military organization and because of the weakness of the other contenders for political power. Consequently, the regime had to create its support by its actions and policies. They were able to do this not only by their actions in the international arena, where they defended the claims of Egyptian Nationalism, but also domestically, through their land reforms and other social programs. One might argue that the land reform and the expansion of education and the state bureaucracy created the basis of their support among the small peasants and the urban lower-middle class. At the same time, their policies of reform did not go so far as to alienate the rich peasantry, whose "passive" or "enabling" support as a "second stratum" was especially important to such a Bonapartist regime.[23]

Such a regime taking power in such a political economy had several important consequences for its agricultural development policies. The regime was more committed to economic development than any previous Egyptian government. They intended, and in important respects succeeded, in increasing national production, improving distribution, and promoting a variety of long-overdue social reforms. Their "developmentalism," summed up in the slogan "al-kifaiyya w'al-adl,"[24] committed them to industrialization as well as social reform. The funds necessary for such programs had to come from the agricultural sector. Whether listening to Western or to Soviet advice, the message, "squeeze agriculture", was the same. Their means were a system of agricultural cooperatives and a variety of crop price and area restrictions and regulations. But here the regime's "top down" orientation, their social origins and their need to rely on the richer peasantry created some anomalies and distortions which sowed the seeds of future difficulties.

The same "Bonapartist" legacy had implications for investment policy as well, since large, highly visible projects could be most easily controlled from Cairo. This, I shall argue, is an important part of the explanation for the regime's preference for land reclamation over investment in drainage. (That hardy perennial of Egyptian agricultural political economy!) The regime was committed to limited social reform and to projects which it could control. Such a political legacy led the regime neither to mobilize the peasantry nor to offer the decentralized incentives which any successful agricultural development strategy must provide. The regime's "Arab Socialism" or "state capitalism" fell between two stools, generating problems which the Sadat regime has inherited.[25] Let us turn to a review of the three principal agricultural policy instruments of the regime: land reform and cooperativization, investment decisions, and price and output regulations.

POLICIES: AGRARIAN REFORM[26]

The reform of land tenure was one of the first actions of the new regime. By redistributing all land held by individuals above two hundred feddans (three hundred for one family) the Free Officers eliminated at one stroke the basis of power of their principal rivals, the pashas. As we have seen, there was some precedent for their action: reform bills had been submitted repeatedly to Parliament, only to be defeated each time by the representatives of large landlords. Nasser's reform not only eliminated political rivals, but also generated political support among the small peasantry. At the same time, by not seizing medium-sized properties, the reforms avoided alienating the much needed "passive support" of the rich peasantry. Even when ceilings were lowered and all peasants compelled to join cooperatives, the rich peasants continued to control their villages, just as they always had done, as mediators between the peasantry and the government.

The reform proceeded in three stages, with the permissible ceilings on individual land ownership falling from 200 feddans in 1952 to 100 feddans in 1961, and finally to fifty feddans in 1969. Both major reforms served political goals: the first (1952) aimed to weaken the pashas and to garner support for the regime, while the second phase (1961) was part of the regime's reaction to the failure of the union with Syria.[27]

A summary of the laws, their dates and the areas affected appears in Table 6.3. Several aspects of the reform stand out. First, taken together, only 12.5% of the cultivated area was affected directly, with some 341,982 families or 1.7 million persons receiving land. Second, primarily ex-tenants received land, since only they were presumed to have the necessary farming skills. Landless day laborers, by contrast, acquired comparatively little land. Thus the old division of the agricultural work force of the 'ezbah system between permanent and day-laborers was partially reproduced by the land reform. Third, more land changed hands as a result of the reforms than the above figures show, because of widespread distress sales by pashas. The more well-to-do peasantry, who had the cash to buy such land, were the principal beneficiaries.

In assessing post-Reform land distribution, one must distinguish between the distribution of land ownership and of farms as productive units. In general, the latter are larger than the former, as large ownerships are broken down while very small owned parcels are consolidated via tenancy arrangements. The distribution of ownership and of farms in 1961, the year of the last agricultural census, are given in Table 6.4; the numbers for 1950 and 1975 are given for comparison.

The reform seems to have achieved its two principal goals of eliminating the power of the large landowning class and of promoting agricultural productivity. Although it is difficult to disentangle the impact of the land reform from other agrarian changes of the period, it would appear that not only did output increase but that the land reform itself may have helped to raise yields.[28] This result is consistent with the very large amount of evidence which shows that small farms are more intensively cultivated and have higher yields per unit area than large farms in the Third World.[29] Although the reforms largely accomplished the first goal (elimination of the large landlords) several caveats are in order. First, there

TABLE 6.3
LAND DISTRIBUTED BY AGRARIAN REFORM LAWS, 1953-1969

| | Origin of Land | Area (feddans) | Families Benefitting |
|---|---|---|---|
| 1. | Land Reform Law No. 178, 1952 | 365,147 | 146,496 |
| 2. | Law No. 152, 1937 Law No. 44, 1962 (transfer of waqf lands) | 189,049 | 78,797 |
| 3. | Second land reform Law No. 127, 1961 | 100,745 | 45,823 |
| 4. | Purchase of lands sequestered in 1956 (response to Suez invasion) | 112,641 | 49,390 |
| 5. | Law No. 15, 1963 (excluded foreigners from land ownership) | 30,081 | 14,172 |
| 6. | Other | 98 | 49 |
| | Total | 797,761 | 334,727 |

Source: Abdel-Fadil (1975), 10.

TABLE 6.4
SIZE DISTRIBUTION OF LAND
BY OWNERSHIP AND AREA FARMED,
1950, 1961, and 1975

| Size | %Area Owned | | %Area Farmed | | |
|---|---|---|---|---|---|
| | 1950 | 1961 | 1950 | 1961 | 1975 |
| 0-5 Feddans | 35.4 | 52.1 | 23.2 | 37.8 | 65.9 |
| 5-50 Feddans | 29.4 | 32.7 | 37.7 | 40.7 | 33.3 |
| 50+ Feddans | 34.2 | 15.2 | 39.1 | 21.5 | 1.8 |

Source:
    1950 and 1961: Abdel-Fadil (1975),
    1975: Ministry of Agriculture.

were a variety of ways of evading the land reform.[30] This was especially true for the first reform. Yet, such ruses apparently continued: even today one hears of families holding much more than the legal land ceilings.[31] Second, although the pashas' power in the countryside was, indeed, drastically weakened, they were not eliminated as a class. They were compensated for their land in (non-negotiable) government bonds for seventy times the value of the land tax.[32] To be sure, much of their industrial property was also expropriated in the sequestrations of the early 1960s. But their real estate was not, and they often managed to acquire important posts in the army and elsewhere in the government bureaucracy.[33] Consequently, they have been able to make a limited comeback in the more liberal climate of the Sadat regime. The pashas had largely disappeared from agriculture, but not from the political economy as a whole.

The consolidation of the rich peasants' position in the country-side was perhaps the most important aspect of the reform. Their purchases of land in distress sales by larger landlords, the elimination of the pashas, and the absence of mobilization of the poor and land-less peasants made them the dominant force in the countryside. It is true that the government greatly increased its role in the rural areas, coming to replace many of the functions of the large landowners on their 'ezbahs. (See below) However, direct government influence was certainly much stronger in land reform areas than elsewhere, even after the extension of the cooperative system to the entire country in 1962. As we have seen, such areas of firm government control comprised a relatively small percentage of the total cultivated area. In the other villages the cooperatives were easily dominated by the rich peasants and the village headmen.[34]

Even in land reform areas, the rich peasants exerted consider-able influence. First, in the initial phases of the land reform, the mushrif or government manager was often the same person as the old landlord's supervisor: the managerial personnel, usually from the stratum of well-to-do peasants, continued to act as representatives of the government, just as they had done for the absentee landlords. Even when qualified government personnel appeared in the Land Reform areas, these men were themselves often of rural middle-class origins; their origins, training, and inclinations (as well as their low pay and lack of incentives) often led them to rely on the more successful local farmers for guidance.[35]

We may note an important historical continuity here. The Nasser regime continued to rely on the well-to-do peasants as a "mediator" between the government and the mass of the peasantry. In this the Nasserites resembled every previous administration of rural Egypt, despite the former's very different ideology and social base. Government regulations certainly did little to weaken the strength of the rural middle-class. Tenancy regulations were enacted, in which leases had to be in writing and for a minimum of three years. Rents were not to exceed seven times the basic land tax assessed in 1949. If sharecropping was used (infrequent on larger estates and less than 20% of rental arrangements on 10-20 feddans), the crops and the costs were to be divided 50:50. Yet such regulations had to be enforced at the local level; in the non-land reform areas, it is unlikely that they would have been faithfully observed. Landowners could simply evade the laws, or could shift to direct exploitation of their lands using wage-labor: the percentage of the cultivated area which was leased declined from roughly sixty percent to fifty percent. Minimum wage regulations were similarly evaded.[36] One could argue that although some of the government's regulations irritated the rich peasants, the regime strengthened them more than any previous modern Egyptian government.

It is evident that the land reforms offered no solution to the problem of landlessness. They did, however, reduce the numbers of landless and their percentage in the population. In 1950, some 1.2 million families (44% of agricultural families) were landless. In 1961, their number had declined to perhaps one million families (40% of agricultural families). But given the small total areas affected by land reform and the continued growth in population, by 1972 some 1.5 million families or 45% of agricultural families, were landless.[37] It should be noted that these numbers can only be rough approximations, since they are obtained as a residual (number of agricultural families minus the number of landowning families). Abdel-Fadil gives a rather different, somewhat lower, estimate for the number of landless. In any event, somewhere between 30 and 45% of the rural population was landless by 1970.

It is doubtful that land reform could have solved the problem of landlessness. First, even if all land held over five feddans (2.912 million feddans) had been redistributed to the roughly one million landless families, the resulting land per family would have been below the subsistence minimum of three feddans. Available land per family would, of course, have fallen further over time as the population grew and Muslim inheritance law was applied. Second, such a radical land reform program would very likely have created considerable production problems, problems which Egypt could ill-afford. Insofar as landless day laborers had had little experience in making the day-to-day decisions of farmers (an inheritance of the 'ezbah system), it seems likely that a massive education and extensions program would have had to accompany any more radical land reform if output were not to fall. There is some evidence that such production problems were encountered by landless day-laborers who did receive land.[38]

One could suggest that the regime might have tried to form more truly cooperative production systems, such as, perhaps, collective farms of some sort. Now, in addition to the generally unimpressive output performance of such arrangements, such a course of action, or the alternative radical land distribution in small plots, was politically impossible. Only if the regime had been willing to take on the rich peasants and to endure substantial rural social and economic dislocations would such an attempt have been feasible. The regime had neither the ideological orientation, social origins, nor political resources to undertake such a strategy. The Nasserist regime was not about to embark on a course of mass mobilization.

A critical component of the Land Reforms and the consolidation of dominance of the rich peasants was the establishment of government cooperatives. All peasants who received land were required to join such organizations; they were extended to the non-land reform areas in 1963. Although the planners' model was the Sudanese Gezira scheme, the parallels between the cooperatives and the pre-War 'ezbah system are striking. As we have seen, there was considerable continuity of personnel initially, with the old supervisory staff of the estates continuing on the job, now paid by the government instead of by the pashas. We have also noted how the recipients of land tended to be renters, rather than day laborers and how the land reforms did not eliminate the landless class. The resulting "two-tiered" nature of the labor force in agriculture is also reminiscent of the 'ezbah system.[39] Just as the pashas could evict a peasant for failure to comply with his wishes and production instructions, so could the government: Law 554 of 1955 made it possible to evict land reform beneficiaries for mismanagement of their holdings (as judged by the supervisory staff).

The cooperatives also took over most of the managerial functions which the pasha's nazir had formerly fulfilled. The cooperative specified the crop rotation, imposing the three-year system. It also took over crop marketing and input supply. It is interesting to note here that the cooperatives, like the pashas before them, were primarily concerned with marketing cotton and wheat; disposal of maize and birsim, on the other hand, were left to the peasants. Although the peasants' plots did not rotate with the crops as was true under the 'ezbah system, they were assigned separate strips of land within consolidated blocks of cotton, maize, birsim, etc.

The cooperatives also supplied inputs, especially short-term credit in the form of cheap fertilizer, pesticides, and seeds. As we shall see, the credit system was the mechanism used to extend the cooperative system to the non-land reform areas in the early 1960s. Finally, the cooperatives took over the maintenance of irrigation and drainage. It would not be a great distortion to say that Land Reform Cooperatives were "Government 'Izab."

The decision to extend the cooperative system to the whole country in 1963 had several roots. First, the experiment with consolidated crop rotation and application of pesticides to the cotton crop had proved quite successful. The devastating attack of the cotton-leaf worm in 1961 made arguments for the extension of this system compelling.[40] Second, the government had committed itself to economic planning in 1958. In order to acquire the resources needed for industrialization, the government needed to control the agriculture surplus. Third, increasing the government's role in agriculture was simply part of the larger trend toward a "socialist" economy, a trend stimulated by Nasser's belief that Syrian capitalists had aborted the union with Egypt, and that Egyptian capitalists would be equally obstructive.[41]

The consequences of the cooperative systems' spread to the whole country were mixed. On the one hand, the consolidated crop rotation and the increased supply of fertilizer and pesticides were surely positive steps. The elimination of the two-year rotation removed one set of ecological problems which had long plagued Egyptian agriculture. Similarly, as discussed in more detail below, the system certainly helped to increase fertilizer consumption and, therefore, crop yields.

On the other hand, the cooperatives created problems for both efficiency and equity. As we shall see, the government's system of controlling prices and areas of some crops but not others had unfortunate consequences for efficiency. It would appear that there were other problems with the incentive structure as well: supervisory personnel were underpaid, and tended to neglect the maintenance of drainage networks.[42] On the equity side, it is clear that the rich peasants dominated the cooperatives, especially those set up in non-land reform areas after 1963. The government simply did not have the cadre to carry out such a massive extension without relying extensively on the local power structure. As Baker has put it, "The attitude of these more prosperous peasants toward the cooperative movement has been not so much one of opposition but one of subverting the service offered by the cooperative to their own exclusive use."[43]

Since inputs were not rationed by price, an alternative allocation mechanism emerged. The government subsidized the price of inputs and allocated them on a quantitative basis. This mechanism excluded the poor as systematically as a price system in an environment of unequal resource endowments would have done. On the one hand, an extensive black market emerged, supplied by the farmers themselves and from other sources in the distribution channels (i.e., the bureaucracy). Often small farmers would sell their quota of fertilizer on the black market to get cash; rich peasants would then purchase it.[44] Insofar as the cooperative society rationed the inputs directly, there is every reason to suppose that such a system has favored the rich peasants.

There is little doubt that rich peasants controlled the board of cooperatives. Before 1969, 80% of the Board members were supposed to be owners of less than five feddans. Yet it is hardly likely that small peasants would have gone against the interests of their more powerful neighbors. They often were tenants of richer peasants (some 50% of the land area was rented in 1966) and depended upon them for cash loans.[45] In 1969 the ceiling for Board members was raised to 15 feddans and illiterates (some 75% of the rural male population) were excluded. Such regulations guaranteed rich peasant dominance. One recent study of cooperatives estimates that perhaps forty percent of village councils were "reactionary."[46] This is surely a lower bound, since the author assumes that the "anti-feudal" campaign which was launched in the wake of the "Kamshish incident" was successful. This seems highly unlikely. It is true that dismissals, and even some arrests, of individuals followed. But six months after the formation of the Committee to Liquidate Feudalism, Nasser directed it to turn "its attention to the problems of corruption in the public sector" (i.e., away from agriculture).[47] Given the character of the national regime, the outlook of the agricultural technocrats by then acting as local managers, and the balance of forces in the villages, such an outcome is hardly surprising.

Rich peasants benefitted from the cooperatives in other ways than by merely standing first in line for inputs. The consolidation of the crop rotation meant that small peasants were required to plant all of their land in cotton one year, all in clover the next, and all in cereals the next, rather than have a little cash, a little fodder, and a little food each year. This increased their need for credit, which the rich peasants supplied. Further, only those owning at least five work animals could participate in government and animal insurance schemes, and therefore qualify for one-hundred fifty kilograms of forrage at state-subsidized prices. Similarly, only those with more than fifteen feddans could acquire selected seed.[48]

Finally, regulations on the output side also favored the well-to-do peasantry. Farmers with less than five feddans were prohibited from planting highly profitable fruit trees, unless they could get enough of their small neighbors to cooperate with them. Richer farmers could evade the regulations on planting of price-controlled crops and shift into the more profitable ones. They could also simply pay the fine for failing to plant the required crops, something which was more difficult for their poorer neighbors. The output consequences of such behavior are discussed below. On the equity side, such a phenomenon widens the gap between rich and poor peasants both directly and indirectly, since rich peasants make higher profits than poor ones,[49] and indirectly, since the crops produced (meat, milk, fruit, vegetables) are primarily consumed by the well-to-do, whether urban or rural. However, the effects on the landless laborers may be more beneficial, since fruit, vegetables, and flowers are labor-intensive crops.[50]

In summary, the land reforms present a mixed picture. On the one hand they certainly increased both efficiency and equity among farmers by eliminating the "bimodalism" characteristic of Egyptian agriculture since the days of Muhammad 'Ali. Over 300,000 poor peasant families benefitted directly as recipients of land, and many others were helped by more secure tenure rights. This was done without losing any of the efficiency advantages of larger holdings with respect to crop rotation, irrigation, and drainage. Indeed, the two-year rotation was eliminated, fertilizer use was stimulated, and pest control was facilitated through the consolidation of crop production. Yet at the same time, some old inequalities were reinforced. The position of the rich peasants was enhanced and the system of selective price control, in the absence of adequate enforcement, engendered allocative anomalies while favoring the well-to-do. Let us now look more closely at the allocation problems of this system.

POLICIES: PRICE AND AREA REGULATIONS

The Egyptian government used cooperatives to market crops and to supply inputs to farmers. The regime's problem was the classic one of how to extract resources for industrialization and military and social spending from agriculture without at the same time undermining production incentives for farmers. The government's goals were 1) to acquire foreign exchange, which would have to come largely from agricultural exports and 2) to provide cheap food for the cities both to prevent food-price inflation (which would undermine its industrialization program) and to avoid the political instability which urban food shortages would engender. The government chose not to use direct taxation to accomplish these goals, but rather sought to purchase certain crops at prices below the international levels and then either to sell them on the international market, or, for foodstuffs, to subsidize urban food prices. Here they followed the precedents of previous regimes. To ensure that adequate quantities of the price-regulated crops were grown, the government imposed area restrictions, in effect requiring farmers to grow certain crops. The government tried to sweeten the pill and to maintain incentives by subsidizing the prices of inputs, such as fertilizer, insecticides, and (to a lesser extent) farm machinery.

The results of this system were mixed. While it may have solved some problems (e.g., the problem of cheap food for the cities), it created others. It should be remembered that a full-scale evaluation of the impact of agricultural price and area policies is an extremely difficult undertaking. As C. Peter Timmer has emphasized, straight-forward answers on price-policy impacts are highly elusive in a multi-class, multi-crop economy, in which there are substitution pos-sibilities in both production and consumption, and in which different income classes consume different foods.[51] Let us first turn to an inventory of the policies which were adopted, and then look at the probable consequences for input use, cropping patterns, foreign exchange, and the inter-sectoral terms of trade.[52]

Government intervention in cotton production has a long history in Egyptian agriculture, as we have seen. In the 1950s cooperative marketing of cotton was largely limited to the land reform areas, but in the 1960s the system was extended throughout the country. Since cotton gins are characterized by economies of scale, taking over the gins ensured that the farmers would use government marketing chan-nels. When farmers deliver their cotton to the gin, they are paid after the prices of inputs supplied by the cooperative are deducted. Similar systems were extended to other export crops such as rice, onions, peanuts, potatos, lentils, sesame and beans in the early 1960s. However, unlike cotton, not all of the output of these crops was sold through the (price-regulated) government channels. A simi-lar system of compulsory deliveries at government-determined prices was established for the principal urban food-stuffs of wheat and sugar.[53]

The government enforced and extended this regulatory-cooperative system by controlling the allocation of inputs. This was especially crucial for fertilizer use. By establishing a monopoly over fertil-izer importation, production, and distribution, the government could exert strong influence over what was grown. The Ministry of Agricul-ture, through the cooperative system, issues to each farmer a "farm holding card," which states the amount of his land and the crop rota-tion to be followed. After 1964 the government specified the amount of fertilizer to be used on each crop; this information, combined with the data on the farm holding card, established each farmer's quota of fertilizer. Farmers could not buy less than this; it was all advanced on credit, with this and other expenses deducted when the crops were marketed through the cooperative. Up to certain specified limits, farmers could acquire more than their quota, but only by paying cash. Most (e.g., over 90% of all nitrogenous fertilizers in 1966/67) was acquired on credit.[54]

The government also promoted fertilizer use by expanding local production facilities; domestic production of nitrogen fertilizer increased nearly ten times from 1952 to 1966; by the latter year, 62% of nitrogen fertilizer consumption was produced domestically.[55] From 1948 to 1967, fertilizer consumption grew at an average annual rate of 68,000 tons per year.[56] (See Table 6.5) Even before the extension of the semi-compulsory fertilizer quota system, the government had held down the price of fertilizer, starting in 1956.

---

**TABLE 6.5**
**FERTILIZER CONSUMPTION**

| Year | Nitrogenous (000 M. Tons) | | Phosphate | |
|------|-----------|--------|--------|------|
| | CAPMAS | OECD | CAPMAS | OECD |
| 1949-50 | | 569 | | 79 |
| 1951 | | 667 | | 113 |
| 1952 | | 696 | | 122 |
| 1953 | 658 | 685 | 92 | 127 |
| 1954 | 679 | 683 | 105 | 68 |
| 1955 | 645 | 764 | 113 | 118 |
| 1956 | 632 | 724 | 156 | 137 |
| 1957 | 698 | 655 | 181 | 157 |
| 1958 | 694 | 849 | 179 | 177 |
| 1959 | 734 | 886 | 171 | 177 |
| 1960 | 763 | 993 | 177 | 168 |
| 1961 | 913 | 1,166 | 230 | 218 |
| 1962 | 939 | 1,336 | 245 | 250 |
| 1963 | 984 | 1,339 | 254 | 257 |
| 1964 | 1157 | 1,458 | 293 | 268 |
| 1965 | 1231 | 1,550 | 322 | 305 |
| 1966 | 1294 | 1,725 | 349 | 350 |
| 1967 | 1100 | 1,482 | 290 | 286 |
| 1968 | n.a. | | n.a. | |
| 1969 | 1083 | | 251 | |
| 1970 | 936 | | 237 | |
| 1971 | 966 | | 272 | |
| 1972 | 975 | | 312 | |
| 1973 | 1061 | | 363 | |
| 1974 | 1125 | | 290 | |
| 1975 | 2578 | | 303 | |
| 1976 | 2646 | | 382 | |
| 1977 | 2797 | | 441 | |
| 1978 | 3135 | | 606 | |

Such a system of controls left (and leaves) rather little room for market forces to influence fertilizer use. Consequently, it is probable that the "inducement" hypothesis, which received considerable attention in Chapter 4, has much less relevance during the Nasser period.[57] However, in view of the tendency of farmers to try to evade controls, it is nevertheless instructive to find that the general trend of prices should have encouraged fertilizer use. The relative price of land to fertilizer seems to have risen: land rents were stable (at least legally), being pegged to the land tax which was readjusted only once every ten years. Further, there is evidence that rent controls were evaded.[58] Fertilizer prices, on the other hand, declined some 30%. Additional support for the inducement hypothesis may be found in the trend of the relative price of crops to fertilizer (See Table 6.6). For all of the major crops (which accounted for 79% of N-fertilizer use in 1961/63), there is a pronounced upward trend, taking the decade as a whole.

However, the government's system of price controls quite probably worked in the other direction in the 1960s. While farmers received less than world prices for crops such as cotton and rice, they paid some 87% more for fertilizer than international prices.[59] Since local prices were tied to local production costs, Egyptian farmers did not benefit as technological change and entry of new producers drove down international fertilizer prices by more than 40% from 1967 to 1971. However, farmers were also insulated from the fertilizer price explosion of the mid-1970s, paying only some 40 to 60% of international prices after 1973. Government policies thus served to insulate farmers from the impact of international price

TABLE 6.6
INDEX OF RELATIVE PRICE OF CROPS TO FERTILIZER,
1961-1970
(1960 = 100)

| Date | Cotton | Rice | Wheat | Maize | Onions | Sugarcane |
|------|--------|------|-------|-------|--------|-----------|
| 1961 | 107 | 110 | 109 | 107 | 154 | 110 |
| 1962 | 113 | 114 | 114 | 99 | 188 | 114 |
| 1963 | 109 | 108 | 109 | 110 | 124 | 108 |
| 1964 | 123 | 116 | 113 | 116 | 129 | 110 |
| 1965 | 111 | 123 | 109 | 106 | 135 | 104 |
| 1966 | 106 | 148 | 113 | 120 | 129 | 124 |
| 1967 | 155 | 229 | 178 | 192 | 195 | 171 |
| 1968 | 163 | 246 | 159 | 155 | 200 | 176 |
| 1969 | 169 | 242 | 161 | 175 | 179 | 176 |
| 1970 | 173 | 226 | 193 | 181 | 224 | 179 |

Source:
Calculated from Abdel-Fadil (1975), 138.

changes.

Further, it will be noticed that the price ratio is relatively stable before 1966, and then increases sharply during the next four years, with most of the change coming in one year, 1967. This is not the pattern of fertilizer consumption, as a glance at Table 6.5 shows. There the main jump in consumption during the Nasser years comes in 1964, the year in which the compulsory system of fertilizer quotas was introduced. We may conclude that although price policies did not interfere with the diffusion of fertilizers, they were of secondary importance in stimulating the use of such inputs.

It is possible that some of the increase in fertilizer use may be the result of peasants' neglecting the major crops and either selling it on the black market or using it themselves on vegetable crops. There is considerable evidence that such practices were (and are) quite common. Yet, although the area in vegetables did expand both absolutely and relatively throughout the period, the low percentage of the land area planted in vegetables could not account for more than a limited proportion of the increase in fertilizer use. Also, unless evasions of cooperative regulations were on a very large scale, the fact that over 90% of fertilizer was allocated through the cooperative system indicates that such a phenomenon probably did not affect aggregate fertilizer use significantly. Nevertheless, such practices may have been one response to the peculiarities of Egyptian agricultural prices.

The major problems which the price and area restrictions caused were on the output side. It is clear that a reallocation of land took place during this period. Not all of this was due to demand forces; the increased supply of water played a central role, as well. (See below) It is noticeable that the major field crops with controlled prices declined as a percentage of the cropped area in favor of higher value products like birsim, fruits, and vegetables. (See Table 6.7) We have already noted that such reallocations primarily benefitted the well-to-do peasantry, due to their greater ability to avoid government regulations. In macro-economic terms, the policies do not seem to have been especially successful. Hansen and Nashashibi[60] have estimated the losses due to the regulations' impact on cropping patterns. They found: 1) that price and area regulations usually reinforced rather than offset both each other and other market imperfections (see Table 6.8) and 2) that the value of these losses was roughly equivalent to the balance of payments deficit for the 1960s.

Nor do these price policies seem to have transferred many resources from agriculture to industry. Several different studies of the trend in the sectoral terms of trade exist. Broadly speaking, although the terms of trade moved in favor of industry in the early 1960s, the trend was reversed at mid-decade, with overall relative stability.[61] Of course, the government did tax agriculture using the price and area systems, and did acquire revenue which they then used for industrial, military and, especially, social spending.[62] But the particular system used had considerable costs, costs which, as we shall see, increased in the 1970s. The government did manage

TABLE 6.7
AREA OF MAIN CROPS BY SEASON DURING THE 1950—78 PERIOD
( 000's feddans )

| Season & Crops | 1950–1954 | 1955–1959 | 1960–1964 | 1965–1969 | 1970–1974 |
|---|---|---|---|---|---|
| A. Winter crops; | | | | | |
| Clover | 2,184 | 2,362 | 2,444 | 2,630 | 2,801 |
| Wheat | 1,571 | 1,501 | 1,387 | 1,268 | 1,302 |
| Broad Bean | 328 | 353 | 365 | 349 | 283 |
| Barley | 122 | 135 | 128 | 110 | 81 |
| Lentil | 74 | 80 | 77 | 65 | 64 |
| Onion | 26 | 36 | 44 | 45 | 33 |
| Fenugreek | 53 | 60 | 55 | 38 | 28 |
| Flax | 8 | 14 | 27 | 30 | 33 |
| Other | 42 | 66 | 83 | 78 | 94 |
| Vegetables | 70 | 104 | 49 | 170 | 189 |
| Sub—Total | 4,478 | 4,711 | 4,759 | 4,783 | 4,908 |
| B. Summer crops: | | | | | |
| Cotton | 1,765 | 1,791 | 1,751 | 1,694 | 1,551 |
| Rice | 505 | 641 | 791 | 1,028 | 1,093 |
| Maize | 29 | 56 | 271 | 1,078 | 1,245 |
| Sorghum | 386 | 393 | 414 | 462 | 465 |
| Sugarcane | 96 | 111 | 122 | 145 | 197 |
| Sesame | 37 | 43 | 45 | 32 | 37 |
| Groundnut | 28 | 36 | 46 | 47 | 35 |
| Other | 13 | 14 | 46 | 54 | 86 |
| Vegetables | 120 | 200 | 260 | 328 | 356 |
| Sub—Total | 2,979 | 3,285 | 3,716 | 4,868 | 5,064 |
| C. Nili | | | | | |
| Maize | 1,717 | 1,794 | 1,456 | 432 | 348 |
| Sorghum | 52 | 58 | 55 | 45 | 29 |
| Other | 23 | 24 | 18 | 31 | 33 |
| Vegetables | 69 | 91 | 138 | 170 | 216 |
| Sub—Total | 1,861 | 1,967 | 1,667 | 678 | 627 |
| D. Totals by Season: | | | | | |
| Winter | 4,478 | 4,711 | 4,759 | 4,783 | 4,908 |
| Summer | 2,979 | 3,285 | 3,716 | 4,868 | 5,065 |
| Nili | 1,861 | 1,967 | 1,667 | 678 | 627 |
| Orchards | 94 | 114 | 147 | 208 | 255 |
| Total Cropped Area | 9,412 | 10,077 | 10,289 | 10,537 | 10,855 |

TABLE 6.7 CONT.

| Season & Crops | 1975 | 1976 | 1977 | 1978 |
|---|---|---|---|---|
| A. Winter crops: | | | | |
| Clover | 2,812 | 2,787 | 2,854 | 2,882 |
| Wheat | 1,394 | 1,396 | 1,207 | 1,381 |
| Broad Bean | 246 | 260 | 292 | 239 |
| Barley | 100 | 104 | 95 | 114 |
| Lentil | 58 | 64 | 48 | 36 |
| Onion | 27 | 31 | 37 | 29 |
| Fenugreek | 32 | 33 | 24 | 26 |
| Flax | 54 | 47 | 59 | 60 |
| Other | 146* | 104* | 128* | 28 |
| Vegetables | 203 | 215 | 214 | 225 |
| Sub-Total | 5,072 | 5,041 | 4,958 | 5,020 |
| B. Summer crops: | | | | |
| Cotton | 1,326 | 1,248 | 1,423 | 1,188 |
| Rice | 1,053 | 1,078 | 1,040 | 996 |
| Maize | 1,426 | 1,490 | 1,323 | 1,403 |
| Sorghum | 468 | 446 | 409 | 413 |
| Sugarcane | 220 | 243 | 250 | 280 |
| Sesame | 33 | 31 | 40 | 40 |
| Groundnut | 32 | 42 | 36 | 31 |
| Other | 105* | 90* | 107* | 188 |
| Vegetables | 419 | 443 | 452 | 461 |
| Sub-Total | 5,083 | 5,111 | 5,080 | 5,000 |
| C. Nili | | | | |
| Maize | 368 | 404 | 401 | 505 |
| Sorghum | 20 | 28 | 15 | 20 |
| Other | 72 | 42 | 86* | 25* |
| Vegetables | 263 | 260 | 248 | 253 |
| Sub-Total | 723 | 734 | 750 | 803 |
| D. Totals by Season: | | | | |
| Winter | 5,072 | 5,041 | 4,958 | 5,020* |
| Summer | 5,083 | 5,111 | 5,080 | 5,000* |
| Nili | 723 | 734 | 750 | 803* |
| Orchards | 285 | 311 | 321 | 325 |
| Total Cropped Area | 11,163 | 11,198 | 11,111 | 11,148 |

Source:
   1950-74 data from El Tobgy (1976), 1975-77 Ministry of Agriculture; 1978 Ministry of Agriculture and U.S. Agricultural Attaché estimates, presented in "Egypt: Agricultural Situation Report," Cairo: U.S. Embassy, 1980.
* Agricultural Attaché estimate.

TABLE 6.8
MISALLOCATION OF LAND, AVERAGE 1962-1968
(percent of optimal acreage)

Deviation from Optimal Acreage

|  | Total | Due to Imperfect Market Forces | Due to Government Interference | | |
|---|---|---|---|---|---|
|  |  |  | Total | Price Distortion | Other Intervention |
| Cotton | 12.1 | 5.4 | 6.7 | 2.3 | 4.4 |
| Rice | 21.8 | 4.7 | 17.1 | 8.6 | 8.5 |
| Corn | 5.3 | 6.1 | -0.8 | -2.1 | 1.3 |
| Millet | 8.9 | 14.0 | -5.1 | -2.2 | -2.9 |
| Wheat | 16.4 | 2.1 | 14.3 | 0.9 | 13.4 |
| Beans | 78.3 | 37.4 | 40.9 | 36.3 | 4.6 |
| Onions | 29.5 | 25.7 | 3.8 | -12.5 | 16.3 |
| Lentils | 10.3 | 11.3 | -1.0 | 2.2 | -3.2 |
| Cane | 93.8 | 42.5 | 51.3 | 0.0 | 51.3 |
| Total | 7.9 | 3.5 | 4.4 | 1.2 | 3.2 |

Source:  Hansen and Nashashibi (1975), 191.

to hold down urban food prices, thus avoiding wage inflation and urban political instability, as well as placing a "subsistence floor" under the urban poor. From a distributional point of view however, they were 1) subsidizing wheat consumption, the preferred food grain and 2) subsidizing urban consumption more effectively than rural. We shall see how this stimulated important changes in the rural political economy during the 1970s. But before turning to these problems, and before summarizing the technical and social changes of the 1952-1970 period, we need to look at the Nasser regime's investment policies.

POLICIES:  INVESTMENT

Three agricultural investment policies under Nasser stand out: 1) the construction of the Aswan High Dam, 2) the relative neglect of drainage, and 3) the emphasis on land reclamation. Agriculture as a whole was not neglected under Nasser: the share of agriculture, irrigation, and drainage investment in total gross fixed investment rose from 10.9% in 1955-56 to 20.2% in 1968-69.[63] Some sources estimate agriculture's share in public sector investment to have been as high as 25% in the mid 1960s.[64] We have seen how fertilizer use expanded rapidly; pesticide application and the number of livestock also increased. (See Tables 6.9 and 6.10) However, most capital formation was in the hydraulic system, which comprised some 75% of total capital investment in 1966, in comparison with 70% in 1950.[65] This is hardly surprising, given that the Aswan Dam accounted for roughly one third of all capital formation during this period. The problem was not that agriculture received _no_ investment; the problem was the <u>kind</u> of investment which took place.  Like the land reform and the cooperative system, the investment policies scored some noticeable successes.  Yet, at the same time, they sowed the seeds of later difficulties.  Let us examine the major investments in turn.

Few major public works projects in the post-war era have been the subject of as much controversy as the Aswan High Dam; few have received as much attention from partisan advocates and from more detached analysts alike.[66] We may summarize this vast debate under two general headings: the reasons underlying the decision to build the dam, and the probable consequences for agriculture of the dam's construction.

It had been clear since the 1930s that some form of "over year" storage was necessary for Egyptian agriculture.  Not only would such a scheme prevent excessively low and high floods, but it would also make possible a dramatic increase in the supply of water during the summer, the growing season for Egypts' major exports, rice and cotton.  It was clear that the growth of population and the hopes for agricultural development in the Sudan and in Egypt required hydraulic systems which could capture more of the hydrological potential of the Nile river for economic uses.  There were really only _two_ such schemes: 1) the "Century Storage Scheme" and 2) the High Dam at Aswan.  The former had originated with British engineers of the Egyptian Irrigation Department.  As Waterbury has described it, "the basic notion...required storing several successive annual floods...a series of reservoirs could be used to bank water against low years or to hold excess water if the flood promised to be especially high."[67] The Century Storage Scheme planned to do this with a series of regulators and dams, which would use the natural lakes of the Nile system (Lake Tana, Lake Albert, Lake Victoria), as reservoirs.  Although there were technical problems, especially the silting up of the reservoirs, the principal difficulties were political.  Such a scheme required the cooperation and agreement of all the states of the Nile watershed.  This would have been difficult enough under colonial rule, when only the British, Belgian, and (during the occupation of Ethiopia) Italian governments were involved.  It was even more problematical in the 1950s, when the Sudan and East Africa remained under British rule, while an anti-British, nationalist regime held sway in Cairo.

It is therefore hardly surprising that the decision to construct the High Dam at Aswan was made very quickly after the Free Officers seized power.  The advantages from their point of view were obvious: the Dam would solve the water supply and uncertainty problems; it would be entirely within Egypt's borders; and it was highly visible.  The latter was an important consideration for a regime of relative unknowns, with no clear political base outside of the officer corps itself.  Furthermore, the problems with the Dam which later emerged (discussed below), were foreseen by very few people.  It is true that voices continued to be raised in favor of the Century Storage Scheme.  But in the political climate of the 1950s, this was not an option for the Nasser regime.  Neither the Consortium of Western Engineers nor

TABLE 6.9
NUMBER OF CATTLE AND BUFFALOES,
1952-1978

| Year | Cattle (000 head) | Buffalos |
|------|---------|----------|
| 1952 | 1,356 | 1,212 |
| 1953 | | |
| 1954 | 1,344 | 1,262 |
| 1955 | 1,362 | 1,323 |
| 1956 | | |
| 1957 | | |
| 1958 | 1,390 | 1,395 |
| 1959 | | |
| 1960[a] | 1,502 | 1,472 |
| 1961 | 1,523 | 1,501 |
| 1962 | 1,544 | 1,530 |
| 1963 | 1,566 | 1,559 |
| 1964 | 1,587 | 1,588 |
| 1965 | 1,608 | 1,617 |
| 1966 | 1,630 | 1,646 |
| 1967 | 1,651 | 1,675 |
| 1968 | 2,058 | 1,943 |
| 1969 | 2,036 | 2,015 |
| 1970 | 2,115 | 2,009 |
| 1971 | 2,112 | 2,058 |
| 1972 | 2,129 | 2,098 |
| 1973 | 2,127 | 2,135 |
| 1974 | 2,119 | 2,170 |
| 1975 | 2,102 | 2,204 |
| 1976 | 2,079 | 2,236 |
| 1977 | 2,048 | 2,266 |
| 1978 | 2,587 | 2,542 |

Source: CAPMAS.

a. Earlier series exclude cattle
in all urban areas and the Cairo governorate.

their Soviet replacements voiced the kind of complaints which later came to dominate the discussion. If the emperor had no clothes, no one was saying so.

Or rather, no one was saying so to whom the regime would be willing to listen.[68] The project became a political symbol almost immediately: those who supported the Nasser regime backed the Dam, while the regime's detractors attacked it as folly. The politics of funding the Dam began with Israel's Gaza raid of 1955, and the ensuing Egyptian-Czech arms deal. The World Bank's financial conditions

TABLE 6.10
PESTICIDE USE

| Year | Quantity Consumed (metric tons) |
|------|------------------|
| 1953 | 2,143 |
| 1954 | 1,627 |
| 1955 | 8,871 |
| 1956 | 9,188 |
| 1957 | 10,489 |
| 1958 | 8,075 |
| 1959 | 15,078 |
| 1960 | 11,062 |
| 1961 | 23,398 |
| 1962 | 7,447 |
| 1963 | 12,552 |
| 1964 | 20,916 |
| 1965 | 20,450 |
| 1966 | 28,639 |
| 1967 | 30,699 |
| 1968 | n.a. |
| 1969 | 25,668 |
| 1970 | 24,664 |
| 1971 | 20,851 |
| 1972 | 35,259 |
| 1973 | 26,344 |
| 1974 | 20,910 |
| 1975 | 27,055 |
| 1976 | 25,593 |
| 1977 | 28,344 |
| 1978 | 26,079 |

Source: CAPMAS, Statistical Indicators.

were already quite stringent. Further, they required the participation of the United States and provided for funding to proceed in two
stages. Nasser, understandably, felt that this second feature would
leave him open for blackmail once construction commenced. As Waterbury puts it "The terms were not only patronizing but ominously reminiscent of the... Caise de la Dette ...Nasser reacted with indigna-
tion."[69] Since the terms required the participation of one government whose imperial interests Nasser opposed (U.K.) and another whose
desire for anti-Soviet military alliances overrode all other regional
objectives (the U.S.), it is hardly surprising that Western financing
for the Dam evaporated. Nor is it surprising that the regime then
turned to the only other available source of funds, the Soviet Union.
The financing of the High Dam illustrates very clearly once again the
importance of the international state system for Egyptian agricultural development.

Not the least of the problems which such politicization created was the stifling of technical criticism of the High Dam. Once the political climate in Egypt shifted in the 1970s, more critical voices have been raised. There have been a number of assessments of the Dam's impact, which have usually followed a "pro" or "anti" Dam/Nasser regime line. It would not be very instructive to attempt a formal cost-benefit analysis of such a gigantic project, with such pervasive impact on the entire national economy, which generated so many different positive and negative externalities, and whose costs and benefits accrue at such different times. No reasonable weighting system exists which could aggregate such disparate factors and uncertain effects into a single index. The best that can be done is to highlight the principal positive and negative impacts of the Dam for agriculture.

On the positive side, the Dam eliminated the threat of either excessively high or low floods from Egyptian agriculture. Just how important this is may be seen by the low flood of 1972: had the Dam not been built, the lack of summer water would have meant the loss of perhaps one third of both the cotton and the rice crops. The increased water supply also made possible the conversion of some 800,000 feddans of Upper Egyptian basin land to perennial cultivation and made possible the increase of the cropped areas from 9,412,000 feddans in 1950-54 to 10,855,000 feddans in 1970-74.

Further, the Dam permitted some significant crop shifts, which increased land productivity and agricultural output. First, it allowed the expansion of the area planted in rice by over 500,000 feddans, bringing the total rice area to over one million feddans in 1970. Second, it led to an increase of over 85,000 feddans in the sugar cane area. Third, it enabled farmers to shift the planting date of the maize crop from July or August (the nili season) to May and June. This resulted in an immediate increase in maize yields of over 40% from 1960-64 to 1965-70, both because of the more favorable sunlight and heat conditions and also because summer maize escaped the ravages of the maize borer.[70] Finally, in view of the well-known complementarity of fertilizer use and water supply, it is reasonable to attribute much of the increase in fertilizer use to increased water availability.

However, some benefits which were hoped for during the planning stages were not only not realized, but may have been adversely affected by the Dam. It was believed that the additional water would make it possible to reclaim over 1.2 million feddans and that drainage would be improved. As we shall see below, land reclamation has been much more limited and more costly than was anticipated, while the neglect of drainage has been disastrous. In general, one may say that the Dam created or exacerbated a set of ecological problems which were either not correctly anticipated or which were ignored because of the severe budget constraint imposed by such a massive spending project as the Dam. The latter theme, of course, is strongly reminiscent of British behavior before World War I. (See Chapter 3)

Three ecological problems stand out: evaporation and seepage, the absence of silt and consequent "scouring" problems, and drainage difficulties. The first problem appears to have been the result of miscalculation. Because of the high temperatures and wind velocities over Lake Nasser behind the High Dam, evaporation losses exceeded anticipations. Similarly, sediment appears to have been deposited largely at the southern end of the reservoir, not uniformly as had been anticipated. Consequently an unknown amount of water seeps out of the reservoir through the Nubian sandstone. These are serious problems:

> when one is dealing with a net benefit to Egypt of only 7.5 billion cubic meters per year, any miscalculation in evaporation or seepage rates could change credits to debits. ...for at least a decade of operation at optimal levels, Egypt's benefit may be nearly cancelled by storage losses...under the best of circumstances, more than half of Egypt's incremental gain from construction of the High Dam will be lost in storage.[71]

The problem of silt and scouring have several aspects. First, there is the often exaggerated deprivation of nutrients due to the absence of the silt from the Nile. As we have seen, most of this silt remained in the canals under the system of perennial irrigation: it was not deposited on the fields by the flood as in the basin system. However, the canals were cleaned out annually, and the silt was then spread on the fields. Since the silt, like all Nile soils, was poor in nitrogen in any case, the main problem was the deprivation of trace elements. There is considerable difference of opinion on the gravity of this problem.

More clearly detrimental, and indeed, one of the more dangerous phenomena in rural Egypt today, is the "robbing" of top-soil to make baladi bricks, which were formerly made from the Nile silt gathered from the canals. Despite prohibitions against this practice, the scarcity of alternative building materials and the consequent high prices offered for top-soil ensure that such activities continue. When one adds the loss of arable land to encroachment from urban areas and the construction of military bases, and the very high cost and disappointing progress of land reclamation, the seriousness of this problem is obvious.[72]

Yet by far the most serious problem for long-run land fertility was the second major feature of the regime's investment policies, the relative neglect of drainage. This story is a depressingly familiar one. We have seen in Chapter 3 how the combination of the increased supply of water, the shift into more water intensive crops, and the lack of investment in drainage created severe constraints to the increase in land productivity in succeeding decades. We have also seen (Chapter 4) how such failures then required a "repair phase," in which investments had to be made in the drainage system to undo the damage, while the population and the demand for agricultural production kept growing. Plus ça change, plus c'est la même chose: the neglect of drainage from 1952 to 1970 required large scale investments in the 1970s under Sadat, and retarded the growth of agricultural supply while demand increased.

Investment in drainage decelerated from 1939 to 1952. Despite World War II, kilometers of drainage canals in Egypt grew at a compound rate of roughly 2% per year from 1939 to 1947 (whereas the rate during the decade 1929-39 had been 3% per year). But the uncertainty, government crises, and political instability after the War led to a dramatic slowing down of drainage construction: from 1947 to 1954 only some 252 additional kilometers were constructed (an annual growth rate of about .3% per year). Although drainage construction picked up somewhat between 1954 and 1960 (kilometers of drains growing at 1.3% per year),[73] the Nasser regime had other priorities, primarily agrarian reform and preparations for the Aswan Dam. The cooperative societies themselves had difficulty maintaining the existing drainage networks. The main drains suffered from shoddy maintenance, either because of dishonest contractors or, if done by the local peasants, because of the usual "free-rider" problem.[74] Since the main drains were frequently in poor condition, peasants often refused to maintain their field drains. As a result of these disparate factors, by 1960 some two million feddans were suffering from salinity problems.[75]

The failure to invest adequate sums in drainage continued in the 1960s. Only 13% of government agricultural investment was directed to increasing yields, including drainage investment.[76] The problem may be divided into the problems of lands converted from basin to perennial irrigation in Upper Egypt as a result of the High Dam, and the difficulties facing lands already cropped year-round. The first problem was straightforward: as a result of shortages of funds and engineers, the government simply ignored drainage construction in those areas, postponing such works in their plans to 1970-71.

Investments in public drains and pumping stations in the Delta did occur. However, such efforts were inadequate, in view of the large increase in water use. We have already seen how the rice area increased, and how the bulk of the maize crop was shifted to the summer. The increased use of water in these months, with their high temperatures and evaporation rates, aggravated salinity problems. In addition to the social problems of maintaining field drains, peasants continued to overwater all of their crops. Their accustomed techniques, developed in the days of periodic summer water shortages, were slow to change, and the governments' extension service had little success in changing these practices. Since water was a free good, this is hardly surprising.[77]

The results of this drainage failure became apparent in the late 1960s, and continued to grow over time. By the mid-1970s the FAO estimated that some 35% of the cultivated area was suffering from salinity problems, while some 90% of the area had problems with the water table. The joint Ministry of Agriculture-U.S.D.A.-U.S.A.I.D. team reported in 1976 that some 80% of the cultivated area would be affected unless tile drainage installation was accelerated.[78] It hardly seems an exaggeration to describe this as a national disaster. The impact on potential yields varies with the crop and, of course, with the degree of deterioration which the soil has undergone. In general, yields seem to have been reduced some 20 to 40% by poor drainage.[79]

The government recognized the gravity of the situation in the late 1960s. Since then, considerable progress has been made, but the problem is far from solved even today. Two programs with World Bank financing were launched to install tile drainage both in the Delta (950,000 feddans) and Upper Egypt (300,000 feddans). Since most of this spending took place under the Sadat regime, its progress and problems are examined below.

At the same time that drainage was neglected, the regime stressed land reclamation. Such a concentration is perfectly under-standable, given the high cropping ratios and rural population densi-ties. The regime's technicians hoped that Egypt could escape the confines of the Nile Valley by such policies. A major study of this effort identifies three periods: 1952-59, "a time of million-feddan proposals and haphazard implementation"; 1960-67, "a period of con-certed effort and rapid expansion;" and post-1967, "an era of conso-lidation and retrenchment."[80] Of L.E. 208 million allocated for agricultural investment under the Five Year Plan of 1960-65, some L.E. 154 million was given over to land reclamation.

The results have been disappointing. The government has estimated the cost of reclaiming land at approximately L.E. 150 per feddan. This is surely too low: others have estimated the cost at closer to 400, 900, and even 1,000 L.E. per feddan.[81] One should compare this to the per feddan cost of tile drainage installation of roughly L.E. 46-50. Returns have been limited by low crop yields, in turn the result of salinity (drainage neglect, once again!), and poor soil structure. Unlike the arid U.S. Southwest or the Negev, "when you're talking about deserts in Egypt, you're talking about blowing sand."[82] Although the government has claimed that between 1952 and 1975 some 912,000 feddans have been reclaimed, in fact only some 309,000 feddans have reached "marginal productivity."[83] This should be compared to the estimates of high quality land which are lost every year to urban encroachment, roads, military bases, and brick making.[84]

One may ask why such egregious blunders were made. At one level, the explanation is simple: miscalculation. For instance, it was thought that the Aswan Dam would cause the water table to fall; it was believed that the new lands would be as fertile as the old. But such an explanation can be only part of the story. We should also note the extreme scarcity of funds, especially after the June War imposed a rising burden of military expenditures on the regime. In the face of severe capital constraints, with very pressing current problems, it is perhaps natural that drainage should take a back seat.[85] But land reclamation also has a very long pay-off period, so such myopia cannot explain the choice of land reclamation over drainage.

It seems likely that these failures were partly the result of the nature of the regime, of its social base, and of its

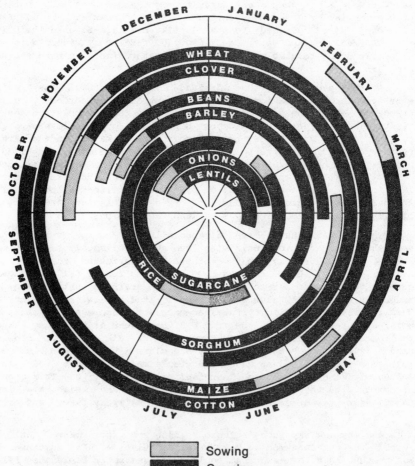

Sowing
Growing
Harvesting

CROP SEASONS IN EGYPT, 1965-1980

bureaucratic/military mentality. Such a regime was highly unlikely to approach the deep-seated, complex, and intractable problems of production on the old lands by launching a thoroughgoing social transformation of the countryside: it was not about to take the Chinese route to agricultural development. Nor was a bureaucratic and highly centralized military regime likely to rely on providing decentralized incentive mechanisms, for instance, to induce farmers to install their own field drains. The engineers of the bureaucracy naturally tended to think of the fundamentally social problems of Egyptian agriculture as essentially technical ones. Given the overall government structure, highly centralized responses were most attractive. New lands could be planned from the top down; the labour was to be imported; one did not have to deal with a pre-existing and complex balance of local power, etc.

Further, it is perhaps understandable that decision-makers sought panaceas for the intractable problems of Egyptian agriculture. In the face of serious social and technical problems, there was a great temptation to deny the existence of such constraints. This "sky is the limit" approach was reinforced by the symbolism of the High Dam, which came to represent Egyptian nationalism and resistance to foreign domination. The regime could not involve the peasantry and could not provide adequate incentives, leaving a centralized, technocratic, "new lands" approach as the only real option.

## A SUMMARY OF TECHNICAL AND SOCIAL CHANGE UNDER NASSER

Technical change in agriculture during this period may be summarized as follows: 1) massive investments in irrigation and land reclamation, combined with under-investment in drainage; 2) a series of shifts in cropping patterns; and 3) large increases in the use of fertilizers, pesticides, improved seeds and, to a lesser extent, agricultural machinery. This section will focus on the latter two types of technical changes, since the problems and progress in the hydraulic system have already been discussed. The interrelated, package aspect of technical change familiar from the interwar period emerges here as well: the shift in cropping patterns and the increase in fertilizer use was contingent upon the large increases in summer water supply. At the end of the section we shall summarize the probable impacts of these technical changes and government policy measures on the relative positions of the three social classes of well-to-do peasants, small peasants, and landless laborers.

The crop shifts are illustrated in Table 6.7. The decline in the relative share of cotton in the cultivated area had two sources. First, the government imposed the three-year crop rotation in the agrarian reform areas and extended it to most cooperativized areas in the early 1960s. Thus the two-year rotation, which played such an important role in Egypt's rural history, came to an end.[86] A second feature leading to the relative demise of cotton was, of course, the low profitability of the crop due to government price controls.[87] The wheat area showed a similar decline. This may have been as much the result of comparative advantage forces as of government policy.[88]

By contrast, the areas planted in rice, summer maize, vegetables, fruits, and birsim showed marked increases. The increased supply of summer water due to the Aswan Dam largely accounts for the first two changes, although rice continued to be a profitable crop. Rising urban incomes, the uncontrolled market for horticultural products, and the subsidy for meat of over 100% promised high profits to those who could grow fruits, vegetables, or birsim.[89] The dramatic increase in the numbers of animals (Table 6.9), both for work and for milk and meat, increased the demand for this fodder crop. Area expansion was especially necessary to increase birsim output, since hardly any plant breeding was undertaken for birsim, in contrast to the other principal field crops.[90]

All major inputs showed important increases in the period 1952 to 1970. Two of the most important, fertilizer and water, have been discussed above. In cereals, plant breeding of wheat and rice made significant gains.[91] Tetraploid wheats were replaced by hexaploids, and Giza 139, an Egyptian bred rust resistant variety, was diffused during the 1960s. Similarly, a rice locally bred from Japanese stock, Nahda, was released in 1954 and occupied 98% of the rice area in 1958. By contrast, plant breeding played little role in maize production. Hybrids were introduced, but they proved susceptible to late wilt. Further, farmers preferred local varieties to meet their combined purpose of home food production and supplementary fodder supplies. Maize yields rose because farmers used more fertilizer and changed the crop's planting date. Summer maize enjoyed better climatic conditions than did Nili, and also escaped from the maize borer.

The long history of cotton breeding in Egypt continued under Nasser. As before, one finds a cycle of continuous breeding of new varieties, their diffusion, and their demise, only to be replaced by new varieties, etc. Of the five types grown in 1950 only one was produced on even a limited area in 1972.[92] In general the new varieties were earlier ripening, thus (potentially) escaping from the cotton bollworm. However, the fact that cotton was generally much less profitable than birsim (which precedes 80% of the cotton crop) led many farmers to take extra cuts of birsim and delay cotton planting--just as small peasants had done between the World Wars.

As in the earlier period, such practices put a brake on yield increases. Delayed planting, and the increase in the birsim area which provided a breeding ground, contributed to serious pest problems for cotton. Indeed, in 1961 nearly one-third of the crop was lost to pests. The principal response has been to apply ever-heavier doses of pesticides. Egypt in 1967 was using nearly four times as many tons of pesticide per cultivated area as the United States.[93] Most of these chemicals were used on the cotton crop; rice and wheat have no significant insect predators in Egypt, while the corn borer has been controlled primarily through the change in planting date.[94] One might note here a potentially vicious circle of increased planting of birsim, increased plant density per acre, increased insect attacks, and increased pesticide application. As is well known from the American experience, such a circle is unlikely to

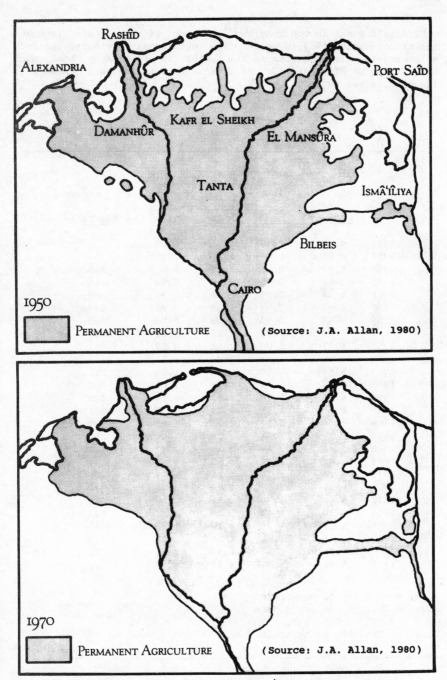

EXTENT OF THE CULTIVATED AREA 1950 & 1970

control the pests in the long run, because of the rapid rate of insect evolution.[95] Given the low educational levels and the lack of protective clothing worn in the fields, it is likely that such heavy levels of pesticide use constitute a public health hazard in the rural areas as well.

TABLE 6.11
CROP YIELDS, 1952-1977

| Year | Cotton[a] | Wheat[b] | Rice[c] | Maize[d] | Sugar Cane[e] |
|------|-----------|----------|---------|----------|---------------|
| 1952 | 4.53 | 0.78 | 1.38 | 0.88 | |
| 1953 | 4.81 | 0.86 | 1.54 | 0.92 | |
| 1954 | 4.41 | 0.96 | 1.84 | 0.92 | 845 |
| 1955 | 3.68 | 0.95 | 2.18 | 0.94 | 865 |
| 1956 | 3.93 | 0.99 | 2.28 | 0.90 | 863 |
| 1957 | 4.46 | 0.97 | 2.34 | 0.85 | 877 |
| 1958 | 4.68 | 0.99 | 2.09 | 0.90 | 860 |
| 1959 | 5.19 | 0.98 | 2.22 | 0.81 | 897 |
| 1960 | 5.11 | 1.03 | 2.11 | 0.93 | 952 |
| 1961 | 3.38 | 1.04 | 2.13 | 1.01 | 853 |
| 1962 | 5.52 | 1.10 | 2.46 | 1.09 | 911 |
| 1963 | 5.43 | 1.11 | 2.31 | 1.09 | 880 |
| 1964 | 6.26 | 1.16 | 2.12 | 1.17 | 823 |
| 1965 | 5.48 | 1.11 | 2.11 | 1.48 | 841 |
| 1966 | 4.89 | 1.14 | 1.99 | 1.51 | 897 |
| 1967 | 5.37 | 1.04 | 2.12 | 1.46 | 882 |
| 1968 | 5.96 | 1.07 | 2.15 | 1.47 | 902 |
| 1969 | 6.68 | 1.06 | 2.15 | 1.59 | 933 |
| 1970 | 6.25 | 1.16 | 2.28 | 1.59 | 851 |
| 1971 | 6.68 | 1.28 | 2.24 | 1.54 | 879 |
| 1972 | 6.62 | 1.30 | 2.19 | 1.58 | 862 |
| 1973 | 6.12 | 1.47 | 2.28 | 1.52 | 837 |
| 1974 | 6.08 | 1.38 | 2.13 | 1.51 | 756 |
| 1975 | 5.68 | 1.42 | 2.30 | 1.52 | 818 |
| 1976 | 6.35 | 1.40 | 2.13 | 1.61 | 782 |
| 1977 | 5.60 | 1.41 | 2.19 | 1.54 | |

a, e: Metric Cantar per feddan
b, c, d: Metric Ton per feddan

Source: Ministry of Agriculture.

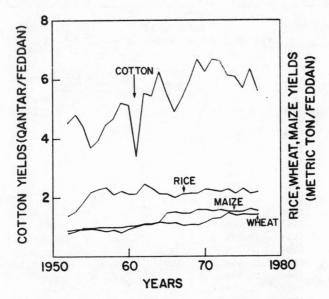

Fig. 6.2 Crop Yields 1952-1977 (Source: CAPMAS)

Nevertheless, it is clear that the expanded use of fertilizers, water, improved seeds and pesticides (for some crops at least) led to substantial increases in crop yields. (Table 6.11) The Nasser years saw a continuation of the land-saving technical change which characterized the interwar years. This is hardly surprising: although rent and price controls make a relative price comparison of limited usefulness, one can simply point to the increasing numbers of persons per cultivated area as the underlying rationale for this type of technical change. However, we have seen that such changes are not the result of a simple market process, because of the extensive involvement of government in agricultural resource allocation during this period.

A very different type of technical change, agricultural mechanization, made some beginnings in this period, and was a goal of the government. There was indeed some increase in the number of tractors (Table 6.12). Some of these were held by the cooperative societies, but most were privately owned. The government nationalized the Behera Machinery Company outside of Alexandria (a maker of agricultural implements and threshers) and established the Nasr Motor Works in Helwan, which assembled tractors. However, the pace of mechanization was fairly slow in this period; we shall see a markedly different picture when we turn to the 1970s.

The combined impact of these technical changes and the government's policies on the relative position of the major rural

TABLE 6.12
TRACTOR STOCKS, FLOWS, 1948-1978

| Year | Tractor Park | Tractor Imports | Domestic Tractor Production |
|------|------|------|------|
| 1948 | 5,400 | | |
| 1949 | | | |
| 1950 | 9,972 | | |
| 1951 | | | |
| 1952 | | | |
| 1953 | 8,850 | | |
| 1954 | 10,355 | | |
| 1955 | 10,750 | | |
| 1956 | 10,753 | | |
| 1957 | 12,086 | | |
| 1958 | 10,994 | | |
| 1959 | 10,994 | | |
| 1960 | 10,994 | | |
| 1961 | 10,994 | | |
| 1962 | 10,999 | | |
| 1963 | | | |
| 1964 | | | |
| 1965 | 14,500 | | |
| 1966 | 15,000 | | |
| 1967 | 15,400 | | 895 |
| 1968 | 15,572 | 3,719 | 731 |
| 1969 | 16,962 | 3,182 | 484 |
| 1970 | 17,500 | 1,901 | 1,071 |
| 1971 | 18,500 | 1,632 | 1,050 |
| 1972 | 18,500 | 1,670 | 937 |
| 1973 | 20,036 | 1,500 | 1,143 |
| 1974 | 21,000 | 1,952 | 1,259 |
| 1975 | 21,500 | 2,850 | 1,435 |
| 1967 | | 3,398 | 2,500 |
| 1977 | | 6,061 | 1,000 |
| 1978 | 29,784 | | 1,260 |

Sources:
Stock (1948); Clawson, et. al. (1970), 40; 1950: Donald C. Mead,
Growth and Structural Change in the Egyptian Economy, (Homewood,
Ill, 1967), 334; 1953-1975: FAO Production Yearbook; 1978: Cal-
culated from imports and domestic production, assuming a 10%
depreciation rate; Imports and Domestic Production: Al-Hosary's
Report (1979).

social classes may now be summarized.  First, it is obvious that  one group, the pashas, was eliminated from the rural scene, so they will not concern us here. There is considerable reason to suppose that the principal beneficiaries of both the land reforms and the government's price and input allocation policies were the rich peasants. They acquired land in distress sales, controlled the rural cooperative societies, had the best access to inputs, and were favored in a number of other special ways. Abdel-Fadil has shown how the richer peasantry's crop mixes differed from those of small peasants; the rich peasants grew more fruit, vegetables, and other high value crops, and were able to avoid the price-controlled crops more successfully than their smaller neighbors. He also shows how richer farmers often acquired tractors and other machinery and rented their services to the smaller fellahin. Radwan has constructed two different terms of trade indices, one of small farmers and one for the well-to-do. That of the latter improved much more than the former (see Table 6.13). The evidence points overwhelmingly to the conclusion that rich peasants' incomes rose relative to those of the smaller fellahin. Abdel-Fadil estimates that the share of agricultural income of peasants owning more than five feddans rose from 25% in 1950 to 32% in 1961.[96]

The position of the landless class appears to have fluctuated considerably during the period, with some long-run improvement. A variety of different real wage indicators are available. Despite differences among these measures, the overall trends are quite similar. Real wages rose sharply during the cotton boom of the late 1940s and early 1950s, fell markedly in the early and mid-1950s, recovered and advanced again in the early and middle 1960s, only to fall again after 1966. On balance one may conclude: 1) that the overall trend of real wages was upward for the period 1938-1974 or for 1948-1974; and 2) that real wages probably fell below their previous trough (1938) in the mid-1950s, but did not fall to such low levels in the second downturn in the late 1960s. The various problems and measures underlying these assertions are discussed in Appendix 6A.

There is a large debate on the theoretical explanation of the movements of wages in Egypt.[97] On balance, it would appear that the supply-and-demand explanation is the more robust. If real wages rose, one would suppose that the demand for labor increased more rapidly than the supply. This appears to be the case, although some anomalies appear. The demand for labor for crops certainly increased during the period. If one looks only at the effects of changes in the cropping pattern (i.e., increase in the cropped area and shifts among crops), there is a fairly steady increase in the demand for labor. From 1952 to 1970, labor use for crops may have increased by about 20% (assuming stable labor input/feddan/crop). (See Appendix 6B)

However, it should be stressed that the fact that the demand for labor increased does not automatically mean that the demand for hired labor rose at a corresponding rate. It is usually argued that the reason for the fall in wages in the immediate post-1952 period is that the Land Reform reduced the demand for hired labor because the

TABLE 6.13
INDICES OF TERMS OF TRADE FOR THE
AGRICULTURAL SECTOR, 1960-75
(1960 = 100)

| Year | A. All Farmers | | | B. Poor farmers | | | C. Rich farmers | | |
|------|-----|-----|-----|-----|-----|-----|-----|-----|-----|
|      | (1) | (2) | (3) | (1) | (2) | (3) | (1) | (2) | (3) |
| 1960 | 100 | 100 | 100 | 100 | 100 | 100 | 100 | 100 | 100 |
| 1961 | 99  | 99  | 100 | 99  | 99  | 101 | 100 | 100 | 101 |
| 1962 | 88  | 96  | 68  | 88  | 96  | 68  | 90  | 99  | 70  |
| 1963 | 86  | 94  | 71  | 87  | 94  | 70  | 90  | 97  | 73  |
| 1964 | 94  | 100 | 78  | 94  | 100 | 78  | 100 | 107 | 83  |
| 1965 | 85  | 88  | 79  | 86  | 88  | 79  | 92  | 95  | 86  |
| 1966 | 93  | 94  | 90  | 93  | 94  | 89  | 105 | 106 | 101 |
| 1967 | 99  | 96  | 112 | 99  | 95  | 115 | 109 | 107 | 121 |
| 1968 | 97  | 93  | 116 | 98  | 92  | 120 | 106 | 102 | 124 |
| 1969 | 99  | 95  | 118 | 106 | 94  | 123 | 108 | 104 | 126 |
| 1970 | 98  | 98  | 101 | 99  | 98  | 104 | 111 | 111 | 111 |
| 1971 | 97  | 96  | 100 | 98  | 96  | 103 | 109 | 109 | 111 |
| 1972 | 97  | 95  | 105 | 98  | 95  | 109 | 113 | 111 | 120 |
| 1973 | 95  | 91  | 106 | 95  | 91  | 109 | 114 | 111 | 125 |
| 1974 | 97  | 93  | 112 | 98  | 93  | 116 | 118 | 115 | 132 |
| 1975 | 102 | 97  | 120 | 102 | 96  | 124 | 127 | 122 | 146 |

Notes:
(1) Over-all index of terms of trade between agricultural output and
all manufactured commodities.
(2) Terms of trade between agricultural output and manufactured con-
sumer goods.
(3) Terms of trade between agricultural output and manufactured inputs.

Source: Radwan (1977), 74.

small peasants who received land tended to hire less labor than the
large estates.[98] There is survey evidence which shows that fewer
small farms hire outside labor than large farms.[99] Since the size
of farms appears to have declined during the period, such an effect
would reduce the demand for hired labor.

The impact of technical change on the demand for labor is also
ambiguous. On the one hand, the same labor-using technical changes
which characterized the interwar years also occurred during the
Nasser era: increased fertilizer use, increased water supply,
increased use of pesticides, and increased land yields. At the same
time, however, agricultural mechanization may have reduced the demand

for labor.[100] The relative weight of these forces is uncertain, although it seems likely that the labor using changes outweighed any adverse effects of mechanization. Despite increased use of tractors, the total tractor park remained fairly small. The other labor-using changes were far more widespread and pervasive.

However, the motor driving agricultural real wages lay outside of the agricultural sector. Increased demand for labor for urban industry, for the Aswan Dam, and for land reclamation may well have driven up wages in the 1960s. As Abdel-Fadil has pointed out, it is very difficult to quantify these factors, since, for example, the number of workers employed on the Aswan Dam is given as a simple number in official sources: 300,000-500,000.[101] We do not know how many were employed in individual years. Nevertheless, it is certainly plausible that such large-scale public works and the growth of demand for labor in the cities drew enough labor out of agriculture that the wages of those remaining increased. Indeed, one can see a fairly sharp deceleration in the growth of the agricultural labor force, which increased by 8% from 1947 to 1960, but only by 1% from 1960 to 1971. (See Table 6.14) It is plausible that once started, the trend toward the cities was irreversible, and that the large increases in the size of the army also drew considerable labor out of the rural areas.[102]

Further, the shift of the basin lands of Upper Egypt to perennial cultivation, as well as the location of the High Dam, would indicate that the increase in the demand for labor would have been especially strong in Upper Egypt. The expansion of the sugar cane area, a labor intensive crop, had the same effect. Since this area typically supplied migrant labor to the rest of the country, such changes probably reduced the supply of tarahil labor.[103]

In summary, the following effects of the demand for hired labor on rural wages may be noted during the period:

1) increase, due to changes in cropping patterns;
2) increase, due to land-saving, labor-using technical changes;
3) increase, due to public works and increased demand for urban industry;
4) decrease, due to land reform and to fragmentation of farms over time;
5) decrease; due to mechanization.

It is not possible to quantify and sum up these different effects; suffice it to say that the third effect was probably largely responsible for the major increase in wages in the early 1960s, while the decline in the early 1950s was the result of both the collapse of the cotton boom and the Land Reform. The combination of these effects with the slow growth of the agricultural labor force plausibly account for the slow growth in agricultural wages.[104] We shall see in the next section that the rural labor market has undergone a transformation in the 1970s, largely as a result of the increased demand for labor outside of agriculture, and, indeed, outside of

TABLE 6.14
AGRICULTURAL LABOR FORCE

| Year | Number[a] (World Bank) | ('000's) (Ministry of Agriculture) |
|------|------------------------|------------------------------------|
| 1946 | 4,086 | |
| 1960 | 4,406 | |
| 1966 | 4,447 | |
| 1971 | 4,471 | 4,056 |
| 1972 | | 4,122.8 |
| 1973 | | 4,162.4 |
| 1974 | 4,212 | 4,212.4 |
| 1976 | 4,224 | 4,222.9 |

a.  These numbers grossly undercount the number of women working  in agriculture.

Source:
World Bank (1978), VI, 11; Ministry  of  Agriculture,  Institute for Research in Agricultural Economics and Statistics.

Egypt itself.

## AGRICULTURE UNDER SADAT, 1970-1980:  THE PROBLEM

The Sadat regime and its policies represent a  curious  amalgalm of  the  old  and the new, in agriculture no less than in foreign and other domestic policies.  Baker has aptly described it as  "Nasserism with  a  Liberal  Face."[105]  On the one hand, Sadat himself was, of course, a Free Officer and a member of the same class as most of  his compatriots.  In many important respects, his regime is a lineal des- cendant of Nasser's.  This is especially so when one  remembers  that the  general  shift  to  the right in social policy had already begun before Nasser's death, in the aftermath of the defeat of 1967.  Sadat has  continued  to rely on many of the same sources of support as the earlier regime, especially in the countryside.[106] He has  inherited the  bureaucratic  structures  and  many  of  the problems of the old regime, such as inadequate drainage, the  cooperative  structure  and price  policies,  the  steady increase in the demand for agricultural commodities, and the burden of massive military spending.

At the same time, the regime faced new problems and  new  oppor- tunities, and responded to both in novel ways.  The sea-change in the

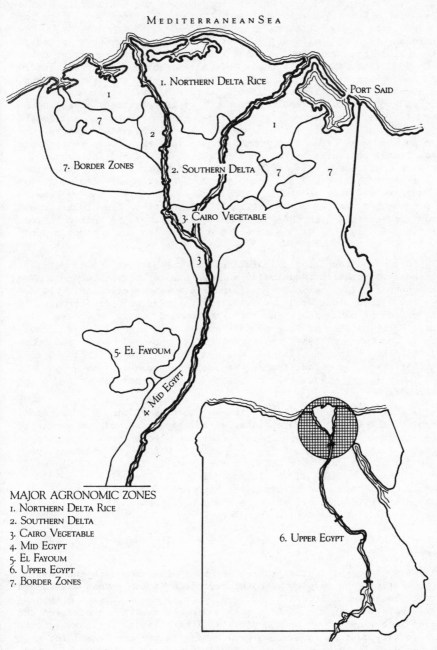

MEDITERRANEAN SEA

1. NORTHERN DELTA RICE

PORT SAID

I

I

7

2

7. BORDER ZONES

2. SOUTHERN DELTA

7

7

3. CAIRO VEGETABLE

3

5. EL FAYOUM

4. MID EGYPT

6. UPPER EGYPT

MAJOR AGRONOMIC ZONES
1. NORTHERN DELTA RICE
2. SOUTHERN DELTA
3. CAIRO VEGETABLE
4. MID EGYPT
5. EL FAYOUM
6. UPPER EGYPT
7. BORDER ZONES

5. **Agronomic Zones, 1975**

regional (and global) political economy due to the conjuncture of the rise in both oil and wheat prices in 1973 accelerated the search for new policies and new escapes from the dilemmas of the 1960s. The increase in oil wealth not only strengthened conservative forces in the region, thus reinforcing Sadat's moves away from the Soviet Union and towards a more open economy, but also created a powerful, external demand for labor of all sorts. From the point of view of technical change in Egyptian agriculture, this was the most important single effect of OPEC's transformation of the international energy economy in 1973. The increase in wheat prices contributed to a dramatic worsening in the balance of payments. This situation forced the regime to seek financing abroad, financing which could really only come from conservative Arabian oil-surplus nations and the West. The lenders, of course, favored and reinforced Sadat's own predilections for a liberalization of the economy. Perhaps even more important was Sadat's belief that no stable, long-term economic progress was possible without a settlement with Israel, and that this, in turn, could only be achieved with the help and cooperation of the United States. Needless to say, the U.S. also urged a loosening of economic controls.

But the resulting liberalization was partial and halting. Further, it actually exacerbated some of the country's problems. The limits to foreign-inspired liberalization were dramatically revealed by the Cairo riots of January 1977 in the aftermath of increases in the prices of basic foods and kerosene imposed under pressure from the I.M.F.[107] In the face of the largest popular uprising since the burning of Cairo in 1952, the regime quickly backed down and rescinded the price increases. Nor was the government likely to risk further unrest by discontinuing the policy of guaranteeing a job to all university graduates, a policy which feeds an ever-growing bureaucracy.

Liberalization itself led to an increased demand for food. Members of the urban middle class, especially those able to engage in the export-import trade, increased their income and, therefore, demand for food and clothing. In effect, middle class demand, pent up under Nasser, was released by Sadat. Rural to urban migration (and thus urban demand) was stimulated by the liberalization-induced construction boom, and by the fact that food and other consumption subsidies were more effective in the cities than in the countryside. Yet the multiple claims on resources, the inherited bureaucratic structures, the price policies, and existing rural class relations meant that supply did not respond to liberalization as rapidly as did demand.

Egypt's agricultural problem in the 1970s, then, was the inability of supply to keep up with the growth in demand. This manifested itself as an agricultural trade problem, with the following elements. 1) Population growth and rural to urban migration have raised the domestic demand for wheat and rice; 2) liberalization, urban growth and other sources of income growth have raised the demand for high income elasticity products such as milk and meat; 3) the price explosion for wheat in 1972-74, greatly increased the per unit costs of imports (see Table 6.15); 4) the system of selective price controls and corruption in cooperative societies lowered cotton production and slowed the growth of wheat and rice output; and 5) the problems of drainage and the need for animal fodder prevented crop yields from achieving their potential increases.

That the ensuing imbalance could be sustained was due to 1) massive concessional financing from abroad, especially from the United States,[108] 2) the explosion of remittances from Egyptian workers abroad, especially from those in the oil states and 3) some "new" sources of revenue, primarily petroleum exports, tourism, and Suez Canal tolls. Workers' remittances replaced cotton as the most important single source of Egyptian foreign exchange.[109] Indeed, by the late 1970s, the problems of financing appeared to have been overcome: with a balance of payments surplus of $700 million in 1980, Egypt was no longer staggering from one repayment crisis to another.[110] Let us look more closely at the two forces underlying increased demand, and then turn to the sluggish supply response. I shall conclude with a discussion of the major technical change of the 1970s, the

TABLE 6.15
FOOD IMPORTS, 1970-1980[a]

| Year | Cereal Imports MT | Value (million $U.S.) | Balance of Agricultural Trade (Agr.x – Agr.m) (U.S. $million) |
|------|------|------|------|
| 1970 | 1,305,730 | 70.7 | +3,027.6 |
| 1971 | 2,447,840 | 162.56 | 2,478.9 |
| 1972 | 1,773,620 | 119.16 | 2,263.1 |
| 1973 | 1,872,210 | 171.43 | 3,998.5 |
| 1974 | 2,997,320 | 738.91 | −2,247.5 |
| 1975 | 3,822,370 | 732.66 | −6,466.8 |
| 1976 | 4,347,270 | 701.63 | −6,357.3 |
| 1977 | 4,935,590 | 681.51 | −7,202.6 |
| 1978 | 5,961,600 | 809.46 | −13,326.9 |
| 1979 | | | |
| 1980 | | | |

Source: FAO Trade Yearbooks.

a. Egyptian trade data, especially for agricultural trade, must be treated with caution. Customs data, the primary source for all compilations, understate imports, because "priority items" are given rapid clearance and are often unrecorded. The above figures should be treated as estimates.

acceleration of farm mechanization.

Some seven million Egyptians were born in the 1970s. Of course, the increase of population had long been of concern to observers of Egypt,[111] for all of the usual reasons. As Table 6.16 shows, the death rate showed a fairly steady fall, while the birth rate rose slightly. The rise in fertility may reflect rising incomes, given the evidence of a positive correlation between family size and income. If true, this would be a rather disturbing trend. A possible counteracting force has been rural-to-urban migration: urban families tend to be smaller than rural.[112] This difference may reflect housing and settlement patterns more than fertility, however. A population growth rate of 2 1/2% per year obviously poses an enormous challenge to agricultural production possibilities in Egypt, whose land yields are already high by international standards.

Although rural to urban migration may help to lower fertility, in the short-run it raises the demand for food, especially for wheat, meat, vegetables and other high income-elasticity goods. Between the two census years of 1966 and 1976, the urban proportion of the population rose from 40 to 44%; there is every reason to suppose that this trend accelerated in the last half of the 1970s. If we assume that urban and rural populations have the same rate of natural increase,[113] then just under one and one-half million Egyptians moved to the cities during those ten years. Lance Taylor[114] has estimated an expected mean household income of LE 716 in the cities and LE 386 in the countryside. For those lucky enough to get a job, the differences are likely to be even larger: unskilled construction wages are roughly three times as high as rural wages. The construction boom in the cities, coupled with the out-migration of most skilled Egyptian construction workers to the Gulf and to Libya, helped to stimulate this movement toward the cities after 1973.[115] Food subsidies were also more effective in the cities. The upsurge in professional and elite incomes as a result of the liberalization of the economy added further impetus to the upsurge in the demand for food, as did the revival of tourism.[116] The liberalization of the economy combined with underlying longer-run trends to drive up the demand for food, especially the more preferred items, such as wheat bread, meat, fruits, and vegetables.

Unfortunately, the supply response has been sluggish. (See Table 6.17) Several factors were at work here: price policies reduced the areas devoted to grains and cotton; land reclamation proceeded quite slowly; yields stagnated. Let us examine these in more detail. First, price policies continued to penalize wheat, rice, and cotton producers; farmers consequently shifted into other crops wherever possible.[117] Although the resulting area shifts reduced the supply of exports and cereals, they increased that of meat, milk, and horticultural products. (See Table 6.7 and 6.17) Second, land reclamation proceeded very slowly, being limited to completion of projects already initiated before 1967. Meanwhile urban encroachment and the use of top soil for bricks steadily nibbled away at the cultivated area. Many specialists came to question the wisdom and feasibility of continued efforts to expand the cultivated area as ever-more

TABLE 6.16
POPULATION, RATES OF BIRTH, DEATH AND
NATURAL INCREASE, AND AGE STRUCTURE

| Year | Population (thousands) | Birth Rates | Death Rates | Rate of Natural Increase |
|------|------|------|------|------|
| | | (Rates per thousand population) | | |
| 1952 | 21437 | 45.2 | 17.8 | 27.4 |
| 1953 | 21943 | 42.6 | 19.0 | 23.0 |
| 1954 | 22460 | 42.6 | 17.9 | 24.7 |
| 1955 | 22990 | 40.3 | 17.6 | 22.7 |
| 1956 | 23532 | 40.7 | 16.4 | 24.3 |
| 1957 | 24087 | 38.0 | 17.8 | 20.2 |
| 1958 | 24655 | 41.1 | 16.6 | 24.5 |
| 1959 | 25237 | 42.8 | 16.3 | 26.5 |
| 1960 | 25832 | 43.1 | 16.9 | 26.2 |
| 1961 | 26579 | 44.1 | 15.8 | 28.3 |
| 1962 | 27257 | 41.5 | 17.9 | 23.6 |
| 1963 | 27947 | 43.0 | 15.5 | 27.5 |
| 1964 | 28659 | 42.3 | 15.7 | 26.6 |
| 1965 | 29389 | 41.7 | 14.1 | 27.6 |
| 1966 | 30139 | 41.2 | 15.9 | 25.3 |
| 1967 | 30907 | 39.2 | 14.2 | 25.0 |
| 1968 | 31693 | 38.2 | 16.1 | 22.1 |
| 1969 | 32501 | 36.8 | 14.4 | 22.4 |
| 1970 | 33329 | 35.6 | 15.0 | 20.6 |
| 1971 | 34076 | 35.1 | 13.2 | 21.9 |
| 1972 | 34839 | 34.4 | 14.5 | 19.9 |
| 1973 | 35619 | 85.7 | 13.1 | 22.6 |
| 1974 | 36417 | 35.7 | 12.7 | 23.0 |
| 1975 | 37233 | 37.7 | 12.2 | 25.5 |
| 1976 | 38228 | 37.6 | 11.7 | 25.8 |
| 1977 | 39860 | 37.6 | 12.0 | 25.8 |

AGE STRUCTURE (Population in Thousands)

| Population Aged: | 1960 | 1965 | 1970 | % distribution in 1960 | % distribution in 1970 |
|------|------|------|------|------|------|
| 0-14 | 11013 | 12458 | 14019 | 42.6 | 42.1 |
| 15-24 | 4915 | 5576 | 6350 | 19.0 | 19.1 |
| 25-49 | 7175 | 8143 | 9276 | 27.8 | 27.8 |
| 50-64 | 2014 | 2288 | 2611 | 7.8 | 7.8 |
| 65-80 | 715 | 924 | 1073 | 2.8 | 3.2 |

Source: CAPMAS

difficult soil conditions and soaring energy prices drove up costs.[118]

**TABLE 6.17**
**PRODUCTION OF MAJOR FOOD STUFFS, 1970-78**

| Crop | 1970 | 1971 | 1972 | 1973 | 1974 |
|---|---|---|---|---|---|
| Cotton, | | | | | |
| unginned | 1,404 | 1,418 | 1,422 | 1,368 | 1,204 |
| Maize | 2,393 | 2,342 | 2,417 | 2,507 | 2,640 |
| Sorghum | 874 | 854 | 831 | 853 | 824 |
| Beans | 277 | 256 | 361 | 273 | 234 |
| Rice (paddy) | 2,605 | 2,534 | 2,507 | 2,274 | 2,247 |
| Wheat | 1,516 | 1,729 | 1,616 | 1,837 | 1,884 |
| Vegetables | 5,159 | 5,232 | 5,415 | 5,688 | 6,006 |
| Sugarcane | 6,945 | 7,498 | 7,713 | 7,349 | 7,018 |
| Citrus Fruit | 706 | 883 | 825 | 923 | 963 |
| Dates | 294 | 340 | 396 | 380 | 396 |
| Other Fruit | 384 | 441 | 522 | 515 | 612 |
| Livestock products | | | | | |
| Meat | 284 | 288 | 295 | 299 | 302 |
| Milk | 1,589 | 1,614 | 1,640 | 1,666 | 1,692 |
| Poultry | 90 | 98 | 102 | 102 | 112 |
| Eggs | 50 | 53 | 54 | 58 | 56 |

| Crop | 1975 | 1976 | 1977 | 1978 |
|---|---|---|---|---|
| Cotton, | | | | |
| unginned | 1,061 | 1,084 | 1,260 | 1,381 |
| Maize | 2,781 | 3,047 | 2,724 | 3,117 |
| Sorghum | 775 | 759 | 648 | 681 |
| Beans | 234 | 254 | 270 | 231 |
| Rice (paddy) | 2,423 | 2,300 | 2,272 | 2,351 |
| Wheat | 2,033 | 1,960 | 1,697 | 1,933 |
| Vegetables | 6,520 | 6,922 | 6,750 | 7,746 |
| Sugarcane | 7,902 | 8,446 | 8,379 | 8,296 |
| Citrus Fruit | 1,013 | 889 | 797 | 990 |
| Dates | 415 | 417 | 461 | 377 |
| Other Fruit | 640 | 705 | 644 | 718 |
| Livestock Products | | | | |
| Meat | n.a. | 313 | 321 | 324 |
| Milk | n.a. | 1,750 | 1,780 | 1,801 |
| Poultry | n.a. | 129 | 121 | 115 |
| Eggs | n.a. | 76 | n.a. | n.a. |

n.a. Not available.

Source: Ministry of Agriculture.

Finally, and most importantly, yields stagnated during the decade, principally for two reasons. First, the failure to invest adequately in drainage in the previous decade bore its insidious fruits in the 1970s. A soil survey carried out in 1973 revealed that only 6% was classified as excellent: the rest was either "good" (45%) or "medium to poor" (49%).[119] It is clear that this situation resulted from inadequate drainage, which led to the familiar problems of a rising water table, soil salinity, and the formation of hard pan. Indeed, there is a consensus among students of Egyptian agriculture that the lack of adequate drainage is _the_ major constraint to increasing agricultural supply.[120] As we have seen, the problem was already recognized in the late 1960s, and some important remedial steps have been taken. A ten-year drainage plan was completed in July, 1972. Due to the large investments needed and due to the very tight capital constraint after the June War, the government sought, and obtained, World Bank loans for drainage installation. The Bank undertook financing and technical assistance for two major projects to cover 950,000 feddans in the Delta and 300,000 feddans in the Sa'id with tile drains. Additional projects covering some 1.8 million feddans have also been approved. Spending on drainage rose from 9% of total agricultural investment in 1969 to 29% in 1975.[121] This represents considerable progress, and is certainly an encouraging start.

However, some problems have been encountered. There is evidence of corruption in both maintenance and construction of drains. The maintenance problems are familiar from the Nasser period. In construction, it appears that some engineering specifications were ignored. The installation system was based upon tender offers. There was a large gap between the low, subsidized prices of inputs which the contractors had to pay to the government and the very high black-market prices (in turn stimulated by the soaring demand for construction materials as urban construction boomed).[122] Consequently, some contractors skimped on materials, installing drains at less than optimal depth and at wider than optimal intervals, so that they could sell their surplus materials on the illegal market. It also proved difficult to obtain enough labor for drainage installation, due to the apparent shift in labor to the urban areas. As the World Bank put it, "Drainage projects have been seriously delayed by administrative constraints (especially a shortage of engineering staff), the capability of the local civil works-contracting industry, and the lack of local funds."[123] Egyptian crop yields thus continued to fail to achieve the 20 to 50% increase which improved field drainage could provide.

The impact on crop yields from the diffusion of higher-yielding varieties has also been limited. (See Table 6.18 on the areas planted in new varieties.) Several factors are at work here. Policies which hold down the price of wheat and rice provide little incentive for farmers to adopt the new varieties. It is possible that the absence of adequate drainage helps to retard the diffusion of new seeds.[124] If the benefits of the new strains are to be realized, more fertilizer must be used. But here farmers confront the problem of the correlation between good drainage and responsiveness to fertilizer. Nevertheless, fertilizer use continued to increase during the decade, despite the stagnation of yields, because of the

TABLE 6.18
AREA PLANTED IN HYV CEREALS, 1970-71

| Year | Wheat Area | | Maize Area | | |
| | (a) | (b) | (c) | (d) | (e) |
|---|---|---|---|---|---|
| 1970/71 | 400 | | | | |
| 1972 | 4,500 | | | | |
| 1973 | 49,700 | | | | |
| 1974 | 525,900 | | | | |
| 1975 | 194,300 | 187,000 | 166,577 | 13,417 | 152,621 |
| 1976 | 183,500 | 177,000 | 202,730 | 26,126 | 125,466 |
| 1977 | 310,100 | 299,000 | | | |
| 1978 | | 288,000 | | | |

Sources:
(a) Acres: Dana G. Dalrymple, Development and Spread of High-Yielding Varieties of Wheat and Rice in the Less Developed Nations. Sixth Edn., U.S.D.A. Foregin Agricultural Report, No. 95 (Washington, D.C., 1978), 47.
(b) Feddans, Semi-dwarf Egyptian Major Cereals Improvement Project, (1979), 7.
(c) Feddans, American Early.
(d) Feddans, Shedwan -3.
(e) Feddans, Hybrids: All from Cereals Project (1978), 11.

government's system of fertilizer distribution.

It has also proved difficult to breed varieties of short-stemmed wheat which are resistant to rust under Egyptian conditions.[125] Such strains also tend to shatter easily and to yield less straw and flour than local varieties. Progress with rice and maize has been even more limited. Rice yields have shown little improvement during the 1970s. So far the strains introduced from the IRRI have shown more susceptibility to the stem borer than the dominant Nahda variety. Social and technical constraints have also limited the diffusion of hybrid maize seed. Farmers grow maize both as a food and as a fodder crop. Consequently, "it is no use growing hybrid maize under the same conditions and the same practices as the local open-pollinated varieties."[126] Farmers tend to strip off the leaves of the standing plant, which may reduce yields. Finally, the hybrids have proved susceptible to late wilt diseases. The "Green Revolution" has had only a limited impact so far on Egyptian agriculture.

AGRICULTURE UNDER SADAT, 1970-1980: RESPONSES

The government's response to this situation of rising demand and relatively stagnant supply seems so far to be one of (relative) "benign neglect." Such a position seems to be due to the intractability of many of the problems, the current surplus in the balance of payments, and the recognition that long-run solutions to Egypt's economic problems must come from industry, not agriculture. This is not to say that positive steps have not been taken. For example, drainage installation continues to be emphasized. Further, the prices to urban consumers and farm gate prices have been substantially "delinked," as the latter have risen closer to international prices. However, the share of investment in agriculture declined from approximately 25% of total public sector investment in 1965 to only 7% in 1975. Nor does private sector investment seem to have made up for this: private investment in agriculture may have been as low as L.E. 2-3 million in the mid 1970s, only 4 to 5% of gross fixed investment. It is widely believed that the government's price policies have been responsible for this lack of investment.[127]

Most private agricultural investment has gone toward the expansion of the fruit and vegetable area.[128] This is consistent with the demand patterns discussed above. Indeed, such a concentration on horticultural production has long been advocated as a rational response to the long-run problems of rising population on relatively fixed, yet highly fertile, land resources.[129] Not only would such a policy make the most intensive use of land, in some cases permitting triple cropping, but it would also provide a great deal of employment. Indeed, if the calculations in Appendix 6B are at all accurate, vegetable and fruit production now accounts for roughly 1/5 to 1/4 of all field crop labor in agriculture. Such products can be, and are, not only marketed at home, but also exported, especially to the Gulf and to other Arab countries.[130] Whether at home or abroad, such a production strategy essentially caters to the demands of wealthier, largely Arab, consumers.

The "horticultural strategy" faces very severe marketing problems, however. The current physical infrastructure is utterly inadequate even for the current level of vegetable output, much less for a greatly expanded volume of fresh produce. For example, Cairo now has one principal wholesale vegetable market (at Rod al-Farrag).[131] Built in the late 1940s, when the city's population was some 2.5 million, it is surrounded by high walls and has two main entrances, each just wide enough to let in two trucks abreast. Packing and handling are entirely unmechanized; as a result, considerable damage is done to the produce, especially to the fresh vegetables which come in on donkey carts packed in handicrafted palm or bamboo crates. Such woeful inadequacies are repeated at every level, from the overcrowded road network to the jammed ports. Tomato losses, due to such marketing problems for example, are estimated at at least 40% of field output.[132]

If vegetable production is to expand markedly, large-scale investments in the marketing system will be necessary. The currently rosy foreign exchange picture may make some such investments possible, although of course there are numerous competitors for the scarce funds. Further, the response of foreign agribusiness investors has been largely limited to New Lands or to input supply.[133] Although some small scale (in relation to need) projects have been initiated with World Bank funding (the Nubariyya cannery), relatively little has been done to date in this critical area.

A second common recommendation is to reduce both full-season clover and the number of work animals. The hope here is to release the land planted in clover (some 25% of the cultivated area) for other uses, while simultaneously orienting livestock production toward meat and dairy production. The (reduced number of) animals would be fattened on maize and sorghum. Mechanization of land preparation, threshing, and irrigation would make all of this possi+ ble. The government is committed to mechanization: The Five Year Plan calls for the mechanization of all farm operations by 1990. Major external aid agencies, such as the World Bank and USAID are assisting this effort: the Bank, for example, has a plan to mechanize completely the agriculture of Minufiyya and Sohag governorates. Nor is the current government alone in backing mechanization; the Left opposition likewise strongly supports the diffusion of tractors, threshers, and water pumps.[134] Mechanization, also long advocated for Egypt, is central to current and alternative agricultural development plans.

Private farmers have been in the lead here. Although national tractor censuses are notoriously unreliable because of the problems of treating old machines, current official estimates place the number of four-wheeled tractors probably exceeded 28,000 in 1978. Some 80% of these are privately owned. Over 1,000 tractors have been manufac- tured at the Nasr works in Helwan every year since 1973; the number of tractor imports has quadrupled in the same period. (See Table 6.12.) In Sharqiyya, land preparation and threshing are almost entirely mechanized; plowing is over 85% mechanized and irrigation pumping nearly 75% so in Behera, where plans are now being made to mechanize the harvest of rice, wheat, and birsim. Mechanization is also spreading in Middle Egypt.[135] Like increased vegetable and horticultural production, mechanization is increasingly becoming a reality rather than a planner's dream in Egypt.

As with the expansion of horticultural production, however, there are reasons for skepticism as to the social benefits of mechan- ization. There can be no question that such technical change brings important private benefits: greater leisure for the farm family, less exhausting and debilitating work, improved local transportation, and (perhaps) increased incomes. But if mechanization displaces labor, workers may simply trade one kind of low productivity job for another. Insofar as mechanization stimulates rural to urban migra- tion, it will exacerbate the already severe housing and transporta- tion crises. We know that mechanization reduces labor per unit of output in agriculture. The social question then becomes: will it raise production enough to prevent a net decline in employment? There are some additional distributional questions, since most trac- tors are owned by the well-to-do peasants, and since it is normally thought that mechanization will reduce grain supplies but increase milk and meat production. Let us look at the production side of mechanization first, and then turn to questions of labor supply, the key to the rapid diffusion of tractors after 1973.

Those hoping for the release of fodder land for other uses usually call for the mechanization of three specific operations: 1) water pumping, 2) land preparation, and 3) grain threshing. They stress that since the aim is to reduce animal numbers, these three functions should be viewed as a package.[136]

At first glance, this argument is appealing. Roughly half of the fodder fed to animals is clover; perhaps 25% of the cropped area is devoted to animal fodder. Clearly, such land has alternative uses. The chopped straw and other crop residues which are used to feed the animals also have a significant opportunity cost, as we have seen. Since birsim provides enough fodder for only about half of the year, fellahin strip green leaves off of maize plants and shun Mexipak wheat because of its low straw yield. These practices are assumed to reduce yields substantially.

However, there are several problems with this viewpoint. First, a recent study of the practice of stripping maize stalks found, somewhat surprisingly, no relation between such a practice and maize yields. The reason appears to be that, as usual, fellahin are clever farmers: they only strip after flowering, never take the ear or above-ear leaf, and seldom take even the next leaf.[137] Second, some authorities estimate that birsim yields could be raised by as much as 95%. As usual, drainage appears to be the main problem here. Also, as noted above, very little research on breeding higher yielding types of birsim has been done.[138] If this additional fodder could be stored, the current animal population could be fed on clover year-round, thereby eliminating the need to strip maize stalks at all. Depending on the relative costs of machinery and improved storage facilities, investing in the latter plus raising birsim yields might be the optimal response to a fodder shortage.

The opportunity cost of the land planted in birsim remains, of course. But here several additional points are in order. First, since birsim is a nitrogen-fixing crop, its removal from the crop rotation could have deleterious consequences for the yields of succeeding crops, usually cotton. (Birsim precedes about 80% of the cotton crop). Birsim not only supplies nitrogen to the soil, but also contributes to the formation of a tilth and to the permeability, and hence drainage, of the soil. More importantly, birsim is a highly profitable crop, both because of liberalization and the high income elasticity of milk and meat, and because government import quotas for meat raise domestic meat prices far above international levels.[139] Given these policies and assuming a growth in real incomes for Egyptians, it is highly unlikely that there will be any substantial reduction in the birsim area.

The multiple functions which water buffaloes and cattle play for small farmers further reduces the likelihood that mechanization will eliminate them and thus free clover land for other uses. Most of these animals are owned by small holders. In addition to traction

power, the animals fulfill four functions for small farmers. First
and most important, they provide cheese and milk for home consumption; such dairy products constitute the main (often the only) source
of animal protein for the fellahin. Second, any surplus can be sold.
Third, the animals constitute a kind of insurance. If farmers have a
bad year, or are faced with a major financial crisis, they can sell
the animals. Fourth, it is likely that owning at least a share of
such animals is necessary if a farmer wishes to lease in additional
land. Clearly, peasant-owned work animals are deeply imbedded in the
current social structure. This militates against any large reduction
in their numbers and release of fodder land.

Such a conclusion is reinforced by the lack of spare parts,
skilled mechanics, and repair shops in the rural areas. There is a
general shortage of adequate spare parts; local manufacturers do not
maintain an inventory of parts, but rather, manufacture them on
demand. Further, some twenty to thirty signatures are needed for
spare parts delivery. There is an acute shortage of rural repair
mechanics; this is hardly surprising since they can make "at least
300% more in industry or abroad." The result is that farmers often
must wait three to six months for repairs.[140] In such an environment, farmers retain their work animals as "spare parts on the hoof."

Several recent studies of Egyptian farm mechanization support
the view that such technical change is unlikely to release much land
for other uses. Both the World Bank study of their Minufiyya-Sohag
mechanization project and the ERA 2000 report argue that any gains
from released land are at best long-run gains. In the near term they
see the benefits of farm mechanization coming from increased
quantity/improved quality of milk and meat production if animals do
little or no farm work. The World Bank estimates that animals who
stop work can double their milk yields.[141] It should be pointed out
here that some 80% of farm animals' work time is devoted to irrigation work. Since this work uses relatively little adult labor (small
boys normally tend the animals turning saqiyas), and less adult hired
labor (tambour and shaduf labor being normally a family task), the
social benefits of mechanizing irrigation should be particularly
high.

Indeed, the fact that mechanization of irrigation would displace
primarily child labor, increase milk yields, increase the precision
of irrigation, and have some potential health benefits indicates that
water pumps should have the highest priority in mechanization. Curi-
ously, what one finds instead is the combination of 65 h.p. tractors
and saqiyas. One wonders what role government policy played in generating such a pattern of mechanization. Further, some current
cost-benefit analyses purport to show that the internal rate of
return on mechanical pumps is substantially lower than that of
threshers and tractors, precisely because pumps save relatively little labor.[142] This raises the crucial question of the social opportunity cost of labor, a point taken up below.

The second principal argument advanced for mechanization is that
it will raise yields per unit land. Such gains may be decomposed
into three parts: 1) the grain saved by mechanical threshing, 2) the

benefits of deeper plowing, and 3) the benefits of more timely planting. It might be noted that the first benefit is thought to derive
not merely from the technical superiority of mechanical drum threshers over nurags, but also from reduced transportation losses when the
grain is moved by tractor-drawn trailers rather than by camels. The
gains from better tillage are variously estimated at between 5 and
15%. Roger Garrett estimates that rotary tillers can raise maize
yields by 13%, cotton yields by 27% and beans by 14.6%. He reports
that chisel ploughs would raise maize yields by 5%. In 1968 the
agricultural engineer Hilmi Abdal-Ghani Sa'ad asserted that rotary
ploughs would raise maize yields 8%, cotton yields 18.2% and bean
yields by 6.8%. Yet in this same symposium these figures were ques-
tioned by the economist Amr Mohi-el-Din, who retorted that there was
little hard evidence that mechanization increased crop yields.[143]
However, further support for the "pro-tractor" argument may be found
in both the World Bank study and one chapter of the ERA 2000 report.
Both place the yield gains from better tillage at approximately 15%.
In a fine example of a book written by a committee, a later chapter
of the ERA 2000 report contradicts the earlier assertions, claiming
that neither cotton nor wheat yields are affected by deeper ploughing, while only maize showed some response.[144]

All of this evidence is based on agricultural experiment station
trials. The World Bank report adds that "evidence from other countries substantiates yield increases in the range of 5 to 15%." This
is highly debatable. Consider a recent summary of the Indian evidence:

> For all regions combined we are at best left with five or six
> instances out of 118 where large yield differences remain in
> the absence of equally large or larger differences in fertil-
> izer use. These studies fail to provide much support for the
> yield increasing effect of tractor cultivation.[145]

The problems inherent in generalizing from agricultural experiment
station data to the field are well-known. Indeed, the World Bank
report explicitly recognized this and points to evidence which suggests that farmers use tractors just as they would buffaloes:

> Tractors with chisel ploughs can cultivate without prior ir-
> rigation, and when used twice can effectively cultivate to a
> depth of twenty-five or thirty centimeters. In practice,
> however, tractor ploughing does not at present result in an
> improvement in work quality and only in some cases does it
> improve timeliness as farmers delay cultivation to await the
> tractor.[146]

At present farmers plough no more deeply with chisel ploughs than
they do with baladi ploughs. This means that "under existing conditions, the only advantage in using mechanical ploughs as opposed to
animal ploughing is the lower cost."[147]

It is possible that such problems could be remedied with time
and with an adequate extension service to diffuse the proper use of
the machines in the rural areas. But the current extension program
is understaffed and starved of resources and materials. The system

offers few incentives for agents to live in the countryside. Peasants commonly perceive agents as regulators who enforce government decrees rather than as technical experts who dispense useful information. The agents' training leave much to be desired; very little of it takes place in the countryside. If mechanization were to spread as the government plans, then a thoroughgoing reform of the extension service would be required. This is an important area for policy action, and is recognized as such by the Ministry of Agriculture. However, the magnitude and complexity of the problems do not engender optimism about the outcome.

Finally, there is the timeliness factor. It should be stressed that although increased timeliness may raise crop yields, it is unlikely to increase the cropped area itself. With a cropping intensity of about 2.0 on non-orchard lands, most authorities agree that the maximum increase in the cropped area would be to 2.1 over the next five to ten years. Perhaps mechanization will increase the cropping intensity of rich farmers. These crop less intensively than their smaller neighbors because of the problems of supervising and controlling a large hired labor force. Mechanization might increase their cropping intensity; if so, by definition, such a change will not generate any additional employment. A recent study has found no relationship between tractor use and cropping intensity in the Fayyum.[148]

Turning to the effect of timeliness on yields, the World Bank suggests that at an aggregate level, crop production could increase by some 5%. They argue that "even a saving of a few days can be expected to increase yields by one or two percent."[149] Using unmechanized techniques, it is apparently very difficult to plant at the optimal time, in part due to a lack of labor, in part because the native plough cannot be used for soil preparation unless the land is irrigated for five to seven days before ploughing. The problem is most acute for planting the summer crops of maize, cotton and rice when they are preceded by wheat or birsim. We have seen that current price policies contribute to the practice of taking extra cuts of birsim and delaying the planting of cotton; some 25% of cotton is planted late. One wonders whether mechanization would change this much if price policies remain the same. Maize and rice, on the other hand, usually follow wheat. Here mechanization of wheat threshing and land preparation would indeed permit earlier planting. Perhaps some 50% of both maize and rice is planted late. As a result, total rice and maize output per unit land is some 84% and 92%, respectively, of what it would be if there were no late planting.[150]

Although these results are plausible, three caveats are in order. First, as for the arguments on depth of ploughing, the data are based on agricultural experiment station results, whose relevance to actual field conditions is questionable. Indeed, one recent study finds no effect of timeliness on maize yields when looking at actual practices of famers; nor did it find any evidence that farmers using tractors plant maize earlier than those not doing so.[151] Second, the ERA 2000 report implicitly assumes that there is one optimal planting date for maize for the whole of Middle and Lower Egypt.

This seems unlikely, given the temperature differences in these areas. Third, historical evidence indicates that only complete mechanization eliminates, or even ameliorates, seasonal bottlenecks. In the U.S. cotton South, for instance, the initial effect of tractor mechanization was to increase the peak season demand for labor by nearly fifty percent.[152] This is encouraging from the point of view of labor displacement due to partial mechanization. The problem, however, is that farmers continue to press their mechanization effort forward, which then eliminates the new, sharper seasonal bottlenecks—and most of the jobs. As noted above, there are already signs of such a process in Behera governorate.

In summary, current studies do not support the notion that mechanization will increase cropping intensity substantially or release fodder land for other uses. The primary benefits of mechanizing threshing and land preparation appear to be grain saved during threshing and transportation, increased milk and meat production, and labor savings. Although there is experimental evidence that mechanization raises yields, this finding is not supported by any actual field studies of which I am aware. In short, there is no evidence that tractor farms have higher yields or cropping intensities than unmechanized farms.

Augmented meat and milk supplies are another alleged benefit of mechanization. The increase in meat might simply be that of which is gained by retaining former work animals and then holding their feed rations constant. However, mechanization advocates more commonly envision an enlarged cattle industry, which, in turn, would require additional resources. This raises some serious distributional questions on the commodity demand side. As one group of mechanization advocates notes, "Per capita supplies of maize and sorghum grains for human consumption would be reduced, but dietary levels would be improved by the projected increase in the supply of livestock products."[153] If livestock production is to be expanded, the former assertion seems likely. But the latter seems highly improbable. Both the urban and the rural poor consume proportionally much more maize than do the urban middle classes and foreign tourists. It is not plausible that the poor will be able to substitute meat for the grain which has now become more expensive because of reduced supply. A much more likely scenario would be: 1) the reduction in grain supplies increases grain prices which 2) leads to a reduction in the poor's consumption of grain and 3) to a proportionally larger reduction in their consumption of meat because of a) the large income effect which an increase in grain prices represents for such consumers and b) the higher income elasticity of meat and milk products.[154] The high grain:meat conversion ratio makes it very likely that increased meat supplies will fail to offset such a dynamic. The problem here is especially severe if, as many experts believe, the primary nutrition problem in Egypt is under nutrition (i.e. not enough calories) rather than malnutrition (although pellagra does continue to be a problem in some areas).

Given the importance of labor savings in the sum of benefits of agricultural mechanization, and given the constant refrain of "labor

shortage" which one hears in the rural areas and in government planning offices, it is essential to look at the conditions of labor supply. There are, indeed, indications of increasingly tight labor
markets. As the data in Appendix 6A shows, there has been a rapid
rise in both money and real wages for agricultural labor. Further,
these data understate the extent of the increase, because they are
given as units per day; there is considerable evidence that the working day in agriculture has been shortened from its former "sunrise to
sunset" pattern, to a day which ends around noon or 2:00 P.M. Workers often demand and receive such "fringes" as meals, tea, and
cigarettes.

It is also clear that labor costs have been rising more rapidly
than other production costs (see Table 6.19). This is not surprising, since the use of many other inputs is subsidized.[155] This is
certainly true for the capital and operating costs of mechanized
techniques. A variety of loan sources and terms is available. In
general the rate of interest on tractor loans does not exceed 12%,
and may be as low as 5%. With an inflation rate of at least 20% per
year, the real rate of interest on such loans is negative. The
government removed the 35% ad valorem tariff on tractors in 1980,
thus further lowering the cost of tractors. Also, some 68% of the
international price of diesel fuel is subsidized to Egyptian farmers;
electric power for pumps is also priced well below international levels.

Government policy may have favored mechanization in a second,
rather inadvertent, way. Since the government sets prices for the
major crops, labor costs have risen much more rapidly than prices;
farmers are then caught in a "profit squeeze." This appears to be the
case for wheat, maize, and cotton, but not for rice.[156] However, if
this problem were truly acute, we would expect to observe falling
land rents and a bearish market for agricultural land. In fact, of
course, we observe the opposite. Nevertheless, government policy on
both the input and output side has clearly promoted mechanization.

Let us look at the forces underlying the principle impetus in
inducing mechanization, the increase in rural wages. It would appear
that the action here is on the supply side: Although there has been a
fairly steady increase in the demand for labor in agriculture since
1950, the rate of increase in demand in the period after 1973 (when
real wages shot upward at unprecedented rates) was no more rapid than
during previous periods when wages showed no such marked upward
trend. Yet as Table 6.11 shows, the rate of growth of the rural
labor force has markedly slowed down. It does not appear that there
is a net decline in the number of workers; rather, the construction
boom in the cities, the subsidized costs of urban living, and the
departure of some 40% of urban construction workers to the Gulf
created openings which rural migrants have hoped to fill. There is
some impressionistic evidence that rural workers themselves go overseas.[157] Although it appears that only the additions to the labor
force are being drawn off, it may be that such workers, usually young
males, are the most productive agricultural workers. Insofar as
rural labor markets are sharply segmented by age and sex, the impact

TABLE 6.19
PRODUCTION COSTS FOR FIELD CROPS, 1972-1976

| Year: | 1972 | 1973 | 1974 | 1975 | 1976 |
|---|---|---|---|---|---|
| Labor Costs as Percent of Variable Costs per Feddan for: | | | | | |
| Wheat | 26% | 27% | 26% | 31% | 32% |
| Rice | 33 | 34 | 36 | 40 | 43 |
| Maize | 33 | 35 | 36 | 41 | 44 |
| Cotton | 46 | 48 | 51 | 53 | 55 |
| Rent (L.E. per feddan) | | | | | |
| Wheat | 15.46 | 15.52 | 16.45 | 18.07 | 18.37 |
| Rice | 10.95 | 11.90 | 13.43 | 14.82 | 17.45 |
| Maize | 12.08 | 12.58 | 13.29 | 15.02 | 18.91 |
| Cotton | 25.14 | 25.27 | 26.1 | 26.11 | 32.00 |

Source:
Calculated from Al Iqtisad Al Zira'i, Vol. 2, Section 35, 1978.
Cairo: Ministry of Agriculture, Center for Agricultural Economic
Research and Statistics.

of any migration would be exacerbated. However, the increased
numbers of women visible at work on rural construction sites gen-
erates skepticism as to the importance of this force.

The key to rising rural wages appears to be migration, which in
turn is the result of 1) the boom in the Gulf and in Libya and 2) the
construction boom in the Egyptian cities, itself the result of
Sadat's liberalization policies. Liberalization also made possible
the dramatically increased outflow of workers. The "Open Door"
swings both ways.

Several other forces may have also changed labor supply. The
experience of military service, the guarantee of government jobs to
veterans, and expanding rural education may have also reduced the
rate of growth of the agricultural labor force. The numbers of men
in the armed forces have risen by almost one-third (from 298,000 to
395,000) between 1973 and 1978.[158] Since all of those who serve are
guaranteed government jobs, their labor was indeed withdrawn from the
private labor market. Further, Egyptians, like soldiers everywhere,

are often reluctant to return to village life after a tour of duty. Spreading education also reduces the number of workers. Obviously, more educated young people are reluctant to engage in the back-breaking work of unmechanized agriculture. However, some skepticism is in order here. Even the Ministry of Education does not claim that all rural male (much less female) children are enrolled in primary school. At best, seventy-five percent are enrolled, and some experts place the figure nearer to sixty-five percent.[159] The number of illiterate entrants to the labor force thus rises yearly.

The diffusion of mechanization appears to be a classic case of induced technical change. While real wages have risen rapidly, the cost of capital has been held down through subsidies, and may have even declined in real terms. The "profit squeeze" has also served as a "focusing" device, helping to induce farmers to invest in tractors, threshers, and irrigation pumps. It is interesting to note the extent to which this pattern of induced technical change is due to international, in this case largely regional, forces. The outflow of workers to the Gulf not only helped to raise wages in agriculture, but has also provided the principal source of foreign exchange in recent years, thus loosening the capital constraint for the economy as a whole. There is no information on the use of remittances in the rural areas; insofar as these funds do flow into rural areas, they would, of course, increase the supply of capital and lower its cost.

There are grounds for disquiet about the future, however. It is not clear that the currently relatively rosy employment and capital flow pictures will be long-lived. While it is unlikely that large numbers of workers abroad will actually be forced to return (barring a major war, or a drastic change in Saudi Arabia's development plan), the rate of migration abroad, and its corollary, the inflow of remittances, are likely to slow down. Most of the Gulf countries have completed their major infrastructural investments; Saudi Arabia seeks to employ (less politically dangerous) Asian labor wherever possible. Indeed, in 1979 Korean construction firms won all new Saudi construction contracts. Egyptians also face competition from Indian, Pakistani, and Yemeni labor in that country.[160] Since mechanization is typically an irreversible phenomenon, it is quite possible that the current pattern of technical change is sowing the seeds of later difficulties. The risks of a decreasing rate of migration abroad for employment, the fairly high incremental capital: output ratio, the labor-displacing potential of mechanization, and the inexorable growth in the domestic labor force should make one cautious about labor-displacing technical change.[161]

Clearly, planners are caught in a cruel dilemma. Who could deny the desirability of eliminating the exhausting work of plowing with gamusa and mihrat, or laboring with a fas? Who could deny that freeing children from field labor or from tending saqiyas would be an enormous stride forward for rural Egypt? Who would deny that they, too, would seek work in the cities if they were given the same choices as discharged rural soldiers? But the removal of jobs from landless, hired workers, the poorest of the poor, should give us pause, as should the other probable distributional consequences of mechanization. If agriculture does not at least retain the current number of jobs, the already horrendous difficulties of the cities will be exacerbated. Father Ayrout may have been correct to say that "to mechanize agriculture will...overthrow a way of life which has grown up as the only one suitable for an abundant population on a limited soil."[162] But as we have seen throughout this book, that same way of life imposed (and imposes) terrible costs on the fellahin. As so often in Egypt's history, it is the peasants who will pay the costs of whatever the next round of policy decisions and technical changes may be.

APPENDIX A:  RURAL REAL WAGES: 1938-1980

Several real wage indices are available. The best and most comprehensive is probably that of Samir Radwan. (Table 6A-1). Other indices are those of Koval and Bahgat, Hansen, and several different indices calculated by Abdel Fadil.[163] An examination of Figures 6A-1 and 6A-2 shows that although there are some minor variations, the overall trends are roughly similar: increase in the early 1950s, sharp fall in the mid-1950s, recovery to the mid-1960s, a second decline from roughly 1966 to 1973, and then a dramatic upsurge after 1974. These calculations deflate money wages by the index for rural consumption constructed by F. Zaghloul.[164] Maize receives a weight of 45% in this index. Koval and Bahgat's index is identical to that of Radwan down to 1966: thereafter, their calculations for money wages diverge from his, for reasons which I am unable to discover. It might be mentioned, however, that their figure for 1974 seems rather more plausible than Radwan's. The dislocation of the rural labor market after 1973 must have pushed wages up because 1) all veterans of the October War were offered government jobs, and 2) the oil price boom, the liberalization of the Egyptian economy, and migration for employment to the oil states accelerated in the aftermath of that war.

Of course, conclusions as to the long-term trend of wages depends upon the choice of beginning and end year. It is obvious that measures from trough to peak (e.g. 1955-1978, or 1938-1978) will show considerable advance, while measuring from peak to trough (e.g. 1951-1970) may show decline. Several points are in order here: 1) the increasing trend obtained when measuring from trough to peak is greater in absolute value than the decreasing trend when measuring from peak to trough and 2) the overall trend from 1938 to 1974 in Radwan's own index is positive. His conclusion that "with the exception of a brief period in the late 1960s, the standard of living of agricultural laborers has more or less remained unchanged over the last 25 years"[165] must be rejected even if one accepts his somewhat suspect calculation for the real wage in the end year, 1974, rather than the higher estimate of Koval and Bahgat. A simple regression of the real wage index on time for 1938 to 1974 yields the following:

Index = 102.92 + 1.468 t, which is significant at the five percent confidence level, implying that the real wage index increased by at least one percent, on average, every year. Radwan goes on to state that "the real wage index in 1974 declined to the 1948 level," (Ibid.) Again, even accepting his rather low figure for 1974,

TABLE 6A-1
MOVEMENT IN REAL WAGES IN RURAL EGYPT, 1938-74
(1938 = 100)

Average Daily

| Year | Money wage | | Cost-of-living index | Real wage index |
|------|-----|-------|--------|--------|
| | PT | Index | | |
| 1938 | 3.0 | 100 | 100 | 100 |
| 1939 | 3.5 | 117 | 101 | 116 |
| 1941 | 3.6 | 120 | 132 | 90 |
| 1942 | 5.0 | 167 | 198 | 83 |
| 1943 | 6.3 | 210 | 238 | 87 |
| 1944 | 9.3 | 310 | 262 | 117 |
| 1945 | 9.3 | 310 | 262 | 117 |
| 1946 | 9.5 | 317 | 297 | 107 |
| 1948 | 10.0 | 333 | 271 | 123 |
| 1949 | 10.0 | 333 | 259 | 130 |
| 1950 | 11.6 | 387 | 264 | 147 |
| 1951 | 12.6 | 420 | 263 | 160 |
| 1952 | 12.0 | 400 | 265 | 151 |
| 1953 | 12.0 | 400 | 269 | 150 |
| 1955 | 7.6 | 253 | 294 | 87 |
| 1956 | 10.0 | 333 | 342 | 97 |
| 1959 | 12.5 | 417 | 334 | 124 |
| 1960 | 12.5 | 417 | 337 | 123 |
| 1961 | 12.3 | 410 | 358 | 113 |
| 1962 | 14.0 | 450 | 367 | 122 |
| 1963 | 15.0 | 480 | 377 | 127 |
| 1964 | 19.0 | 609 | 438 | 138 |
| 1965 | 22.0 | 704 | 519 | 135 |
| 1966 | 25.0 | 801 | 468 | 170 |
| 1967 | 24.5 | 784 | 479 | 162 |
| 1968 | 24.5 | 784 | 499 | 156 |
| 1969 | 25.5 | 817 | 536 | 151 |
| 1970 | 25.0 | 801 | 576 | 138 |
| 1971 | 25.5 | 817 | 580 | 140 |
| 1972 | 27.5 | 880 | 613 | 143 |
| 1973 | 29.2 | 930 | 661 | 140 |
| 1974 | 32.2 | 1,001 | 792 | 125 |

Source:  Samir Radwan (1977), 31.

TABLE 6A-2
DAILY WAGE RATE IN RURAL EGYPT, 1966-1978

|  |  |  | Average Daily Cost of Living/Rural Consumer Index | Real Wage Index |
|---|---|---|---|---|
|  | Money Wage |  |  |  |
| Year | P.T. | Index |  |  |
| 1966 | 25.5 | 850 | 468 | 181 |
| 1967 | 25.0 | 833 | 479 | 174 |
| 1968 | 24.0 | 800 | 499 | 160 |
| 1969 | 25.0 | 833 | 536 | 155 |
| 1970 | 25.5 | 850 | 576 | 147 |
| 1971 | 25.8 | 860 | 580 | 148 |
| 1972 | 26.5 | 883 | 613 | 144 |
| 1973 | 28.5 | 950 | 661 | 144 |
| 1974 | 35.1 | 1170 | 753 | 155 |
| 1975 | 46.5 | 1550 | 845 | 183 |
| 1976 | 61.6 | 2053 | 947 | 217 |
| 1977 | 76.0 | 2533 | 1039 | 246 |
| 1978 | 90.0 | 3000 | 1199 | 250 |

Source:  Koval and Bahgat (1980), 7.

repeating the above procedure for the period 1948-1974, we obtain:

Index = 127.54 + .63652 t, again significant at the five percent level: we may reject the hypothesis that there was no upward trend in rural real wages from 1948 to 1974.

The argument that the overall trend in real wages was upward would of course be reinforced if one used Koval and Bahgat's figures. Given the evidence of considerable increase in the demand for labor taking the period as a whole and the slow growth in the agricultural labor force, it would be surprising if real wages had shown no increase for the period.

One should remember, however, that the above calculations have used 1938 as a base year. As the tables show, that is near the "trough" of 1934. Average "maize wages" exceeded those of the first decade of the twentieth century (about 6.5 kg. of maize per day) only in the early 1950s and again after 1964 or so. If one looks at the twentieth century as a whole, only after the mid-1960s do real wages rise above the pre- World War I levels. Hansen's index shows a somewhat different picture, with peaks in 1928 and 1951.

One should also remember that these wages are at very low levels. Egypt's agricultural workers remain extremely poor; since it is well known that they do not work for a whole year due to seasonal fluctuations in demand, their standard of living remains an insult to human dignity. This is especially so for the tarahil workers, who

TABLE 6A-3
REAL WAGE INDEX (HANSEN)

| Year | Index of Real Wages |
|------|---------------------|
| 1914 | 100 |
| 1920 | 109 |
| 1928 | 164 |
| 1929 | 125 |
| 1933 | 147 |
| 1934 | 71 |
| 1937-39 | 110 |
| 1941 | 101 |
| 1943 | 98 |
| 1945 | 130 |
| 1950 | 144 |
| 1951 | 188 |
| 1955 | 116 |
| 1956 | 131 |
| 1959 | 141 |
| 1960 | 138 |
| 1961 | 127 |

Source:

> Bent Hansen, "Marginal Productivity Wage Theory and Subsistence Wage Theory in Egyptian Agriculture." Journal of Development Studies, 2 (July, 1966), 407.

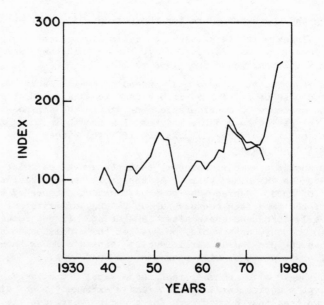

Fig. 6.3 Indices of Rural Real Wages
1938-1978

TABLE 6A-4
MOVEMENT IN RURAL REAL WAGES, 1950-71

| Year | (1) Index of average daily money-wages for men (1950 = 100) | (2) Index of money cost of the 'grain equi- valent' of rural wages in terms of maize (1950 = 100) | (3) Official wholesale price index for cereals (1950 = 100) | (4) Special cost-of-living index for agricultural labourers (1950 = 100) |
|------|------|------|------|------|
| 1950 | 100 | 100 | 100 | 100.0 |
| 1951 | 100 | 100 | 106 | 100.0 |
| 1952 | 100 | 110 | 102 | 100.5 |
| 1953 | 100 | 110 | 112 | 101.7 |
| 1954 | 100 | 108 | 118 | 104.3 |
| 1955 | 100 | 154 | 125 | 112.0 |
| 1956 | 100 | 171 | 136 | 129.7 |
| 1957 | 100 | 143 | 134 | 128.6 |
| 1958 | 100 | 143 | 127 | 131.5 |
| 1959 | 109 | 143 | 128 | 126.0 |
| 1960 | 115 | 143 | 128 | 128.0 |
| 1961 | 115 | 143 | 134 | 136.0 |
| 1962 | 115 | 148 | 133 | 139.3 |
| 1963 | 164 | 148 | 131 | 144.0 |
| 1964 | 164 | 200 | 137 | 167.5 |
| 1965 | 169 | 172 | 149 | 198.1 |
| 1971 | 227 | 229 | 190 | 214.6 |

Source: Abdel-Fadil (1975), 68.

often work less than half the year. Radwan has developed estimates of the numbers of rural Egyptians living below a minimum subsistence level: the percentages of the rural population below the poverty line were 22.5%, 17%, and 28% in 1959, 1965, and 1975, respectively.[166] Although the number for 1975 may be too high, due to the likely underestimate of real wages in that year, such figures still represent at least four million (hungry) people (Radwan gives over five million). Although it is likely that the transformation of the labor market in the middle and late 1970s has ameliorated this grim picture, we do not know by how much, or for how long.

APPENDIX B: THE DEMAND FOR LABOR IN AGRICULTURE

Two problems confront any attempt to estimate the total amount of labor input into field crop production in Egypt. First, there is the problem of which of the many, varied estimates of labor input per crop per feddan to use. The labor input coefficients were taken from

TABLE 6A-5
THE 'GRAIN EQUIVALENT' OF RURAL WAGES, 1950-71

| Year | Price of maize (piastres) Per Ardeb* (1) | Per Kg (2) | Money cost (piastres) of 3000 calories' worth of maize (3) | Index (4) | Daily money-wages (piastres) (5) | (5)-(4) (6) |
|------|------|------|------|------|------|------|
| 1951 | 245 | 1.75 | 1.45 | 100 | 11.0 | 11.0 |
| 1952 | 270 | 1.93 | 1.60 | 110 | 11.0 | 10.0 |
| 1953 | 270 | 1.93 | 1.60 | 110 | 11.0 | 10.0 |
| 1954 | 264 | 1.89 | 1.57 | 108 | 11.0 | 10.2 |
| 1955 | 378 | 2.70 | 2.24 | 154 | 11.0 | 7.1 |
| 1956 | 419 | 2.99 | 2.48 | 171 | 11.0 | 6.4 |
| 1957 | 350 | 2.50 | 2.07 | 143 | 11.0 | 7.7 |
| 1958 | 350 | 2.50 | 2.07 | 143 | 11.0 | 7.7 |
| 1959 | 350 | 2.50 | 2.07 | 143 | 12.0 | 8.4 |
| 1960 | 350 | 2.50 | 2.07 | 143 | n.a. | . |
| 1961 | 350 | 2.50 | 2.07 | 143 | 11.6 | 8.8 |
| 1962 | 360 | 2.57 | 2.14 | 148 | n.a. | . |
| 1964 | 400 | 3.50 | 2.90 | 200 | 18.0 | 9.0 |
| 1965 | 420 | 3.0 | 2.50 | 172 | 18.6 | 10.8 |
| 1971 | 360 | 4.0 | 3.32 | 229 | 25.0 | 10.9 |

Source: Abdel-Fadil (1975), 66.

Mohi el-Din and the I.L.O. Although there are many alternative estimates, the same overall picture emerges regardless of the precise set
of coefficients which are used. The coefficients are given in Table
6B-1.

Obviously, such a figure, when used for the whole country can
only be a rough guess, at best. Differences among regions and among
farmers with different relative resources suggest that there is considerable variation even for one year. They also do not capture
changes in seasonality, as when the shift in maize planting date and
increased rice cropping exacerbated the May-June bottleneck. Calculations over time are also complicated by technical change, which, as
we have seen, had both labor-using and labor-saving elements during
this period. The numbers in Table 6B-2 are, of course, only very
rough estimates of the change in the demand for labor as a result of
1) changes in the cropped area and 2) changes in the crop-mix. As
for technical change, as argued in the text, it seems likely that the
labor-using changes were more important: indeed, the Ministry of
Agriculture estimates the total time utilized in field crop production in 1978 as 12% more than the estimate given here. Of course,

TABLE 6B-1
COMPARATIVE SETS OF LABOR COEFFICIENTS
(DAYS PER FEDDAN)

| | MEDIUM SET | | LOW SET | |
|---|---|---|---|---|
| CROP | Men | Women, Children | Men | Women, Children |
| COTTON | 42 | 87 | 26.33 (c) | 87 |
| MAIZE | 25 | 10 | 17.5 (e) | 10 |
| WHEAT | 27 | 4 | 11.17 (c) | 4 |
| RICE | 35 | 40 | 30 (d) | 37.5 (e) |
| BIRSIM | 15.625* | .5* | 15.625* | .5* |
| MILLET | 42 | 9 | 23 (e) | 9 |
| BEANS | 19 | 5 | 17 (e) | 1 (e) |
| BARLEY | 18.25 | 3 | 10.75 (e) | 1 (e) |
| SUGAR | 98 | 31 | 55 (e) | 31 |
| LENTILS | 21 | 2 | 12 (e) | 1 (e) |
| ONIONS | 33.5 | 70 | 33 (a) | 70 |
| VEGETABLES | 60 | – | 60 | – |
| FRUIT | 68 | – | 68 | – |

| | HIGH SET | |
|---|---|---|
| CROP | Men | Women, Children |
| COTTON | 48.5 (d) | 102 (d) |
| MAIZE | 39.5 (b) | 43.7 (b) |
| WHEAT | 27 | 7.5 (d) |
| RICE | 45 (e) | 51.55 (b) |
| BIRSIM | 44.33 (e) | 26.8 (b) |
| MILLET | 42 | 16 (e) |
| BEANS | 19 | 5 |
| BARLEY | 18.25 | 3 |
| SUGAR | 98 | 59 (e) |
| LENTILS | 21 | 2 |
| ONIONS | 36 (e) | 80 (e) |
| VEGETABLES | 66+ (d) | – |
| FRUIT | 127++ (d) | – |

* average;  + tomatoes only;  ++ oranges only.

Unless footnoted, figures are from Ministry of Agriculture, Department of Economics, Agricultural Economy, Vol. 15, No. 2, Cairo, February, 1964.

Other Sources:
  a) "Debate on Agricultural Mechanization," L'Egypte Contemporaine, 331, Jan., 1968.
  b) ERA 2000, Inc., Further Mechanization of Egyptian Agriculture,

234

1979.

c) Calculated from EWUMP data, kindly supplied by Dr. Gene Quenemeon.

d) IBRD, 1975. IBRD uses Ministry of Agriculture data.

e) M.R. Ghonemy, Resource Use and Income in Egyptian Agriculture, Unpub. Ph.D., North Carolina State University, 1953, 81.

there are many possible reasons for such a disparity; given the crude level of all such calculations, it appears that 1) the numbers are fairly close and 2) there is no evidence that technical change in agriculture has displaced labor, at least as of 1978.

It will be seen that the increase in the demand for labor resulting from such cropping changes appears to have been considerable, rising over one-fourth from 1952 to 1978 (28%). Most (78%) of this change had occurred by 1969. This supports the hypothesis that the changes in rural real wage rates in the late 1970s are fundamentally the result of supply factors.

TABLE 6B-2
LABOR INPUT BY CROP - MAN DAYS (MILLIONS)

| TIME | 1950 | 1951 | 1952 | 1953 | 1954 | 1955 |
|------|------|------|------|------|------|------|
| COTTON | 83.143 | 83.143 | 82.723 | 55.427 | 66.347 | 76.423 |
| MAIZE | 36.242 | 41.492 | 42.595 | 50.382 | 47.610 | 45.845 |
| WHEAT | 36.982 | 40.492 | 37.854 | 48.338 | 48.460 | 41.129 |
| RICE | 24.496 | 17.146 | 13.076 | 14.791 | 21.336 | 20.989 |
| BIRSIM | 34.056 | 33.119 | 34.406 | 33.431 | 35.461 | 36.711 |
| MILLET | 16.376 | 17.636 | 18.182 | 20.399 | 19.207 | 18.358 |
| BEANS | 6.838 | 6.078 | 6.753 | 5.675 | 5.985 | 6.806 |
| BARLEY | 2.190 | 2.190 | 2.493 | 2.121 | 2.228 | 2.477 |
| SUGAR | 7.840 | 8.820 | 9.016 | 9.800 | 11.760 | 10.780 |
| LENTILS | 1.680 | 1.680 | 1.210 | 1.451 | 1.825 | 1.697 |
| ONIONS | 1.340 | 1.340 | 0.871 | 1.340 | 1.340 | 1.675 |
| VEGET | . | . | 15.138 | 17.388 | 18.714 | 20.490 |
| FRUIT | 5.916 | . | 6.372 | 6.650 | 6.895 | 7.385 |
| TOTAL | 257.100 | 253.137 | 270.688 | 267.195 | 287.168 | 290.765 |

TABLE 6B-2 CONT.

| TIME | 1958 | 1959 | 1960 | 1961 | 1962 | 1963 |
|------|------|------|------|------|------|------|
| COTTON | 78.884 | 73.920 | 78.666 | 83.412 | 69.594 | 68.334 |
| MAIZE | 48.740 | 46.450 | 45.525 | 40.075 | 45.800 | 43.025 |
| WHEAT | 38.332 | 39.825 | 39.312 | 37.368 | 39.285 | 36.315 |
| RICE | 18.196 | 25.515 | 24.710 | 18.795 | 29.050 | 33.565 |
| BIRSIM | 37.180 | 37.469 | 37.719 | 38.250 | 38.156 | 38.031 |
| MILLET | 17.640 | 19.614 | 19.068 | 19.236 | 19.068 | 20.328 |
| BEANS | 6.840 | 7.087 | 7.163 | 6.859 | 7.277 | 7.258 |
| BARLEY | 2.555 | 2.573 | 2.701 | 2.208 | 2.391 | 2.208 |
| SUGAR | 10.780 | 10.976 | 10.878 | 10.976 | 11.858 | 13.034 |
| LENTILS | 1.470 | 1.659 | 1.785 | 1.323 | 1.659 | 1.638 |
| ONIONS | 1.675 | 1.641 | 1.641 | 1.943 | 1.675 | 1.742 |
| VEGET | 24.744 | 28.620 | 29.760 | 30.420 | 32.100 | 35.220 |
| FRUIT | 7.834 | 8.568 | 8.908 | 9.384 | 9.928 | 10.676 |
| TOTAL | 294.870 | 303.917 | 307.836 | 300.249 | 307.841 | 311.374 |

TABLE 6B-2 CONT.

| TIME | 1966 | 1967 | 1968 | 1969 | 1970 | 1971 |
|------|------|------|------|------|------|------|
| COTTON | 78.078 | 68.292 | 61.488 | 68.124 | 68.334 | 64.050 |
| MAIZE | 39.375 | 37.125 | 38.850 | 37.100 | 37.575 | 38.050 |
| WHEAT | 34.857 | 33.615 | 38.151 | 33.642 | 35.208 | 36.423 |
| RICE | 29.540 | 37.625 | 42.140 | 41.720 | 40.005 | 50.495 |
| BIRSIM | 39.563 | 42.483 | 41.859 | 42.594 | 42.938 | 43.281 |
| MILLET | 21.756 | 21.924 | 22.344 | 19.908 | 21.042 | 20.748 |
| BEANS | 7.942 | 6.954 | 6.365 | 6.878 | 6.270 | 5.491 |
| BARLEY | 1.788 | 1.953 | 2.135 | 1.880 | 1.515 | 1.277 |
| SUGAR | 13.034 | 13.426 | 15.190 | 16.660 | 18.228 | 18.914 |
| LENTILS | 1.575 | 1.386 | 1.071 | 0.945 | 0.987 | 1.365 |
| ONIONS | 1.943 | 1.407 | 1.306 | 1.876 | 1.139 | 1.239 |
| VEGET | 39.120 | 38.460 | 42.480 | 43.260 | 42.840 | 43.140 |
| FRUIT | 13.260 | 14.076 | 15.300 | 15.776 | . | 16.932 |
| TOTAL | 321.831 | 318.680 | 328.680 | 330.362 | 316.080 | 331.406 |

TABLE 6B-2 CONT.

| TIME | 1972 | 1973 | 1974 | 1975 |
|---|---|---|---|---|
| COTTON | 64.344 | 67.200 | 61.026 | 55.692 |
| MAIZE | 38.275 | 38.850 | 43.875 | 44.850 |
| WHEAT | 33.453 | 33.696 | 36.990 | 37.638 |
| RICE | 40.110 | 34.895 | 36.855 | 36.855 |
| BIRSIM | 44.047 | 44.906 | 43.703 | 43.938 |
| MILLET | 200.328 | 20.454 | 20.958 | 20.538 |
| BEANS | 6.935 | 5.738 | 4.636 | 4.674 |
| BARLEY | 1.661 | 1.533 | 1.405 | 1.825 |
| SUGAR | 19.796 | 19.404 | 20.384 | 21.560 |
| LENTILS | 1.407 | 1.554 | 1.386 | 1.218 |
| ONIONS | 1.038 | 0.837 | 1.306 | 0.904 |
| VEGET | 44.940 | 47.940 | 49.200 | 53.100 |
| FRUIT | 17.204 | 17.544 | 18.564 | 19.380 |
| TOTAL | 333.538 | 340.289 | 340.289 | 342.172 |

TABLE 6B-2 CONT.

| TIME | 1976 | 1977 | 1978 |
|---|---|---|---|
| COTTON | 52.416 | 59.766 | 49.896 |
| MAIZE | 47.350 | 43.100 | 47.700 |
| WHEAT | 37.692 | 32.589 | 37.287 |
| RICE | 37.730 | 36.400 | 34.860 |
| BIRSIM | 43.547 | 44.594 | 45.031 |
| MILLET | 19.908 | 17.808 | 18.186 |
| BEANS | 4.940 | 5.548 | 4.541 |
| BARLEY | 1.898 | 1.734 | 2.080 |
| SUGAR | 23.814 | 24.500 | 27.440 |
| LENTILS | 1.344 | 1.008 | 0.756 |
| ONIONS | 1.038 | 1.239 | 0.971 |
| VEGET | 55.080 | 54.840 | 56.340 |
| FRUIT | 21.148 | 21.828 | 22.100 |
| TOTAL | 347.905 | 344.954 | 347.189 |

Calculated using "Medium Set" of labor input coefficients.

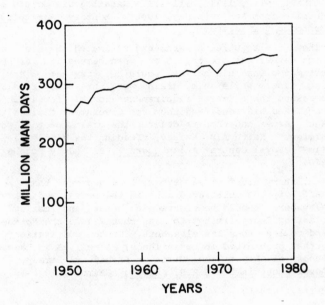

Fig. 6.4 Estimated Total Crop Labor Use
1950-1978 (Source: Table 6.B.2)

## Footnotes

[1]  Montgomenry's army at Alamein alone numbered some 195,000 men.  Peter Calvocoressi and Guy Wint, Total War: Causes and Courses of the Second World War, (N.Y., 1972), 364.

[2]  M.M. Hamdi, "The Nationalization of Cotton", L'Egypte Contemporaine, 265 (July, 1951), 60.

[3]  Martin W. Wilmington, The Middle East Supply Centre. (Albany, N.Y., 1971), 115-116.  Retail grain prices more than doubled from April to May, 1942.  Monthly Agricultural and Economic Statistics.

[4]  "They (the Egyptian government) estimated the area of crops each farmer was growing; then, when harvesting was near, they estimated what his harvest would be; they sat on his threshing floor while his grain was threshed; finally, they assessed the farmer's requirements for his family, his livestock, and his seed, and then the Government claimed the rest. Each farmer had then to deliver the grain to the Government stores."  Keith A.H. Murray, "Feeding the Middle East in War-Time", Royal Central Asian Journal 32 (July-October 1945), 239.

[5]  A military order of September, 1942 decreed that more than half of the cultivated area in the Delta, and sixty percent elsewhere, should be planted in wheat and barley.  Cotton plantings were limited to less than 22% in the Northern Delta and to less than 15% elsewhere.  Cotton cultivation was altogether prohibited in basin lands, as well as in the perennial lands of Asyut and Girga.  Rashed el-Barawy, Economic Development in the U.A.R. (Egypt) (Cairo, 1970), 18.

[6]  Murray, (1945).

[7]  Animal numbers changed as follows:

| Year | Cattle | Buffalos |
|------|--------|----------|
| 1939 | 1,229,879 | 965,826 |
| 1941 | 991,233 | 912,760 |
| 1943 | 1,202,284 | 1,001,124 |
| 1945 | 1,265,185 | 1,064,118 |

Annuaire Statistique du Poche, various issues.

[8]  Hamdi, (1951).

[9]  Peter Mansfield, The British in Egypt (N.Y., 1971), 273-4.

[10]  Agricultural Census, 1939, and 1950.  The demobilization, and the resulting depression of the urban labor market, underlay the slump in real wages in 1946.  The decline in the labor force was not exclusively the result of migration; a malaria epidemic in the Sa'id may have claimed 500,000 lives.

[11]     Department of Agriculture, Bureau of Agricultural Economics,
         Bulletin #6, Vol, &, October, 1949; Agricultural Census,
         1950. Some large farmers retained the 'ezbah system, how-
         ever. See, e.g., Samir Saffa, (1948).

[12]     See, among many others, the discussion in Fellah Department,
         Ministry of Social Affairs, "Notes on Production Relating to
         Agriculture", Cairo, 1948.

[13]     See Ghonemy, (1953) 46ff, for a discussion of possible prob-
         lems in comparing the data from 1939 with those of ten years
         later. He concludes that tenancy "increased substantially".

[14]     See Chapter 3.

[15]     Murray, (1945), discusses the "need" to import tractors to
         replace labor during the War years. By the, end of the war
         the number of tractors had more than doubled. They were
         especially favored by those with farms larger than one hun-
         dred feddans. British Board of Trade: Report of British
         Goodwill Trade Mission to Egypt, November-December, 1945.
         (London, 1946), p. 12. Charles Issawi, Egypt in Revolution
         (London, 1963), 134, gives the Egyptian tractor park as 1,200
         in 1939, 3,000 in 1945, and 7,600 in 1952. See also Table
         6.11. and the discussion on mechanization in the 1970s at
         the end of this chapter.

[16]     See Robert L. Tignor, "Equity in Egypt's Recent Past", Mimeo,
         Princeton University, (1978), 40 ff.

[17]     Anouar Abdel-Malik, (1968), 68.

[18]     Ghonemy, (1953), 146-147.

[19]     Tarek el Bishry, "Aperçu politique et social", La Voie Egyp-
         tienne Vers le Socialisme (Cairo, n.d.) See also Gabriel
         Baer, "Submissiveness and Revolt of the Fellah", in (1969).

[20]     Such a withdrawal and the reduced supervision which it
         implies would also be consistent with the evidence in Chapter
         4 on the further spread of the two-year cotton rotation dur-
         ing the immediate pre-Coup years.

[21]     Republique Francaise, Ministere de Finance et des Affaires
         Economiques, Memento Economique: l'Egypte. (Paris, 1950),
         53-55; Barawy, (1950).

[22]     There is a voluminous literature on the politics of the
         Nasser regime. The following paragraphs rely heavily on Ray-
         mond William Baker, Egypt's Uncertain Revolution under Nasser
         and Sadat. (Cambridge, Mass., 1978).

[23]     On the rich peasantry as a "second stratum" or local elite,
         see Leonard Binder, (1978). Of course, Nasser's enormous
         personal charisma comprised an additional, indispensable ele-
         ment in the regime's legitimacy.

[24]     "Efficiency and justice."

[25]    See the parallel analysis of Samir Radwan and Eddy Lee, "The
        State and Agrarian Change: A Case Study of Egypt, 1952-77" in
        Dharam Ghai, et. al., eds., Agrarian Systems and Rural
        Development (N.Y., 1972). See also Bent Hansen, "Arab
        Socialism in Egypt", World Development, 3,4 (April, 1975).

[26]    There is a large literature on Egyptian land reform. I have
        relied heavily on Mahmoud Abdel-Fadil, (1975); Samir Radwan,
        Agrarian Reform and Rural Poverty: Egypt, 1952-75. (Geneva,
        1978); Gabriel Saab, (1967); Eprime Eshag and M.A. Kamal,
        "Agrarian Reform in the U.A.R. (Egypt)," Bulletin of the
        Oxford University Institute of Economics and Statistics, 30,
        2 (May, 1968).

[27]    See Baker, 64 ff, especially 68.

[28]    See Hossein Askari, John Cummings, and Bassam Harik, "Land
        Reform in the Middle East," International Journal of Middle
        East Studies, 8, 4 (October 1977), 437-451.

[29]    See R. Albert Berry and William R. Cline, Agrarian Structure
        and Productivity in Developing Countries (Baltimore, 1979).

[30]    Saab, (1967), Ch. 2; Baker (1978), 204 ff.

[31]    I have spoken to sharecroppers in Minya Governorate who
        insist that the family who owns their land holds at least
        seven hundred feddans.

[32]    This is only true for the first round of reform. The bonds
        with which former owners were paid in 1961 were negotiable on
        the stock exchange. As Saab puts it, "The preferential
        treatment granted to those affected by the Second Agrarian
        Reform Law may well have been to prevent antagonizing too
        strongly the bourgeoisie and medium-sized farmers, from whose
        ranks much of the Egyptian officer class were drawn." Saab,
        (1967), 184.

[33]    See Anthony Nutting, Nasser (N.Y., 1972), 299-300.

[34]    Binder, (1978), 182; Baker, (1978), Ch. 8; O'Brien, (1966),
        144.

[35]    This may have contributed to the successful production per-
        formance of the reforms. Unlike the experiences in other
        countries, agricultural output showed no decline, indeed,
        rather the reverse, as a result of the reforms. See Saab
        (1967), and fn. 28 above.

[36]    Abdel-Fadil (1975), and Baker (1978).

[37]    Radwan (1978), 23. The World Bank estimates the landless at
        38% of persons active in agriculture. World Bank, Land
        Reform. Sectoral policy paper. (Washington, D.C., 1975).

[38]    Saab (1967).

[39]    However, the demand for hired labor was almost certainly
        lower under the new system. See Appendix 6.1.

[40]    Saab (1967), "Appendix I: The Nawag Experiment," 190-196;
        H.A. El-Tobgy, Contemporary Egyptian Agriculture. (Cairo,
        1976), 56.

[41]    See Abdel-Fadil (1975), Nutting (1972), and O'Brien (1965).

[42]    Saab (1967), 73-77.

[43]    Baker (1978), 205.

[44]    There is a large literature by now on rich peasant dominance
        of cooperatives. See, e.g., James B. Mayfield, Local Insti-
        tutions and Egyptian Rural Development. (Cornell University
        Rural Development Committee Monograph, 1978), 130; Michel
        Kamal, "Feodeaux, Paysans Riches, et Fellahs," Democratie
        Nouvelle, (April, 1968); René Dumont, "Les problèmes agraires
        de la R.A.U.," Politique étrangère, 2 (1968); United States
        Department of Agriculture, cooperating with the United States
        Agency for International Development and the Egyptian Minis-
        try of Agriculture, Egypt: Major Constraints to Increasing
        Agricultural Productivity. (Washington, D.C.: USDA Foreign
        Agricultural Economic Report, #120, 1976); and Baker (1978),
        Ch. 8.

[45]    Rural Employment Problems in the United Arab Republic.
        (Geneva, 1969), p. 36; Baker (1978), Ch. 8.

[46]    Mayfield (1978), 109-110.

[47]    Binder (1978), 342. In the "Kamshish incident" a minor offi-
        cial of the Arab Socialist Union was murdered by agents of a
        large landlord, who had successfully evaded land reform and
        had engaged in a number of other illegal abuses.

[48]    Michel Kamal (1969), and El-Tobgy (1976).

[49]    Farms of less than five feddans currently plant some 37.5% of
        their land in cotton (the last profitable crop), while those
        with more than five feddans plant only 29% of their cul-
        tivated area in cotton. However, since small peasants, as
        usual, crop more intensively than larger farmers, they may be
        able to offset this disadvantage on a per unit land basis.
        ERA 2000, Inc., Further Mechanization of Egyptian Agricul-
        ture. (Gaithersburg, Md., 1979).

[50]    See Appendix 6.2 on labor use by crop. Such highly profit-
        able crops require substantial investment and, for feed lots,
        minimum size for liscensing. Since there was no long-term
        credit source for small farmers, only rich farmers, often
        those with sources of saving from outside the agricultural
        sector, could afford to make the switch. See, e.g. WIlliam
        Cuddihy, "Agricultural Price Management in Egypt" World Bank
        Staff Working Paper No. 388 (Washington, 1980) iv, vii.

[51]    C. Peter Timmer, "Food Prices and Food Policy Analysis in
        LDCs." Food Policy, 5,3 (August 1980), 188-199.

[52]    The most detailed and thorough study to date of which I am
        aware is that of Cuddihy, (1980).

[53]    For cotton 17% of the crop was cooperatively marketed in
        1962-63, 42% in 1963-64, 60% in 1964-65 and 100% thereafter.
        Average marketing ratios for other crops were roughly as fol-
        lows: Rice, 65%; Onions, 60%; Wheat, 40%; Sugar Cane, 90%.
        Abdel Fadil (1975), 135.

[54]    Ezz el Din Hammam Ahmed and Mohammed Galal Abu el Dahab, Fer-
        tilizer Distribution in the Arab Republic of Egypt. (Paris,
        1972).

[55]    Ibid, 94. There is some discrepancy between the OECD figures
        for fertilizer use and those of CAPMAS. The trend toward
        greater domestic share of consumption was reversed following
        the closing of the Suez plant during and after the 1967 war.

[56]    Ibid., 41.

[57]    "Inasmuch as the farmer has no say in determining fertilizer
        application levels (except by purchases of nutrients over and
        above the NPK dosages prescribed on his holding card) and
        furthermore, since prices do not fluctuate with market condi-
        tions, such a situation rules out the possibility of incen-
        tives having any effect on fertilizer sales." Ibid., 89.

[58]    Cf. Robert Mabro, The Egyptian Economy 1952-72. (London
        1974), 67.

[59]    Farmers received some 60% and 70% of the international price
        of cotton and rice, respectively, between 1965 and 1976. See
        Cuddihy, (1980), iii, 57, 80.

[60]    Bent Hansen and Karim Nashashibi, Foreign Trade Regimes and
        Economic Development: Egypt. (New York, 1975), esp. 137-199.

[61]    "The general pattern that emerges from our study suggests
        that the domestic terms of trade have not turned greatly
        against agriculture during the period under consideration.
        The crucial policy variable in this respect has been the
        favorable terms of trade between agricultural outputs and
        manufactured inputs (mainly fertilizers)." Abdel Fadil
        (1975), 103. However, Cuddihy has estimated that cotton,
        rice, and onions were taxed (i.e. nominally protected) at a
        rate of 30%, net of such input subsidies, from 1965 to 1972.
        Cuddihy, (1980), iii.

[62]    William Cuddihy has estimated that of the L.E. 1.2 billion
        total net transfer from farmers due to price policies in
        1974-75, some 85% was transferred to consumers, while the
        government acquired only some 15%, or L.E. 179 million, in
        revenue. Cuddihy, (1980).

[63]    World Bank, Arab Republic of Egypt: Economic Management in a
        Period of Transition (Report No. 1815-EGT, Washington, D.C.,
        1978). Vol. VI: Statistical Appendix, 25. Much of this
        report has just been published: Khalid Ikram, coordinating
        author, Egypt: Economic Management in a Period of Transi-
        tion, (Baltimore, 1980). In current prices, investment rose
        from 18.7 million L.E. to L.E. 67.6 million. The share of

agriculture alone was much lower, however: 4.1% in 1955–56, 7.4% in 1960–61, 8.6% in 1966–67, 7.6% in 1968–69.

[64]  World Bank (1978), Vol. III: Productive Sectors, 44.

[65]  Samir Radwan, Capital Formation in Egyptian Industry and Agriculture, 1882–1967. (London, 1974).

[66]  By far the most dispassionate and convincing analysis is to be found in John Waterbury, Hydropolitics of the Nile Valley (Syracuse, N.Y., 1979). I have relied extensively upon Waterbury's work in the following brief discussion. Readers interested in more detail are urged to consult this excellent book.

[67]  Ibid., 87.

[68]  For example, Dr. Abd al–Aziz Ahmad anticipated many of the post hoc critiques in 1955. However, by presenting his findings at the British Institute of Civil Engineers, just at the time of the Suez crisis, he ensured that his arguments would be ignored. See Waterbury, (1979), 120.

[69]  Waterbury (1979), 105. "The suggested terms...: (1) that Egypt...divert one–third of internal revenues to the project over tén years: (2) that relevant aspects of the country's economy be subject to periodic IBRD review; (3) that contracts be awarded through open bidding with a prohibition on any being awarded to a communist country; (4) that Egypt would have to avoid incurring new foreign obligations (such as acquiring more arms), and any new loans or credits would require prior approval by the IBRD; (5) that no disbursement would take place before Egypt reached a new agreement with the Sudan over the allocation of the Nile's waters." Ibid. The extent to which the whole project was politicized may be seen in Nasser's 1964 speech at the opening of the first stage of the Dam: "There is no spot which represents the great battle of contemporary Arab man better than this site on which we stand. Here the political, social, national and military struggles of the Egyptian people intermingle and combine like the huge blocks of stone which dam up the old course of the Nile." Cited in Nutting (1972), 356. See also the extensive discussion in Tom Little, High Dam at Aswan: The Subjugation of the Nile. (N.Y., 1965).

[70]  El–Tobgy (1976), 102.

[71]  Waterbury (1979), 125.

[72]  Sand brick manufacture had reached only 50 million per year in 1973. Waterbury (1979), 181. Estimates of the total amount of land lost due to all these effects every year vary considerably; some place it at 35,000 or 40,000 feddans per year: Chase World Information Corporation, The Chase World Information Series on Agribusiness Potential in the Middle East and North Africa: Egypt (New York: 1977), 34, and El–Tobgy (1976), 43, respectively. A recent study has placed

the loss at 75,000 feddans annually, largely due to village expansion. Ragaa Rizk, Urban Encroachment on Agricultural Land in Egypt, With Special Reference to Sharkia Governorate. Unpublished M. Sc. Thesis, Department of Agricultural Economics, Zagazig University, 1980.

[73] Kilometers of drains were as follows:

    1939    10,246
    1947    12,064
    1954    12,316
    1960    13,330

Annuaire Statistique, various issues.

[74] Saab (1967), 73-77. As a former under Secretary of Agriculture put it: "...public drains alone were not as effective in draining the areas they were planned to serve because only a small proportion of these areas were covered by the necessary field drains...and hence the benefits of drainage were in reality limited to the narrow strips of land adjacent to the public drains." El-Tobgy (1976), 49.

[75] El-Tobgy (1976), 20.

[76] O'Brien (1966), 116.

[77] Farmers pay no fee for water, receiving it by right of their payment of the land tax.

[78] FAO, Near East Regional Office, Research on Crop Water Use, Salt affected Soil and Drainage in the ARE. (Rome, 1975); USDA-USAID-Egyptian Ministry of Agriculture (1976); The World Bank estimates that about 70% of the cultivated area is afflicted with drainage problems. World Bank (1978), III, 46.

[79] El-Tobgy (1976), 50; World Bank (1978), III, 12; for some cotton farms, improved drainage leads to a doubling of yields. Ibid.

[80] Sarah Potts Voll "Egyptian Land Reclamation Since the Revolution," Middle East Journal, 34, 22 (Spring, 1980), 129.

[81] Ibid; Waterbury (1979), 138; World Bank (1978), 46-47.

[82] Beverley Horsley. "Keeping Up with Demand: The Big Challenge for Egyptian Agriculture," Foreign Agriculture, November 7, 1977. The speaker was H. Reiter Webb, U.S. Agricultural Attache in Cairo.

[83] Voll (1980); World Bank (1978), III, 46-47;

[84] See note 68 above.

[85] This seems to be El-Tobgy's explanation: (1976), 48-49.

[86] The two year rotation is permitted in some very fertile areas. See El-Tobgy (1976), 37-38.

[87]   The Ministry of Agriculture reports the following returns per feddan above variable cost in 1978: Maize, LE 47; Wheat, LE 73; Cotton, LE 66; Rice, LE 84; Birsim, LE 85; Oranges, LE 122; Tomatoes, LE 179. Some survey data from the Egyptian Water Use Management Project are also available: Rice LE 11; Maize, LE 31; Wheat, LE 40; Cotton, LE 60; Birsim, LE 80; Tomatoes, LE 188; Eggplant, LE 324; Artichokes, LE 317. I am indebted to Dr. Gene Quenemeon for the latter set of figures.

[88]   Government price policies did little to distort the international price of food grains before the world-wide price explosion of 1974. Cuddihy (1980), iii.

[89]   See Cuddihy (1980) on meat subsidies.

[90]   "The field is wide-open for real breeding achievements in (birsim)." El-Tobgy (1976), 135. Yields may have increased as a result of increased applications of phosphatic fertilizer. Birsim receives roughly one-third of all the $P_2O_5$ fertilizers used in the country, whose consumption rose from 163,000 tons in 1955/59, to 256,000 in 1960/64, to 315,000 in 1965/69, to 480,000 tons in 1970/75. El-Tobgy (1976), 67. This represents nearly a three-fold increase in fertilizer use per feddan planted in birsim.

[91]   The following discussion relies heavily upon El-Tobgy (1976), 92-158.

[92]   Ibid.

[93]   Marion Clawson, Hans H. Landsbert, and Lyle T. Alexander. The Agricultural Potential of the Middle East. (N.Y., 1971), 45.

[94]   El-Tobgy (1976), 99-105.

[95]   One should be quite cautious about viewing increases in pesticide application as an unmixed blessing or even necessarily a sign of agricultural development. For example, Texas cotton farmers used only one-tenth as much insecticide in 1976 as in 1964, because of the use of Integrated Pest Management. See Allan A. Boraiko, "The Pesticide Dilemma," National Geographic, 157, 2. (February, 1980), 145-183.

[96]   Abdel-Fadil, (1975), 31ff.

[97]   The principal articles are those of Hansen (1966; 1970); Mabro, (1966); Abdel-Fadil (1975); and Amr Mohi el-Din, "Under Employment in Egyptian Agriculture," in Manpower and Employment in Arab Countries: Some Critical Issues (Geneva, 1975).

[98]   Saab (1967), 123; Abdel-Fadil, (1975), 125.

[99]   Robert Mabro, (1971), presents the following evidence

          Size of Farms     % of holdings employing
                            outside wage labor

```
0.5-<2        24%
2.0-<5        36
5.0-<10       53
  <10         85
```

[100]   Some estimates have the number of man-hours per feddan of
        wheat falling from sixty to less than two for wheat, from
        ninety to less than four for cotton. Izz al-Din Kamil, Al-
        Zira'ah Al-Mikanikiyyah, (Mechanized Agriculture) (Cairo,
        1976), 110-118. However, as the enormous debate on the
        Indian experience with mechanization shows, the effect of
        partial mechanization on employment is often not obvious.
        See, e.g., Hans Binswanger The Economics of Tractors in the
        Indian Subcontinent: An Analytical Review. (Hyderabad,
        1977).

[101]   Rural Employment Problems ...(1960), 108; Abdel-Fadil (1975),
        125-127.

[102]   Recall that a similar phenomenon occurred in World War II.
        See above note 9. The number of men in the Egyptian armed
        forces rose from 100,000 in 1960 to 205,000 in 1970. In the
        mid-1960s, some 70,000 Egyptian soldiers were fighting in the
        Yemeni Civil War.

[103]   A similar phenomenon has been observed in Kafr el-Shaykh
        governorate, where the expansion of the rice area served to
        increase the local demand for labor and reduce the supply to
        tarahil workers to neighboring governorates. I am indebted
        to various employees of the Ministry of Agriculture in Ghar-
        biyya governorate, and to Sawsan el-Messiri, for this infor-
        mation.

[104]   Some additional information on income distribution is avail-
        able from personal expenditure surveys. These show: 1) a
        slight improvement in distribution from 1958/59 to 1964/65;
        2) a somewhat larger deterioration from 1964/65 to 1974/75;
        3) that neither movement was large; 4) that the distribution
        of income per person was more nearly equal than the distribu-
        tion per family (the rich tend to have larger families); 5)
        that the distribution of personal income is more equal in the
        rural areas than in the urban; 6) that with a Gini coeffi-
        cient of .27 for rural personal income distribution, the dis-
        tribution of income in rural Egypt is quite egalitarian.
        This should be borne in mind when assessing critiques
        (including those presented here), of the Nasser and Sadat
        regime's policies. World Bank, (1976), I, Annex 1, 82-95.

[105]   Baker (1978), 151. He was referring to individual freedoms.
        However, the same characteristization may also be applied to
        agricultural policies.

[106]   See, e.g., Binder (1978), 398.

[107]   Actually, the price of the basic food stuff of the poor,
        baladi bread, was not increased. The key to the explosion

seems to have been the abruptness with which the changes were announced.

[108]    U.S. aid to Egypt was the largest such peacetime program ever. By 1980, U.S. annual aid reached $2 billion.

[109]    Workers' remittances rose from $6.4 million in 1971, to $81.4 in 1972, to $188.6 million in 1974, and had reached over $1,000 million by 1979. Data from Central Bank of Egypt.

[110]    Many observers have warned that such a happy state of affairs may not persist, however; as will be discussed below, there are reasons to suspect that the rate of growth of remittances will slow down in the early 1980s.

[111]    Early warnings appear in Crouchley, (1938), 239 ff. Such concern has only recently been shared by Egyptian policy makers. See the discussion in John Waterbury, Egypt: Burdens of the Past, Options for the Future (Bloomington, Ind., 1978), 49-66.

[112]    Family size increases with income in rural Egypt: using the 1974/75 expenditure survey data; one obtains a positive correlation coefficient of .235 between expenditure per person and persons per household in rural areas. Wealthy families are roughly three times as large as poor ones. Average rural family size is 6 persons per household, while urban families have an average of 5.6 persons. World Bank (1978), I, Annex 1, 85.

[113]    The evidence here is ambiguous. It seems fairly well established that the rate of natural increase in Cairo, and perhaps other urban centers, was higher than rural areas before 1963. The position may have been reversed since then, however. In the absence of evidence to the contrary, the assumption of equal rural and urban rates of natural increase seems acceptable for the limited purpose here. See, e.g., Waterbury (1978), 67-84; Hansen and Marzouk, (1965), 31.

[114]    Cited in World Bank, (1978), I, Annex 1.1.

[115]    John Waterbury, "Urbanization and Income Distribution in Egypt." Draft, mimeo, Princeton University Income Distribution Project, Princeton, N.J., 1979; Nazli Choucri, Richard S. Eckaus, Amr Mohi el-Din, "Migration and Employment in the Construction Sector: Critical Factors in Egyptian Development," Cairo University/Massachusetts Institute of Technology (1978).

[116]    The number of tourist arrivals rose from 535,000 in 1973 to 1,052,000 in 1978. Central Agency for Public Mobilization and Statistics, Statistical Yearbook, (1979), 203.

[117]    In the late 1970s, farmers received the following percentages of international prices adjusted to the farm gate: Cotton, 49%; Wheat, 69%; Rice, 75%; Sorghum, 78%; Maize, 69%; Birsim, 100%; Milk 100%; Meat, 180%. World Bank, Arab Republic of Egypt: Agricultural Development Project (Minufiyya-Sohag).

Revised report, 1979, 40. The rise in international grain prices after 1973 plus unchanged government prices led to high nominal protection rates for grain crops. The differential narrowed after 1976 as international prices fell and government prices rose. Cuddihy, (1980), iii.

[118]  Voll (1980); J.A. Allan, (1980); Pacific Consultants, New Lands Productivity in Egypt: Technical and Economic Feasibility (Washington, D.C., 1980)

[119]  El-Tobgy (1976), 20.

[120]  See, among many others, El-Tobgy (1976); USDA-USAID-Egyptian Ministry of Agriculture (1976); Sandra Hadler and Maw-Cheng Yang, "Developing Country Foodgrain Projections for 1985," World Bank Staff Working Paper No. 247 (November, 1976), 39.

[121]  World Bank (1978), III, 45.

[122]  The open market price for cement was three times higher than the official price, while that of iron rods was four times larger. Mohammad Khedr, The Impact of Drainage on Agricultural Production in the A.R.E. Unpublished Ph.D. dissertation, Minufiyya University, 1979.

[123]  World Bank, (1978) III, 45-46. The shortage of engineers is largely due to their taking much higher paying jobs in the Oil States.

[124]  See, e.g., Egyptian Major Cereals Improvement Project, Technical Report, Consortium for International Development, Project Design Team. Mimeo (Cairo, 1979), 23.

[125]  The following relies on Ibid., and El-Tobgy (1976), 99-111.

[126]  El-Tobgy (1976), 104.

[127]  World Bank (1978), III, 44. Such figures do not include small peasants' investments in farm animals. See Table 6.9.

[128]  World Bank, (1978), III, 53.

[129]  See, e.g., O'Brien (1966); Saab (1967), 126; Dumont (1968).

[130]  Vegetable exports have risen from a seven year low of 12,743 tons in 1974 to 38,445 tons in 1978. Foreign Agriculture XVII, 10 (October, 1980).

[131]  The following is based on my observations as a consultant to the Ministry of Agriculture in November, 1977.

[132]  Foreign Agriculture, XVII, 10. (October, 1980).

[133]  See, e.g., William Scofield, "Foreign Investment in Egypt: Opportunities and Obstacles," Foreign Agriculture xiii, (July, 1976); Chase World Information Corporation (1977); J.A. Allan (1980) notes the public relations nature of much of this investment. This resembles some of the investment by oil companies in agribusiness in Khuzestan in Iran under the Shah, much of which was apparently done to please the Monarch and to protect petroleum investments. See Helmut Richards,

"Land Reform and Agribusiness in Iran," Middle East Research and Information Project Report, #43, (Dec., 1975).

[134]  e.g. Izz al Din Kamil (1976); Fathi Abd el-Fattah, Al-Qaryah al-Misriyyah bayna al-Islah w'al-thawrah (The Egyptian Village Between Reform and Revolution) (Cairo, 1975).

[135]  Conversations with agricultural economists, at Zagazig University; conversation with Abd al-Rahman Shazli, Agricultural Director, Behera governorate; conversations with Nicholas Hopkins, Director, Social Research Center, American University in Cairo.

[136]  USDA-USAID-Egyptian Ministry of Agriculture (1976); El-Tobgy (1976). The same is argued in an unpublished, untitled Arabic report prepared for the Office of the Undersecretary of Agriculture for Engineering Affairs, apparently prepared in 1979. I am indebted to Dr. Eng. A.M. Al-Hossary, Undersecretary of Agriculture for Engineering Affairs, for showing me this report. It is referred to as "Al Hossary's Report" hereafter.

[137]  James B. Fitch, Ahmad A. Goueli, and Mohammad el-Gabely, "The Cropping System for Maize in Egypt." Paper presented to FAO workshop on "Improved Farming Systems in the Nile Valley." (Cairo, May 8-14, 1979).

[138]  El-Tobgy, 135. This is, of course, a world-wide problem. As Day and Singh argue in their study of the Indian Punjab, "When animals provide the main source of farm power, greater attention must be paid to finding the ways and means of keeping this source cost effective, at least in the short run." Richard H. Day and Inderjit Singh, Economic Development as an Adaptive Process, (Cambridge, 1977), 177.

[139]  World Bank, Arab Republic of Egypt: Agricultural Development Project (Minufiyya-Sohag); Staff Appraisal Report, (1977), 40. See note 83 above on relative crop profitability.

[140]  ERA 2000, Inc., (1979); World Bank (1977), A2,11.

[141]  Al Hossary places the gains at 33%. "Report," (1979).

[142]  IBRD (1977). Diesel Fuel is also subsidized. See below.

[143]  Roger Garrett. "Impressions with respect to Mechanization." USAID mimeo (1978). These numbers are cited repeatedly by students of Egyptian farm mechanization: they also appear, for example, in Izz al-Din Kamil (1976), 115, and Al-Hossary's "Report." They all appear to have the same source: a set of trials conducted at the Cairo University experimental farm. "Nadwah 'an al-mikanikiyyah al-zira'iyyah al-misriyyah." (Debate on Egyptian Farm Mechanization) L'Egypte Contemporaine, 311 (Jan., 1968).

[144]  World Bank (1977); ERA 200, Inc., (1979), Chs. IV and X, respectively.

[145]   Binswanger (1977).

[146]   World Bank (1977), Al, p. 3.

[147]   World Bank (Minufiyya–Sohag, 1978), 16. Al Hosary also recognizes this point. Interview of March 1980, Cairo.

[148]   Preliminary Data from Ford Foundation Farm Management Survey. I, am indebted to Dr. James B. Fitch for this information. If mechanization were to increase rich farmers' cropping intensity, it would also increase inequality between such farmers and their smaller neighbors. See above.

[149]   World Bank (Minufiyya–Sohag, 1978), 38.

[150]   ERA, 200, Inc. (1979), X, 45, 51.

[151]   Fitch, et. al. (1979).

[152]   Richard H. Day, "The Economics of Technological Change and the Demise of the Sharecropper." American Economic Review, 57, 3 (June, 1967); Day and Singh (1977), Ch. 6.

[153]   USDA–USAID–Egyptian Ministry of Agriculture (1976), 15.

[154]   Cf. John W. Mellor. "Food Price Policy and Income Distribution in Low Income Countries." Economic Development and Cultural Change, 27, 1 (Oct., 1978), 1–26. Of course, the poor may be protected from such effects by government subsidies. In that case the government would have to increase its food imports, thereby reducing foreign exchange and raising the social opportunity cost of capital.

[155]   Information on subsidies comes from the ERA, 2000 report, Al-Hossary's report, and from Dr. Gene Quenemeon of Montana State University.

[156]   Data from the Center for Agricultural Economics Research of the Ministry of Agriculture show that while the average yearly percentage rise in price for wheat, maize, rice and cotton in 1975–78 was 6 2/3%, 12 2/3%, 18 2/3%, and 10 2/3%, respectively, costs of the same crops rose 15 1/3%, 20%, 13 1/3%, and 15 2/3%, respectively, during the same period. See Table 6.16.

[157]   Regrettably little is known about the composition of Egyptian overseas migrants. There are a variety of (often contradictory) sources of information. See IBRD, (1978), III; Chourki, Eckhaus and Mohi-el Din (1978); J.S. Birks and Clive Sinclair, International Migration and Development in the Arab Region. (Geneva, 1980).

[158]   International Institute of Strategic Studies, The Military Balance, various issues.

[159]   I am indebted to Professor Saad Gadalla of the American University in Cairo for this information.

[160]   Recent research by over 115 British banks with Middle Eastern branches predicts a deceleration in construction activity in

the principal labor-importing states, simply because "basic infrastructure is now well in place." Gerard Castoriades, "Bouncy Contracting Market Draws Far East Challenge," Middle East Economic Digest, June 6, 1980, 11. See also Birks and Sinclair (1980).

[161]  Perhaps 300,000 new jobs must be created every year just to keep up with the growth of the labor force. Even if all of the savings from mechanization which its advocates hope for were realized, and even if all the savings were reinvested, only if the incremental capital-labor ration were to be as low as L.E. 500 would the new jobs created keep up with such growth. Further, such a calculation assumes that no jobs are lost as a result of mechanization, a highly questionable assumption.

[162]  Ayrout, (1963), 53.

[163]  Samir Radwan (1977); Andrew J. Koval and Ahmed A. Bahgat, "Ten Horsepower Agriculture." Paper presented to Symposium on Appropriate Mechanization of Small Farms in Africa, (Cairo, 1980); Hansen (1966); Abdel Fadil (1975).

[164]  F. Zaghloul, "A Cost-of-Living Index for Rural Labourers, 1913-1961." Egyptian Ministry of Planning, Institute of National Planning Memo # 557. (Cairo, 1965).

[165]  Radwan (1977), 31. See also Radwan and Eddy (1979).

[166]  Radwan (1977), 46.

# 7

# Conclusion

In this book I have tried to trace the broad outlines of technical and social change during more than 180 years of Egyptian agrarian history. The major change was, of course, the transformation of the irrigation system. The step-by-step increase in the quantity and certainty of summer water supply was the crucial factor. Such a change permitted the shift from basin to perennial irrigation, and, within the latter system, made possible still further increases in summer water. These changes were central for three reasons. First, they permitted the growing of new crops (e.g. cotton and maize in the nineteenth century), or the raising of the yields of existing crops (e.g. the shift in the maize planting date and the increased area planted in rice after the completion of the High Dam.) Second, the way in which the irrigation system was changed, along with other factors discussed below, shaped the distribution of productive resources, especially land. That is, the spread of perennial irrigation and cotton cultivation altered the class structure. Of course, this distribution and this class structure in turn have been the fundamental determinants of the distribution of the benefits and costs of growth and technical change in agriculture. Third, the transformation of the hydraulic system created a set of agro-ecological problems. These difficulties both slowed the growth of agricultural output and stimulated additional technical change.

Two other types of technical change were (are) important for Egyptian agriculture: (1) changes in crop rotations and the timing of crop planting, especially for cotton and maize, and (2) increased use of industrial inputs. The spread of the two-year rotation from 1890 to 1952 bulked large in the story, as debt constrained small peasants first adopted the system, with the large landholders following only after drainage problems had been somewhat ameliorated. As for industrial inputs, Egypt seems to have followed the typical pattern of a land-scarce economy: substitutes for land, such as fertilizer and pesticide, were adopted first, while labor-saving implements spread only after forces external to the agricultural sector had driven up wages.[1]

It is clear that these technical changes were shaped by, and in turn moulded, rural social relations. The two-year rotation was first adopted by small farmers who were in debt to village moneylenders. The small size of their farms, the absence of alternative sources of credit, and the special position of Greek and Levantine moneylenders in Egypt were the result of political decisions and the social structure. Unequal access to credit also contributed to the more rapid adoption of fertilizer in the interwar period by the larger farmers. Earlier cotton planting, so useful in avoiding pest attacks, was difficult for small peasants to adopt since they needed to take as many cuts of birsim (the preceding crop) as possible to feed their animals. Similar fodder problems for small farmers have inhibited the diffusion of HYV cereals, especially maize. Rich peasants during the Nasser and Sadat years have been able to alter their cropping patterns more easily than their poorer neighbors. Throughout modern Egyptian agricultural history, the diffusion of new techniques has been shaped and constrained by social inequality.

The distribution of resources is not the only social relationship which has affected the technical change. Market relations are, of course, also social relations, and market forces have played a central role in the diffusion of new agricultural techniques. They appear to be especially important for the spread of fertilizer in the interwar period and for the current enthusiasm for tractors. The latter change appears to be a clear case of "induced technical change" as farmers switch to tractors in the face of rising labor costs. However, we have seen that although the spread of fertilizers was quite responsive to wheat prices in the interwar decades, the explanation for its increased use for cotton is a bit more complex. In particular, drainage quality constrained the effectiveness of fertilizer use, especially for cotton.

Two points should be emphasized here: first, the interwar changes, as an interrelated package, were a response to past production problems. These problems themselves, rather than relative price changes, were the "focusing device"[2] which stimulated the adoption of the new techniques of earlier planting, closer spacing, improved drainage, and increased fertilizer use. Second, in so far as relative prices have stimulated technical changes, such as in the case of fertilizer used for wheat in the interwar period and the current spread of tractors, government policy has played a central role. Wheat prices were supported during the interwar years; capital and operating costs for mechanical technologies are currently subsidized. The spread of fertilizer use under Nasser was also largely the result of explicit government policy. In general, the pattern of technical change in Egyptian agriculture was the product of the interaction rural class forces, market changes, responses to production problems generated by past technical change, and government policy.

The relationship between social classes and technical change was a reciprocal one. The transformation of the irrigation system, as noted above, underlay the creation of the major social classes and the land tenure system. My discussion of the income distribution effects of technical change has had, perforce, to focus upon the distribution of cotton revenues. Changes here may be summarized as follows. In the period from 1880–1914, and especially from 1890 to c. 1900, the small landholding peasants probably improved their position vis-a-vis the pashas, although from c. 1900 to 1914 the earlier gains made by the small peasants were being reversed. However, the relative income position of the landless class deteriorated. The first outcome was largely the result of the rapid climb in yields on peasant lands (for which there are plausible arguments and scanty data), plus their adoption of the two-year crop rotation. They were planting cotton more frequently on land which had now received much more water than before, thanks to British hydraulic works.

It is important to remember, however, that a significant portion of this revenue, at least one-fourth and quite probably considerably more, was going not to the peasants themselves, but to the moneylenders. Further, those peasants who held no land suffered a relative decline. In part, this was a result of the fact that much of the increased demand for labor occasioned by the switch to the two-year rotation was supplied by the small landowning peasant and his family. Rich peasants hired labor, but the outward shift of demand was apparently swamped by outward shifts in supply. The supply of labor increased because of (1) population growth, (2) fragmentation of peasant properties, (3) land loss for debt, and (4) the abolition of the corvee and the rise of the tarhila labor system, relying heavily on the "reserve army of seasonally unemployed" Sa'idis.

The interwar period presents a different picture. Among landowners, the changes favored those with more resources, whether pashas or rich peasants. This is what we would expect given the existence of interrelatedness and indivisibilities in the adoption of the new techniques. The peasants' lands had undergone relatively more deterioration, since they had adopted the "land-wasting" two-year rotation. They consequently had a greater need for drainage. But drainage was an indivisible and costly input, giving rise to conventional economies of scale. Apparently, peasant cooperation was not a viable solution to this problem. Small peasant lands therefore suffered from poorer drainage than did those of the pashas. Had the peasants adopted earlier planting, they would have needed to change birsim's place in the crop rotation. But then their cotton would have required more nitrogen fertilizer. Drainage, fertilizer, and perhaps the additional seed required by the new planting techniques required capital. Yet in the 1920s finance was relatively more expensive for the small peasants, since they had to rely on moneylenders. Even after the creation of the Credit Agricole Egyptien, many small peasants did not borrow from this bank for fertilizer. There is thus every reason to suppose that the pashas' and rich peasants' yields improved more than those of poor peasants. Since fragmentation affected the pashas' and rich peasants' lands relatively less, and since the former were turning to the two-year rotation system, it seems safe to conclude that the pashas improved their cotton revenue position relative to the small peasants. Rich peasants also appear to have done better than the small fellahin.

On the other hand, the technical changes of the interwar period probably helped to prevent the position of the landless laborers from deteriorating any further. The labor-using technical changes helped to absorb the growing agricultural population. The factor share of labor was roughly constant. Since the share of rents was falling, one might infer that the relative position for the landless class vis-a-vis the pashas and the small landholding peasants improved.

Government policy largely determined the structure of technical and social change in the 1950s and 1960s. The state became more directly involved in all aspects of the agricultural economy under the Nasser regime than under any government since Muhammad 'Ali. Nasser redistributed land; but unlike "the founder of Modern Egypt", Nasser's direct social changes clearly promoted equality, especially among landholders. He eliminated the pashas from agriculture and benefitted numerous small-holders. However, the reforms also strengthened the (already secure) position of the rich peasants, who were able to bend the cooperatives to their own purposes. In consequence, the position of the rich peasants rose relative to poor peasants. The position of the landless class fluctuated, however, with some overall improvement. Labor using technical change, and especially the growing demand for labor for public works, offset the decline in the demand for labor due to land reform and mechanization. The crucial role of the demand for labor outside of agriculture for the position of the landless is even clearer during the 1970s, when Egyptian real wages in agriculture have reached historically unprecedented levels. This is a hopeful sign, but one should remember that the international economy can have the opposite effect on rural wages, as it did in the 1930s. The question of the role of the international sphere in Egyptian rural political economy is examined in more detail in the final section of this chapter.

## UNSOLVED PROBLEMS IN EGYPTIAN AGRICULTURE

The various technical changes in Egyptian agriculture have produced some impressive successes. The cultivated area has risen from roughly three million feddans in 1813 to 5.6 million feddans in 1975.[3] Although much of this increase occurred in the 19th century, some 300,000 feddans have been reclaimed since 1952 for productive uses.[4] The expansion of the cropped area is even more impressive. The average ratio of cropped to cultivated area is approximately 1.9; when orchards are eliminated, the ratio is 2.0. Egyptian fertilizer input at roughly 200 kg. per hectare is exceeded only by the developed areas of Europe and Japan, and by the Republic of Korea. Yields of some crops are among the highest in the world: first for extra-long staple cotton, second for sugar cane, third for peanuts. Her yields of maize and wheat are exceeded only by the U.S., France, Italy and Bulgaria (maize), and the U.S., Sweden, West Germany, France and Australia (wheat). Needless to say, her yields far exceed the average for the Third World or for the Arab League.[5]

These are impressive achievements. And yet, serious problems remain. Indeed, one of the reasons why the current problems appear so intractable is that Egypt has already done so much. Simple, easy advances have long since been attained. But despite fluctuations,

per capita consumption of cereals and pulses had not achieved its
pre-World War I levels by the late 1960s.[6] Current levels of per
capita grain consumption are some 25% higher than those of the late
1960s and early 1970s;[7] however, these gains have been achieved
through increased imports, not through increased per capita production: per capita grain production is below the levels of ten years
ago. The present average consumption levels are thus dependent upon
Egypt's ability to finance food imports.

The unsolved problems facing Egyptian agriculture and rural
society may be summarized under four headings: (1) population growth,
(2) ecological difficulties, (3) poverty, and (4) inequality. The
expansion of population has followed the usual Third World pattern:
slow initial growth when the death rate remained high, and then an
acceleration during and after World War II when the death rate fell
precipitously. The level of fertility has been much less responsive,
declining only slowly. By the year 2000 there will be at least 60
million Egyptians and probably more like 70 million. Providing food
and jobs for so many people is a daunting challenge indeed.

Egypt's population problem is exceptionally severe because of
the great difficulties in expanding the cultivated area., As a
result, the current population density exceeds one thousand persons
per square kilometre of cultivated land. With only 0.08 cultivated
hectares per person, Egypt is more crowded than any country in the
world except Japan, South Korea, and Taiwan. Barring some unforeseen
technological breakthrough in desert land use, this problem will only
get worse. Other ecological constraints have impeded Egyptian agricultural development as well. The quality of fertile Delta and Nile
valley soils has been undermined repeatedly by the failure to install
adequate drainage networks. With the level of pesticide use already
far above that of the United States, one may wonder whether some new
"super bugs" are not evolving in Egyptian cotton and clover fields.
And for all difficulties which surround agricultural mechanization in
the social context of modern Egypt, it is surely true that the use of
nearly one-fourth of the very scarce cultivable area for animal
fodder constitutes a serious problem in such a land poor-country.

We have seen how these ecological problems were generated by the
social structure, by the land tenure system, and by public policy
errors. When looking at Egypt's current agricultural problems, one
has a sense of historical dejà vu: difficult problems require sophis-
ticated solutions, but these solutions are impeded by the nature of
the government's base of support, by rural class relations, by the
international environment, and by the illiteracy and poverty of the
mass of the fellahin.[8] It is true that there has been some improvement in the standard of living in rural Egypt, but we have now seen
how food consumption per capita and rural real wages remained below
pre-World War I levels until the last seven years. We have also seen
reasons for disquiet as to the permanence of the recent favorable
trends. It is also clear that even with the current improvements,
basic needs for sanitary water supply and waste removal, electricity,
health care, and education remain a hope rather than a reality for
many rural Egyptians.

For many, but not for all: inequality remains the fourth problem of rural Egypt. Access to the increasingly scarce resource of land remains the key to inequality in rural Egypt, although mechanization, differences in crop mixes, and marketing activities also differentiate the peasantry. We have scrutinized two watersheds here: 1) the creation of the pasha and landless classes and the bimodal land tenure system in the nineteenth century, and 2) the land reforms of Abdel Nasser. It bears repeating that the distribution of income in rural Egypt appears fairly egalitarian by Third World Standards. It is clear, however, that although the pashas have been (largely) removed, a "rural middle class" controls much of the wealth and power in the Egyptian countryside,[9] and that the problems of the landless class remain.

The role and strength of the rural middle class have received considerable attention here. We have seen how their power has deep historical roots. The rich peasants have been able to turn their role of intermediary between the government and the village to their own advantage regardless of the nature and aims of the regime in Cairo. The _forms_ of this role have changed, but never the role itself: now weakened, now strengthened, whether under Mamluks, Muhammad Ali and his successors, the British, the Constitutional Regime, Nasser, or Sadat, rich peasants have used their political position and skill as farmers to enhance their wealth and status. And although future regimes might try to weaken them, the historical record in Egypt (and in those countries which have tried to eliminate similar rural middle groups) suggests that they will survive. Meanwhile the thirty to forty-five percent of the rural population which holds no land at all likewise remains, only recently receiving wages above nineteenth century levels.

INTERNAL AND EXTERNAL FORCES IN EGYPTIAN AGRICULTURAL DEVELOPMENT

How is one to explain the persistence of the relative underdevelopment of Egyptian agriculture? In seeking to explain why Egypt, or any other country remains (or as some would have it, becomes) underdeveloped, one is trying to account for the pattern of technical, structural, and institutional change. The plethora of theories of underdevelopment may be grouped into two very broad groups. Some believe that the basic problems are internally generated; others hold that it is involvement in the international economic and state systems which is to blame. Dependency theorists, for example, believe that it is the international systems of trade and war themselves which retard growth, distort structural change, impede technical progress, and twist local social forces into barriers to development.[10] For such social theorists, the causes of underdevelopment are fundamentally external to the local economy.

Both orthodox economists and orthodox Marxists have attacked this position, arguing that it gives insufficient weight to internal factors. The former stress a wide range of deleterious influences, ranging from ecological problems to lack of entrepreneurs to distorted internal prices.[11] The Marxists, on the other hand, stress the local "semi-feudal" class structure, especially the slow development of wage labor, as the main retarding force.[12] For all of these scholars, the basic factors are <u>internal</u>.

I believe that at least for the Egyptian case neither of these explanations is sufficient. I would like to offer a synthetic explanation for the relative underdevelopment of Egyptian agriculture which draws on some current work in political economy. The evidence for Egypt indicates that international forces shape, but "underdetermine", the path of development.[13] In Egypt the international trade and state systems shaped (1) the class structure and (2) local government policy which (a) had a direct impact on the course of development and (b) had an indirect impact via its influence on the country's agroecology. Thus what are often considered to be "internal" forces are themselves often the product of international pressures. But such international forces were not <u>sufficient</u> to generate the institutional patterns which blocked development. Rather, it was the specific conjuncture and the pattern of interaction of truly indigenous forces with the international trade and state systems which produced the structures which fettered Egyptian agricultural development.

Consider the origins of the bimodal land tenure system and government policy toward agriculture before World War I. The unequal distribution of land and the distinctive land/labor system ("the 'ezbah system") used on large estates had several consequences for agricultural development. First, the fact that many peasants had very little land led them to plant cotton every two years, instead of every three like the larger landlords. This was especially so since many of them were in debt to village moneylenders. This, in turn, led to a rise in the water table, increased insect attacks, and reduced soil fertility. Insofar as peasant (cash or crop share) renters were not closely supervised, such a tenure system would also have contributed to the over-exploitation of the soil. Trying to repair such socially induced agroecological damage absorbed the energies of Egyptian farmers and governments throughout the interwar period. Second, the "'ezbah system" was a kind of "semi-commercialized" agriculture: the estates produced for a market which did not include its own labor force. Because of the low level of cash payments, the low level of income, and the existence of the subsistence plots, there was little market for either food crops or simple industrial goods. The bimodal distribution of land thus impeded Egyptian agricultural development on both the demand and the supply side.

The origins of this land tenure system cannot be explained by an exclusive focus upon either internal or external forces. It logically follows that the same may be said of those features of underdevelopment which derived from such bimodalism. Growing cotton for the international market required the transformation of the irrigation system; this, in turn, necessitated large amounts of labor and capital. The government acquired cash resources through taxation, the corvée, and foreign borrowing. Peasants lost land as they fled from the excesses of such taxation, or because they sought to avoid the military draft. The latter, in turn, was imposed by Muhammad 'Ali as part of his drive to establish himself as an independent regional power. Just as the peasants' loss of land (and therefore, the emergence of the bimodal land system) due to taxation can only be understood by reference to the international economy, so can their losing plots as they fled the draft only be comprehended within the context of the international state system of the time.

Yet, such international forces are not sufficient to explain the resulting pattern of land distribution. The brutality of the corvée, of land taxation, and especially of the military draft can, it seems to me, only be fully understood by considering Muhammad 'Ali and his House as the heirs of what Marshall Hogson called the "Gunpowder Empires."[14] Muhammad 'Ali's tax collection procedures remind one of the Mamluk's brutal methods. The same may be said for those who profitted by the situation. This is obviously so for the Turko-Circassian ruling class; these men gained access to land because of their relationship with the Ruler, in a manner very reminiscent of the Mamluk system. Equally, the village shaykhs owed their good fortune to their position as intermediaries between the state and the peasantry. This crucial location in the taxation system was an inheritance of the old order; in our terms, the rise of the large landlords, whether Turkish mercenaries or village shaykhs, was the result of the conjuncture of both external and internal forces.

The same may be said of the development of the institution of private property in land, without which the dislocations of the period would not necessarily have had such important consequences. After all, only this institution made it possible for someone to appropriate land which the peasants abandoned, and therefore, to generate a bimodal land tenure system. It seems fairly well-established that tendencies toward private property in land were emerging through the indigenous institutions of waqf and the ard al-usya in the eighteenth century, well before the coming of cotton and perennial irrigation.[15] Nor is there evidence of Europeans' pressuring the Egyptian state to establish private property in land. What we know of British behavior in the nineteenth century would lead us to believe that so long as British merchants in the ports were unmolested and their property rights respected, H.M. Government would have been singularly indifferent to the internal system of property rights in land.[16]

In summary, it was the conjuncture of the international trade system (cotton and the transformation of the irrigation system), the international state system (Muhammad 'Ali's attempt to set up his own dynasty against the claims of the Ottoman state, in turn backed by the British, necessitating an army) with the internal forces of (a) local ecology (the nature and timing of the Nile flood which required the shift to perennial irrigation if cotton was to be grown on a large scale), (b) the local class structure, (with its Turko-Circassian landlords and village shaykhs), (c) a set of state policies strongly influenced by the Ottoman/Mamluk tradition, and (d) the growth of private property in land which led to the bimodal distribution of land ownership, and, therefore, to the barriers of development which such bimodalism engendered. The same may be said for the origin of the 'ezbah system. It presupposed the distribution of land

ownership, the growing of cotton for export, and commercially minded,
economically rational land owners. These calculating attitudes were
not adopted from the West. Long distance trade and commercial attitudes had bulked large in Islamic lands for many centuries before the
coming of Western capitalist influences.[17]

A similar pattern of interaction of internal and external forces
appears when one looks at the ecological problems of Egyptian agriculture and the role of British policy in generating such difficulties. As we have seen, cotton yields fell from the 1890s until the
outbreak of World War I. The origins of these problems lay with the
land tenure system and with British credit and irrigation policies.
The land tenure system must be taken as given when explaining the
switch of small peasants to the two-year crop rotation. Small
peasants, but not large estate owners, adopted the two-year system.
The unequal access to information and credit, which led to this
difference in technique, was clearly rooted in disparities in wealth,
in turn fundamentally the result of inequalities in land ownership.
The ecological problems of Egyptian agriculture were thus, in part,
the result of the bimodal land tenure system. And we have already
seen how this bimodalism must be explained as the result of the conjuncture of "external" and "internal" forces.

British policy reinforced the unequal access to credit inherent
in the land tenure system. Initially, they did nothing to promote
the development of lending institutions which would cater to small-
holders. They did establish an Agricultural Credit Bank in 1902 precisely to deal with the problem of peasant credit. However, in
response to a rise in arrears, the government began to restrict the
operation of this bank. The promulgation of the "Five Feddan Law" in
1912, of course, effectively closed the bank. Such policies forced
the peasants into the arms of village moneylenders who charged the
usual high interest rates. These, in turn, encouraged the shift to
the "land destroying" two-year rotation as peasants sought cash to
get out of debt as quickly as possible.

Such credit policies, as well as British irrigation policies
(discussed below), naturally presuppose the presence of the British
in Egypt. Here again, the origins of the forces which ultimately
promoted underdevelopment must be seen as the result of the interplay
of internal and external forces. As by now seems well established,
the British occupied Egypt because of their position in the international state system: their interests in India were perceived as
threatened by internal upheaval in Egypt.[18] This was more important
in 1882 than it would have been in say, 1860, because Britain's
supremacy in the international economy and (especially) state system
was under increasing pressure. The "internal conflict" in Egypt was
what precipitated the British occupation. It, too, must be understood as the result of the international trade and state systems
interacting with indigenous social forces, as argued above in dis-
cussing the origins of the land tenure system.

The crucial, yet underdetermining, influence of the international system is again evident when we look at the role of British
irrigation policy in undermining cotton yields. I have argued that

budget constraints prevented the British from undertaking the large-scale spending which adequate drainage installation required. These constraints in turn derived from the nature of the British occupation. Contemporary international rivalries sharply circumscribed the decision-making powers of the British administration in Egypt. Furthermore, the administration in Egypt had to finance itself because of Parliamentary opposition to the occupation. Nor could the British rely on taxation to raise revenue for drainage, since reducing land taxes was part of British "pacification policy." The British wished not only to forestall peasant unrest but also to secure the cooperation of their chief Egyptian political allies, the large landowners. Accordingly, land-tax rates were lowered, and the burden of taxation fell.

Once again, however, one can argue that while international forces were necessary, they were by themselves insufficient to generate the agronomic problems of inadequate drainage. First, despite a wholly different regime and a transformed international system, the same failure to invest in drainage occurred under the Nasser regime. The same mistake could be made by a very different regime with a very different ideology facing very different internal and external challenges. Second, the drainage problem was exacerbated by the failure of peasants to construct private drains to connect to the (inadequate) government drains (a problem which persists to this day). Here we have a case of the failure of local village institutions to handle an important externality problem. The small scale of many peasant holdings meant that a drain would occupy a substantial portion of any individual's land; the problem then became: Whose lands shall be used? The peasants were, apparently, unable to agree; it is highly likely that the endemic distrust of Egyptian peasant life, surely a "local" force, played a central role in the peasants' failure to construct private village drains. External and internal forces worked together to thwart the maintenance of Egypt's soil fertility.

A similar conjuncture of internal and international forces has more recently moulded the agricultural policies of Nasser and Sadat. The issue of Palestine, the decline of British power, and the context of rural unrest stimulated by the changes of World War II and the Korean War cotton boom set the stage for the July 23 coup. But such international forces, while necessary, were insufficient for this outcome. The internal stalemate of Palace, Wafd, and Ikhwan was also crucial for the Free Officers' success. The political nature of the Nasser regime—its lack of a firm social base, its nationalism, and its hierarchical methods—were the result of the interplay of international and internal forces. The subsequent agricultural policies of this "Bonapartist" regime had the same origin.

The three policies which stand out here are the land reform, the price policies, and the hydraulic investments of the regime. The land reform eliminated bimodalism and increased equality. Trying to develop industry and to strengthen the military, the regime created cooperatives, a sort of government 'ezbah, as the mechanism for a set of price policies which were the principal instrument of rural taxation. Some of the results were salutory: consolidation of crops for better pest control and the diffusion of fertilizer stand out as contributions. But at the same time, the system generated problems. Since the land reform strengthened the rich peasants, and since the regime had neither the cadre nor the political inclination to enforce its decrees equally, many farmers reduced their plantings of cotton, wheat and other controlled crops, thus depriving the government of the very revenues which it sought to obtain. This situation set the stage for the economic trade problems which exploded in the early 1970s. Here again, the interaction of indigenous social structure (the position of the rich peasants being something of a constant of modern Egyptian rural history) with international events shaped the pattern of agricultural development.

The decision to build the Aswan High Dam and the neglect of drainage investment have a similar explanation. The Dam's location entirely within Egypt's borders was an irresistable attraction for a nationalist regime. Further, funding the project became entangled with international conflict almost immediately. The ensuing financial constraints and the regime's "top down" methods led to the disastrous neglect of drainage investment. The resulting (very serious) agroecological brake on supply was left as a legacy for the decade of the 1970s.

The Sadat regime inherited the problems of drainage neglect and of declining area planted in export crops and wheat, as well as the steadily growing population and, of course, the Israeli occupation of the Sinai. The response to these problems, and again, the international conjuncture, generated a new pattern of agricultural development. Two features stand out: the liberalization of the Egyptian economy and the boom in the oil exporting states. Again, the interplay of internal and external forces is clear, especially for the liberalization in Egypt. Sadat's political base, his own perceptions, past developments which had led to relative economic and agricultural stagnation, the problem of importing adequate supplies of food, the strengthening of the rural middle class and their urban cousins—all came together to generate the Infitah. The international conjuncture, in the form of the regional political ascendancy of the Saudis, also contributed to the new policy. Such changes are incomprehensible without an international perspective.

The agricultural consequences of these developments have been dramatic. The resulting urban construction boom and the loosening of controls on external migration of workers have transformed, at least temporarily, rural labor markets. Rural workers in the cities appear to have been filling in behind urban construction workers who have departed for the oil states. Both the outflows of labor and the inflows of remittance income, are the result of the (clearly international) oil boom. Taken together, labor migration and remittances were the basic motor of change in rural Egypt in the 1970s. The rise in labor costs stimulated a fairly rapid spread of farm mechanization, while remittances became the principal source of foreign exchange. The improved balance of payments situation allowed Egypt to finance its growing food import requirements, requirements which the liberalization itself had stimulated.

A complex interaction of internal and external forces, then, has produced the pattern of Egyptian agricultural development. Indeed, as the current scene shows, the internal and external forces are by now so intertwined that it is hard to disentangle them. The oil boom is geographically external to Egypt. Yet the 1973 increase in oil prices was at least in part the result of the Egyptian army's crossing of the Suez Canal and the ensuing October War. Such a decision was, in turn, partly a response to the economic impasse in Egypt's domestic economy, in turn the outcome of past interplay of domestic and international forces, and so on. Looking at Egyptian agricultural development is like looking at two juxtaposed mirrors: one finds an infinite regress of domestic and international causalities. The "Second Image" is reversed, and re-reversed, and reversed yet again. Anyone who has come to know rural Egyptians can only hope that the interplay of international and domestic forces will cease leaving the peasants "as poor as a needle, which clothes others, but remains itself unclad".[19] But the historical record suggests that the path ahead will not be smooth.

Footnotes

[1]     See Hayami and Ruttan, (1971).

[2]     Nathan Rosenberg, "The Direction of Technical Change: Induce-
        ment Mechanisms and Focusing Devices". Economic Development
        and Cultural Change, 18 (Oct., 1969).

[3]     O'Brien, in Holt, (1968), for 1813; CAPMAS, Statistical Year-
        book, for 1975.

[4]     As noted in Chapter 6, the government claims that 900,000
        feddans have been reclaimed, yet it acknowledges that only
        300,000 are agriculturally productive.

[5]     FAO, Production Yearbook, 1978.

[6]     Hansen and Wattleworth, (1978).

[7]     United States Department of Agriculture, Global Food Assess-
        ment, 1980. (Washington D.C., Foreign Agricultural Economic
        Report #159, 1980).

[8]     A perspective repeatedly stressed by John Waterbury. See
        (1979) and "'Aish: Egypt's Growing Food Crisis", in (1978).

[9]     They now constitute some ten percent of farm owners holding
        some one third of the farms. Data from the Egyptian Ministry
        of Agriculture.

[10]    The literature on "dependency" theory is very large. See
        among others, Andre Gunder Frank, Capitalism and Under-
        development in Latin America (N.Y., 1969); Fernando Henrique
        Cardoso and Enzo Faletto, Dependency and Development in Latin
        America, M.M. Urquidi, trans. (Berkeley, 1979). For an
        elegant restatement and thoughtful critique of such theories,
        see Carlos, F. Diaz-Alejandro, "Delinking North and South:
        Unshackled or Unhinged?" in Albert Fishlow, Carlos F. Diaz-
        Alejandro, Richard R. Fagen, and Robert D. Hansen, Rich and
        Poor Nations in the World Economy (N.Y., 1978).

[11]    On ecology, see Andrew Karmack, The Tropics and Economic
        Development: A Provacative Inquiry into the Poverty of
        Nations (Baltimore, 1976); on entrepreneurial failure,
        Russell Stone, "Egypt," in W.Arthur Lewis, ed. Tropical
        Development, 1880-1913: Studies in Economic Progress (Lon-
        don, 1970); on price distortions, T.W. Schultz, ed., Distor-
        tions of Agricultural Incentives (Bloomington, Ind., 1978).

[12]    Ernesto Laclau, "Feudalism and Capitalism in Latin America",
        New Left Review, 67.

[13]    Cf. Gourevitch, (1978).

[14]    Marshall G.S. Hodgson, The Venture of Islam, Volume III: The
        Gunpowder Empires. (Chicago, 1975).

[15]    Cuno, (1980); Abd al Rahim Abd al Rahim, (1973); Peter  Gran,
        Islamic Roots of Capitalism (Austin:  University of Texas
        Press, 1978).

[16]    D.C.M. Platt, Finance, Trade and Politics in British  Foreign
        Policy, 1815-1914 (Oxford, 1968).

[17]    See, e.g., Hodgson, (1975); Maxime Rodinson, Islam and  Capi-
        talism, Brian Pearce, trans. (Austin, Texas, 1974).

[18]    Robinson and Gallagher, (1961).

[19]    Peasant saying, cited by Ayrout, (1963), 58.

# Bibliography

Abbass, M.H. The Role of Banking and Credit Institutions in the Economic Development of Egypt. Unpublished Ph.D. thesis, University of Wisconsin-Madison, 1954.

Abd al-Raheim A. "Hazz al-Quhuf: A New Source for the Study of the Fallahin of Egypt in the XVIIth and XVIIIth Centuries." Journal of the Economic and Social History of the Orient, XVIII, Part III (1975), pp. 245-270.

Abd al-Rahim, Abd al-Rahim. Al-Rif Al-Masri Fi qarn al-thamin 'ashr. (The Egyptian Countryside in the Eighteenth Century). Cairo: 1974.

Abd al-Rahim, Abd al-Rahim and Miki, Wataru. Village Life in Ottoman Egypt and Tokugawa Japan: A Comparative Study. Tokyo: Institute for Studies in the Languages and Cultures of Asia and Africa, 1977.

Abd el-Fattah, Fathi, Al-Qaryah al-Misriyyah bayna al-Islah W'al-thawrah. (The Egyptian Village Between Reform and Revolution). Cairo: 1975.

Abdel-Fadil, Mahmoud. Development, Income Distribution, and Social Change in Rural Egypt; A Study in the Political Economy of Agrarian Transition. Cambridge: Cambridge University Press, 1975.

Abdel-Latif. La Loi des Cing Feddans, Cairo: 1913.

Abdel-Malek, Anouar. Egypt: Military Society. Charles Lam Markham, trans. N.Y.: Random House, 1968.

Abu-Lughod, Janet. "The Transformation of the Egyptian Elite—Prelude to the Urabi Revolt." Middle East Journal, XII (1967).

Ahmed, Ezz el Din Hammam, and Abu el- Danab, Mohammad Galal. Fertilizer Distribution in the Arab Republic of Egypt. Paris: OECD, 1972.

Allan, J.A. "Some Phases in Extending the Cultivated Area in the Nineteenth and Twentieth Centuries in Egypt." Paper presented to the Annual Meeting of the Middle East Studies Association, Washington, D.C., November, 1980.

Amin, Galal. Food Supply in Relation to Economic Development with Special Reference to Egypt. London: F. Cass, 1966.

Amin, Samir. Le Development Inegal. Essai sur les formations sociales du capitalisme périphérique. Paris: Editions de Minuit, 1973.

Ammar, Abbas. A Demographic Study of an Egyptian Province (Shaqiyya). London: P. Lund, Hamphries and Co., Ltd., 1942.

Ammar, Abbas. "Conditions of Life in Rural Sharqiyya." Sociological Review, 1940.

Ammar, Hamad. Growing Up in an Egyptian Village. Silwa, Province of Aswan. London: Routledge and Kegan, 1954.

Amr, Ibrahim. Al-'ard w'al-fellah: al-mas'ilah az-zira' iyya fi misr. (The Land and the Peasant: The Agriculural Problem in Egypt.) Cairo: 1958.

Anhoury, J. "L'économie agricole de l'Egypte." L'Egypte Contemporaine, 32 (1941).

Anhoury, J. "Les engrais chimiques et leur rôle dans l'économie de l'Egypte." L'Egypte Contemporaine, 33 (1942).

Anhoury J. "Les Grandes Lignes de l'économie agricole de l'Egypte." L'Egypte Contemporaine, 33 (1942).

Anhoury, J. "Le Blé en Egypte." L'Egypte Contemporaine, 14 (1925), pp. 194-204.

Anhoury, J. "Le nitrate de soude du Chili et son emploi dans l'agriculture égptienne." L'Egypte Contemporaine, 19 (1928), pp 378-390.

Anis, Mahmoud. "The National Income of Egypt." L'Egypte Contemporaine, 41 (1950), pp. 551-926.

Ar-Rafi'i, 'Abderraham. Thawrah 1919. (The Revolution of 1919.) Cairo: 1955.

Arrow, Kenneth. The Limits of Organization. N.Y.: W.W. Norton, 1974.

Askari, Hossein; Cummings, John; Harik, Bassam. "Land Reform in the Middle East." International Journal of Middle East Studies, 8, 4 (October, 1977).

Avigdor, S. "L'Egypte agricole." L'Egypte Contemporaine, 21 (1930), pp. 72-104.

Ayalon, D. "Studies in al-Jabarti—Notes on the Transformation of Mamluk Egypt under the Ottoman." Journal of the Economic and Social History of the Orient, 3, pp. 148-275.

Ayrout, Henry Habib. The Egyptian Peasant. John Allen Williams, trans. Boston: Beacon Press, 1963.

Azmi, Hamed el-Sayyid. "The Growth of Population in Egypt." L'Egypte Contemporaine, 28 (1937), pp. 267-303.

Azmi, Hamed el-Sayyid. "A Study of Agriculture Land and Present Incidence of the Land Tax." L'Egypte Contemporaine, 25 (1934), pp. 693-717.

Baer, Gabriel. "Dissolutions of the Village Community." Studies in the Social History of Modern Egypt. Chicago: University of Chicago Press, 1969.

Baer, Gabriel. A History of Land Ownership in Modern Egypt, 1800-1950. London, N.Y.: Oxford University Press, 1962.

Baer, Gabriel. "Submissiveness and Revolt of the Fellah." Studies in the Social History of Modern Egypt. Chicago: University of Chicago Press, 1969.

Baer, Gabriel. "The Village Shaykh, 1800-1950." Studies in the Social History of Modern Egypt. Chicago: University of Chicago Press, 1969.

Baker, Raymond William. Egypt's Uncertain Revolution Under Nasser and Sadat. Cambridge, Mass: Harvard University Press, 1978.

Ballou, H.A. "Notes on the Cotton Crops of Egypt." Agricultural Journal of Egypt, IX (1921), pp. 14-48.

Balls, W.L. "The Cotton Plant in Egypt." Studies in Physiology and Genetics. London: Macmillan and Co., 1912.

Balls, Wm. Lawrence. Egypt of the Egyptians. London: Sir I. Pitman and Sons, Ltd., 1915

Balls, Wm. Lawrence. The Yields of a Crop. London: E. & F.N. Spon, 1953.

Balls, Wm. Lawrence, D.S. Gracie, U. Khahil. Egypt, Ministry of Agriculture, Technical and Scientific Bulletin #249. (Cairo, 1936).

Barakat, Ali. Tatawwur al-milkiyyah as-zira'iyyah fi misr, w'atharuh 'ala al-harakah as-siyasiyyah 1813-1914. (The Development of Agricultural Landownership in Egypt and Its Influence Upon Political Development 1813-1914.) Cairo: 1977.

el-Barawy, Rashed. Economic Development in the U.A.R. Cairo: 1970.

Barois, J. Irrigation in Egypt. Major A.M. Millier, trans. U.S. Congress, House of Representatives, 50th Congress, second session, Misc. Doc. #134. Washington, D.C., 1889.

Basheer, Abdel-Mawia. Approaches to the Economic Development of Agriculture in Egypt. Unpublished Ph.D. thesis, University of Wisconsin-Madison, 1961.

Bath, B.H. Slicher van. The Agrarian History of Western Europe, A.D. 500-1800. London: Edward Arnold, Ltd., 1963.

Berry, R. Albert, and Cline, William R. Agrarian Structure and Productivity in Developing Countries. Baltimore: John Hopkins University Press, 1979.

Berque, Jacques. Histoire Sociale d'un Village en Egypte au XXeme Siècle. Paris: Mouton, 1957.

Berque, Jacques. Egypt: Imperialism and Revolution. Jean Stewart, trans. London: Faber, 1972.

Binder, Leonard. In a Moment of Enthusiasm: Political Power and the Second Stratum in Egypt. Chicago: University of Chicago Press, 1978.

Binswanger, Hans. The Economics of Tractors in the Indian Subcontinent: An Analytical Review. Hyderabad: I.C.R.I.S.A.T., 1977.

Birks, J.S., and Sinclair, Clive. International Migration and Development in the Arab Region. Geneva: ILO, 1980.

el-Bishry, Tarek. "Aperçu Politique et Social." La Voie Egyptienne Vers le Socialisme. Cairo: u.d.

Blackman, Winnifred S. The Fellahin of Upper Egypt. London: G.G. Harrap and Co., Ltd., 1927.

Blanchard, G. "Le bien de famille et la loi des cinq feddans." L'Egypte Contemporaine, 4 (1913), pp. 337-355.

Blanchard, G. "La crise en Egypte." L'Egypte Contemporaine, 22 (1931), pp. 313-338.

Bonne, A. Economic Development of the Middle East. London: K. Paul, Trench, Trubner and Co., Ltd., 1947.

Boraiko, Alan A. "The Pesticide Dilemma." National Geographic, 157, 2 (February, 1980): 145-183.

Bosch, Firman van den. Vingt années de l'Egypte. Paris: 1932.

Bowring, John. Report on Egypt and Candia. London: W. Clowes & Sons, 1840.

Braudel, Fernand. The Mediterranean and the Mediterranean World in the Age of Phillip II. Vols. 1 and 2. Sian Reynolds, trans. N.Y.: 1974.

Brinton, J.Y. The Mixed Courts of Egypt. New Haven: Yale University Press, 1930.

Brown, C.H. Egyptian Cotton. London: Leonard Hill, 1955.

Burns, A.C. "Investigations on Raw Cotton. Deterioration of Cotton during Damp Storage." Cairo: Ministry of Agriculture, Technical and Scientific Service Bullitin #71.

Calvocoressi, Peter, and Wint, Guy. Total War: Causes and Courses of the Second World War. New York: Penguin Publishers, 1972.

Cardoso, Fernando Henrique and Faletto, Enzo. Dependency and Development in Latin America. M.M. Urquid, trans. Berkeley: University of California Press, 1979.

Cartwright, W., and Gerald C. Dudgeon. Miscellaneous Publications. Cairo: Ministry of Agriculture, 1922.

Casoria, M. "Chrônique agricole de l'année 1921." L'Egypte Contemporaine, 13 (1922), pp. 44-86.

Casoria, M. "Chrônique agricole de l'année 1922." L'Egypte Contemporaine, 14 (1923), pp. 141-187.

Castoriades, Gerard. "Bouncy Contracting Market Draws Far East Challenge." Middle East Economic Digest, June 6, 1980.

Catzeflis, Emile. "Le drainage des terres humides et salées du delta égyptien." L'Egypte Contemporaine, 6 (1915-16), pp. 307-342.

de Chabrol de Volvic. "Essai sur les moeurs des habitants modernes de l'Egyte," in Description de l'Egypte, Etat Moderne. Vol. 2,

Part 2. Paris: 1822.

de Chamberet, Raoul. Enquête sur la condition du fellah égyptien. Dijon: Imprimerie Darantière, 1909.

Charles-Roux, Francois. La production du coton en Egypte. Paris: A Colin, 1908.

Chase World Information Corporation. The Chase World Information Series on Agribusiness Potential in the Middle East and North Africa: Egypt, New York: 1977.

Cheung, Steven N.S. The Theory of Share Tenancy. Chicago: University of Chicago Press, 1969.

Chirol, Sir Valentine. The Egyptian Problem. London, 1920.

Choucri, Nazli; Eckaus, Richard S.; el-Din, Amr Mohi. "Migration and Employment in the Construction Sector: Critical Factors in Egyptian Development." Cairo University/Massachusetts Institute of Technology, 1978.

Cipolla, Carlo M. Guns, Sails, and Empires: Technological Innovation and the Early Phases of European Expansion, 1400-1700. N.Y.: 1965.

Clark, Colin, and Margaret Haswell. The Economics of Subsistence Agriculture. Second Ed. N.Y.: St Martin's Press, 1966.

Clarke, S. The Unrest in Egypt: A Plea for Justice to the Oppressed Fellahin and Tales of the Omdeh. Cairo: 1919.

Clawson, Marion; Landsbert, Hans H.; Alexander, Lyle T. The Agricultural Potential of the Middle East. New York: American Elsevier, 1971.

Cleland, William. "Egypt's Population Problem." L'Egypte Contemporaine, 28 (1937), pp. 67-87.

Cleland, William. "A Population Plan for Egypt." L'Egypte Contemporaine, 30 (1939), pp. 461-484.

Cleland, William. The Population Problem in Egypte. Lancaster, Pa.: Science Press Printing Co., 1936.

Clough, Shepard B. France: A History of National Economics, 1789-1939. N.Y.: 1939.

Consortium for International Development, Project Design Team. Egyptian Major Cereals Improvement Project. Technical Report. Mimeo, Cairo: 1979.

Coult, Lyman H. An Annotated Research Bibliography of Studies in Arabic, English and French of the Fellah of the Egyptian Nile, 1789-1955. Coral Gables: University of Miami Press, 1958.

Craig, J.I. "Notes on the Cotton Statistics of Egypt." L'Egypte Contemporaine, 2 (1911), pp. 166-198.

Craig, J.I. "Notes on the National Income of Egypt." L'Egypte Contemporaine, 15 (1924), pp. 1-9.

Cressaty, Compte. L'Egypte d'aujourd'hui. Paris: M. Rivière et Cie., 1912.

Cromer, Earl of. Modern Egypt. Two volumes. London: Macmillan, 1908.

Crouchly, A.E. "A Century of Economic Development." L'Egypte Contemporaine, 30 (1939), pp. 133-155.

Crouchley, A.E. The Economic Development of Modern Egypt. London: Longmans, Green, and Co., 1938.

Cuddihy, William. "Agricultural Price Management in Egypt." World Bank Staff Working paper, #388 (1980).

Cuno, Kenneth M. "The Origins of Private Ownership of Land in Egypt: A Reappraisal." International Journal of Middle Eastern Studies. 12, 3 (November, 1980), pp. 245-275.

Dawood, H.A. "Farm Land Aquisition Problems in Egypt." Land Economics, 26 (1950), pp. 305-308.

Day, Richard H. "The Economics of Technological Change and the Demise of the Sharecropper." American Economic Review. 57,3 (June, 1967).

Day, Richard H., and Singh, Inderjit. Economic Development as an Adaptive Process. Cambridge: Cambridge University Press, 1977.

Dessuqi, 'Asim. Kibar Mullak al-'Aradi Al-zira'iyya wa dawruhum Fi al-mujtama' al-masri, 1914-1952. (Large Agricultural Land Owners and their Role in Egyptian Society, 1914-1952). Cairo: 1975.

Diaz-Alejandro, Carlos F. "Delinking North and South: Unshackled or Unhinged?," in Albert Fishlow, Carlos F. Diaz-Alejandro, Richard R. Fagen, and Robert D. Hansen. Rich and Poor Nations in the World Economy. New York: McGraw Hill, 1978.

Doggett, Hugh; Curtis, David L.; Laubscher, F.X.; Webster, Orrin J. "Sorghum in Africa" in Joseph S. Wall and William M. Ross, Sorghum Production and Utilization. Westport, Conn.: AVI Pub. Co., Inc., 1970.

Dorner, Peter. Land Reform and Economic Development. Baltimore: Penguin Books, 1971.

Dudgeon, Gerald C. "Cotton Cultivation in Egypt." L'Egypte Contemporaine, 14 (1923), pp. 517-532.

Dudgeon, Gerald C. "Cotton Demonstration Farms in Egypt, 1914." Agricultural Journal of Egypt, III (1915), pp. 95-114.

Dudgeon, Gerald C. "Cotton Demonstration Farms in Egypt, 1915." Agricultural Journal of Egypt, IV (1916), pp. 42-72.

Dudgeon, Gerald C. "Gossypium Spp. Cotton, Qotn in Egypt: History, Development and Botanical Relationship of Egyptian Cottons." Egypt, Ministry of Agriculture, Cairo: 1917.

Dudgeon, Gerald C. and W. Cartwright. "Treatment of Cotton in the Field." Agricultural Journal of Egypt, VII (1919).

Duff Gordon, Lady L. Letters from Egypt, Gordon Waterfield, ed. London: Routlege and Kegan, Paul, 1969.

Dumont, René. "Les problèmes agraires de la R.A.U." Politique étrangère. 2 (1968).

Egypt, Central Agency for Public Mobilization and Statistics, Statistical Yearbook. Cairo: 1979.

Egypt. New Haven: Human Area Relations Files, 1957.

Egypt, Department of Statistics and Census. Statistics of Wages and Working Hours in Egypt. Cairo: 1946.

Egypt, Ministry of Agriculture. Agricultural Census of Egypt, 1929. Cairo: 1934.

Egypt, Ministry of Agriculture, Agricultural Census of Egypt, 1939. Cairo: 1946.

Egypt, Ministry of Agriculture, La Culture du Coton en Egypte, Cairo, 1950.

Egypt, Ministry of Agriculture. L'Egypte agricole. Cairo: 1936.

Egypt, Ministry of Agriculture. Bureau of Agricultural Economics. Agricultural Census, 1950. Bulletin no. 6. Vol 8. (October, 1949).

Egypt, Ministry of Agriculture. "Preliminary Report on a Field Experiment Upon Cotton Carried Out at Talbia, 1912." Giza: 1913.

Egypt, Ministry of Agriculture. "Report on the the Great Invasion of Locusts in Egypt in 1915 and the Measures Adopted to Deal with It." Cairo: 1916.

Egypt, Ministry of Agriculture. "Report on the Motor Tractor Trials Organized by the Ministry of Agriculture." Cairo: 1921.

Egypt, Ministry of Finance. Egyptian Customs Administration. Annual Statement of Foreign Trade.

Egypt, Ministry of Finance, Statistical Department. Annuaire Statistique de l'Egypte, Various Issues.

Egypt, Ministry of Finance, Statistical Department. The Census of Egypt taken in 1907. Cairo: 1909.

Egypt, Ministry of Finance, Statistical Department. The Census of Egypt taken in 1917. Cairo: 1920-21.

Egypt, Ministry of Finance, Statistical Department. The Census of Egypt, 1227. Cairo: 1931.

Egypt, Ministry of Finance, Statistical Department. The Census of Egypt, 1937. Cairo: 1942.

Egypt, Ministry of Finance, Statistical Department. Monthly Agricultural Statistics. Sept., 1919-Aug., 1934. Renamed Monthly Bulletin of Agricultural and Economic Statistics, Sept., 1934-

April, 1941.

Egypt, Ministry of Justice. <u>Annuaire</u> <u>du</u> <u>Ministère</u> <u>de</u> <u>la</u> <u>Justice</u> <u>pour</u> <u>l'annee</u> <u>1908</u>. Cairo: 1909.

Egypt, Ministry of Justice. <u>La</u> <u>Législation</u> <u>en</u> <u>Matière</u> <u>en</u> <u>Matiere</u> <u>Immobilière</u> <u>en</u> <u>Egypte</u>. Cairo: 1905.

Egypt, Ministry of Justice. <u>La</u> <u>loi</u> <u>sur</u> <u>les</u> <u>digues</u> <u>et</u> <u>canaux</u>. Cairo: 1905

Egypt, Ministry of Justice. <u>Reports</u> <u>1900–1908</u>. Cairo: 1901–1909.

Egypt, Ministry of Public Works. <u>The</u> <u>Agricultural</u> <u>Journal</u> <u>of</u> <u>Egypt</u>. Cairo: 1911–1924.

Egypt, The Ministry of Public Works. <u>Index</u> <u>to</u> <u>Annual</u> <u>Reports</u>, <u>Irri</u>-<u>gation</u> <u>Department</u>. Cairo: 1884–1920.

Egypt, Ministry of Social Affairs. Fellah Department. "Notes on Production Relating to Agriculture." Cairo: 1948.

Egypt, Miscellaneous Official Publications. <u>Reassessment</u> <u>of</u> <u>the</u> <u>Land</u> <u>Tax</u> <u>for</u> <u>various</u> <u>Moudirieh</u>. Cairo: 1900–1907.

Egypt, Miscellaneous Official Publications. <u>Statistical</u> <u>Returns</u>, <u>1881–1897</u>. Cairo: 1898.

<u>L'Egypte</u> <u>Indépendante</u>. Le group d'études d'Islam. Paris: Centre d'etudes de la politique étrangère, 1937.

Elkhaisy, Pasha, M.F. "The Public Security in Egypt in 1927." <u>L'Egypte</u> <u>Contemporaine</u>, 19 (1928), pp. 21–64.

England, Empire Cotton Growing Corporation. "Conference on Cotton Growing Problems." Manchester: 1930.

England, Empire Growing Cotton Corporation. "Reports Received from the Experiment Stations." Manchester: 1925.

ERA 2000, Inc. <u>Further</u> <u>Mechanization</u> <u>of</u> <u>Egyptian</u> <u>Agriculture</u>. Gaithersburg, Md.: 1979.

Eshag, Eprime, and Kamal, M.A. "Agrarian Reform in the U.A.R. (Egypt)." <u>Bulletin</u> <u>of</u> <u>the</u> <u>Oxford</u> <u>University</u> <u>Institute</u> <u>of</u> <u>Econom</u>-<u>ics</u> <u>and</u> <u>Statistics</u>, 30, 2 (May, 1968).

Esteve, Comte. "Mémoire sur les finances de L'Egypte depuis sa con-quete par le sultan Selym Ier, jusqu'à celle du général en chef Bonaparte." <u>Description</u> <u>de</u> <u>l'Egypte</u>, <u>Etat</u> <u>Moderne</u>, Vol. I, Paris: 1809.

Fadil, Mahmoud Abdel. <u>Development</u>, <u>Income</u> <u>Distribution</u>, <u>and</u> <u>Social</u> <u>Change</u> <u>in</u> <u>Rural</u> <u>Egypt</u>: <u>A</u> <u>Study</u> <u>in</u> <u>the</u> <u>Political</u> <u>Economy</u> <u>of</u> <u>Agrarian</u> <u>Transition</u>. Cambridge: Cambridge University Press, 1975.

Fahmi, Abd al-Latif Muhammad. <u>La</u> <u>loi</u> <u>des</u> <u>cinq</u> <u>feddans</u>, <u>ou</u> <u>l'insaisissabilite</u> <u>de</u> <u>la</u> <u>petite</u> <u>propriété</u> <u>agricole</u> <u>en</u> <u>Egypte</u>. Paris: 1914.

F.A.L. des écoles chrétiennes. Géographie du bassin du Nil. Paris: Librarie Delgrave, 1933.

Falcon, Walter P. "The Green Revolution: Generations of Problems." American Journal of Agricultural Economics, 52 (Dec., 1970), pp. 698-710.

F.A.O. Near East Regional Office. Research on Crop Water Use, Salt Affected Soil and Drainage in the A.R.E. Rome: 1975.

F.A.O. Production Yearbook.

Fenn, G. Manville. The Khedive's Country. London: Cassel & Co., Ltd., 1904.

Finley, M.I. The Ancient Economy. Berkeley and Los Angeles: University of California Press, 1973.

Fisher W.B. The Middle East. Fifth ed. London: Methuen & Co., 1963.

Fitch, James B.; Goueli, Ahmad A.; el-Gabely, Mohammad; "The Cropping System for Maize in Egypt." Paper presented to the FAO workshop on "Improved Farming Systems in the Nile Valley," Cairo, May 8-14, 1979.

Foaden, G.P., and F. Fletcher. Textbook of Egyptian Agriculture. Two vols. Cairo: National Printing Dept., 1908, 1910.

Fox, Carl A., ed. Readings in the Economics of Agriculture. Homewood, Ill.: R.D. Irwin, 1969.

France. Ministère des Finances et des Affaires Economiques. Memento Economique: l'Egypte. Paris: 1950.

Frank, Andre Gunder. Capitalism and Underdevelopment in Latin America: Historical Studies of Chile and Brazil. N.Y.: Monthly Review Press, 1969.

Frankel, Francine R. India's Green Revolution: Economic Gains and Political Costs. Princeton: Princeton University Press, 1971.

Freebairn, Donald K. "Income Disparities in the Agricultural Sector: Regional and Institutional Stresses," in Thomas T. Poleman and Donald K. Freebairn, eds. Food, Population, and Employment: The Impact of the Green Revolution. N.Y.: Praeger, 1973, pp. 97-114.

Gali, Kamil. Essai sûr L'agriculture de l'Egypte. Paris: Henri Jouve, 1889.

Gallatoli, Anthony. Egypt in Mid-passage. Cairo: Urwand & Sons, 1950.

Garrett, Roger. "Impressions with respect to Mechanization." USAID mimeo, 1978.

Geertz, Clifford. Agricultural Involution: The Processes of Ecological Change in Indonesia. Berkeley and Los Angeles: University of California Press, 1963.

Genovese, Eugene. Roll, Jordan, Roll: The World the Slaves Made. N.Y.: Random House, 1974.

Ghali, Mirrit Boutrus. The Policy of Tomorrow. Isma'il el-Faruqi, trans. Washington, D.C.: 1953.

Ghonemy, M.R. Resource Use and Income in Egyptian Agriculture. Unpub. Ph.D. thesis, North Carolina State College, 1953.

Gibb, Hamilton A.R., and Harold Bowen. Islamic Society and the West. Vol. I: Islamic Society in the Eighteenth Century. Two parts. London: Oxford University Press, 1950, 1957.

Girad, P.S. "Mémoire sur l'agriculture, l'industrie et le commerce de l'Egypte," Description de l'Egypte, Etat Moderne, Vol. ", Part I. Paris: 1813.

Gourevitch, Peter. "The Second Image Reversed: The International Sources of Domestic Politics." International Organization, 32, 4 (August, 1979.

Gracie D.S. and I. Khalil. Ministry of Agriculture, Technical and Scientific Service Bulletin #152. Cairo: 1935.

Gran, Peter. Islamic Roots of Capitalism. Austin: University of Texas Press, 1978.

Griess, Kamel A. "L'usure en Egypte." L'Egypte Contemporaine, 12 (1921), pp. 92-111.

Griffen, Keith. The Political Economy of Agrarian Change: An Essay on the Green Revolution. Cambridge, Mass.: Harvard University Press, 1974.

Griliches, Zvi. "The Demand for Fertilizer: An Economic Interpretation of Technical Change." Journal of Farm Economics. 40 (August, 1958).

Griliches, Zvi. "Distributed Lags, Disaggregation and Regional Demand Functions for Fertilizer" Journal of Farm Economics. 41 (February, 1959).

Gracie, D.S., Mahfouz Rizk, Ahmed Moukhta, Abdel Hamid I. Moustapha. "The Nature of Soil Deterioration in Egypt." Ministry of Agriculture, Technical and Scientific Service Bulletin #148. Cario: 1934.

Guimei, Mokbel. Le crédit agricole et l'Egypte. Paris: 1931.

Habachi, Marc. L'état économique de l'Egypte sous le régime de la monoculture. Cairo: 1931.

Hadler, Sandra, and Yand, Maw-Cheng. "Developing Country Foodgrain Projections for 1985." World Bank Staff Working Paper No 247. (November, 1976).

Hamdan, G. "Evolution of Irrigation Agriculture in Egypt," in Dudley Stamp, ed. A History of Land Use in Arid Regions. Paris: U.N.E.S.C.O., 1961, pp. 119-142.

Hamdi, M.M. "The Nationalization of Cotton." L'Egypte Contemporaine, #265 (July, 1951).

Hamid, Ra'uf 'Abbas. Al-Nidham al-Ijtima' Fi Misr Fi dhill al-milkiyyat al-zira'iyya al-kabira, 1887-1914. (The Social System in Egypt Under the Influence of Large Agricultural Landowner-ship, 1887-1914). Cairo: 1973.

Hansen, Bent. "Arab Socialism in Egypt." World Development, 3, 4 (April, 1975).

Hansen, Bent. "The Distributive Shares in Egptian Agriculture, 1897-1961. International Economic Review, 9 (June, 1968), pp. 175-194.

Hansen, Bent. "Economic Development in Egypt" in C.A. Copper and S.S. Alexander, eds. Economic Development and Population Growth in the Middle East. New York: Elsevier, 1972.

Hansen, Bent. "An Economic Model for Ottoman Egypt: The Economics of Collective Tax Responsibility." Unpublished MS presented at the Conference on the Economic History of the Near East, Prince-ton University, Princeton, N.J., June 16-20, 1974.

Hansen, Bent. "Employment and Wages in Rural Egypt." American Economic Review, 59 (June, 1969), pp. 298-314.

Hansen, Bent and Marzouk, Girgis A. Development and Economic Policy in the U.A.R. (Egypt). Amsterdam: North Holland, 1965.

Hansen, Bent and Nashashibi, Karim. Foreign Trade Regimes and Economic Development: Egypt. New York: N.B.E.R., 1975.

Hansen, Bent, and Wattleworth, Michael. "Agricultural Output and Consumption of Basic Foods in Egypt, 1886/87-1967/68" Interna-tional Journal of Middle Eastern Studies, 9,4 (November, 1978).

Harari, R.A. "Banking and Financial Business in Egypt." L'Egypte Contemporaine, 27 (1936), pp. 129-149.

Hartmann Fernande. L'agriculture dans l'ancienne Egypte. Paris: 1923.

Hayami, Yujiro, and Vernon W. Ruttan. Agricultural Development, An International Perspective. Baltimore: Johns Hopkins University Press, 1971.

Hayami, Yujiro, and Vernon W. Ruttan. "Agricultural Productivity Differences Among Countries." American Economic Review, 60 (Dec., 1970), pp. 895-911.

Hayami, Yujiro, and Vernon W. Ruttan. "Professor Rosenberg and the Direction of Technical Change." Economic Development and Techni-cal Change, 21 (March, 1973).

Hefnawi, Mahmoud Tawfik. "Manuring of Vegetables." Ministry of Agri-culture, Horticultural Section Leaflet #15. Cairo: 1920.

Herlihy, Patricia. "Odessa and Europe's Grain Trade in the First Half of the Nineteenth Century." Unpub. MS, n.d.

Herodotus. The Histories. Aubrey de Selincourt, trans. Baltimore: Penguin, 1954.

Herold, J. Christopher. Bonaparte in Egypt. London: Hamish Hamilton, 1963.

de Herreros, E.G. "For the Suppression of Usury: Notes on the Spanish Law." L'Egypte Contemporaine, 7 (1915-1916), pp. 86-103.

Hirschman, A.O. The Strategy of Economic Development. New Haven: Yale University Press, 1958.

Al-Hitta, Ahmad Ahmad. Ta'rikh az-zira'iyya al-misriyya fi 'ahd Muhammad 'Ali al-kabir. (The History of Egyptian Agriculture in the Age of Muhammad 'Ali the Great). Cairo: 1950.

Hobsbawm, E.J. Primitive Rebels. Studies in Archaic Forms of Social Movement in the Nineteenth and Twentieth Centuries. N.Y.: W.W. Norton, 1965.

Hobsbawm, E.J. and George Rude. Captain Swing. A Social History of the Great English Agricultural Uprising of 1830. London: 1968.

Hodgson, Marshall G.S. The Venture of Islam, Vol. III: The Gunpowder Empires. Chicago: University of Chicago Press, 1975.

Hollings, M.A., ed. The Life of Colonel Scott-Moncrieff. London: J. Murray, 1917.

Holt, P.M. Egypt and the Fertile Crescent, 1516-1922. London: 1966.

Holt, P.M. The Madhist State in the Sudan, 1881-1898. 2nd edn. London: Oxford University Press, 1970.

Horsely, Beverly. "Keeping up with Demand: The Big Challenge for Egyptian Agriculture," Foreign Agriculture. November 7, 1977.

"Al Hossary's Report." Unpublished, untitled report prepared for the Office of the Undersecretary of Agriculture for Engineering Affairs, 1979.

Hourani, ALbert. "Preface" in Alexander Schoelch, Aegpt den Aegyptern! Die politische und gesellschaftliche Krise de Jahre 1878-1882 in Aegypten. Zurich: 1972, pp. 9-13.

Husayn, Husni. "'Ummal At-Tarahil fi-il-'ard al-jadidah" (Tarahil Laborers on new Lands). At-Ta'liah, Vol. 7, No. 1, Jan., 1971, pp. 20-27.

Husein, Mahmoud. Class Conflict in Egypt, 1945-1970. N.Y.: 1973.

Ikram, Khalid. Egypt: Economic Management in a Period of Transition. Baltimore: Johns Jopkins Univeristy for the World Bank, 1980.

Imlah, A.H. Economic Elements in the Pax Brittanica. Cambridge, Mass.: Harvard University Press, 1958.

Inalcik, Halil. "The Heyday and Decline of the Ottoman Empire," in P.M. Holt, Ann K.S. Lambton, and Bernard Lewis, eds., The Cambridge History of Islam, Vol. 1. The Central Islamic Lands. Cambridge: 1970.

278

International Institute of Strategic Studies. The Military Balance. London: IISS.

Israel, V. "Le problème du blé en Egypte." L'Egypte Contemporaine, 20 (1929), pp. 515-522.

Issa, Hossam M. Capitalisme et sociétés anonymes en Egypte. Paris: Librairie Générale de Droit et du Jurisprudence, 1970.

Issawi, Charles. Egypt at Mid-Century. An Economic Survey. London and N.Y.: Oxford University Press, 1954.

Issawi, Charles. Egypt in Revolution. An Economic Analysis. London and N.Y.: Oxford University Press, 1963

Issawi, Charles. "Egypt Since 1800: A Study in Lopsided Development." Journal of Economic History, xxi 1, March, 1961, pp. 1-25.

James, E. "L'organisation du crédit en Egypte." L'Egypte Contemporaine, 30, (1939), pp. 537-594.

de Janvry, Alain, and Garramou, Carlos. "The Dynamics of Rural Poverty in Latin America." Journal of Peasant Studies, 4.3 (April 1977), 206-207.

Janzen, Daniel H. "Tropical Agroecosystems." Science, 182, pp. 1212-1219.

Johnston, Bruce, and Cownie, John. "The Seed-Fertilizer Revolution and Labor Force Absorption." American Economic Review, 59 (Sept. 1969), pp. 569-582.

Johnston, Bruce F. and Kilby, Peter. Agriculture and Structural Transformation: Economic Strategies in Late-Developing Countries. New York: Oxford University Press, 1975.

Jones, L.A.H. "The Economic Condition of the Fellahin." L'Egypte Contemporaine, 20 (1929), pp. 407-412.

Jullien, L. "Chrônique agricole de l'année 1913." L'Egypte Contemporaine, 4 (1913), pp. 407-412.

Jullien, L. "Chrônique agricole de l'année 1919." L'Egypte Contemporaine, 11 (1920), pp. 47-62.

Jullien, L. "Chrônique agricole de l'année 1920." L'Egypte Contemporaine, 12 (1921), pp. 48-68.

Jullien, L. "Chrônique agricole de l'année 1923." L'Egypte Contemporaine, 15 (1924), pp. 13-31.

Jullien, L. "Chrônique agricole de l'année 1924." L'Egypte Contemporaine, 16 (1925), pp. 486-507.

Jullien, L. "Chrônique agricole de l'année 1927." L'Egypte Contemporaine, 18 (1927), pp. 541-546.

Jullien, L. "Chrônique agricole de l'année 1928." L'Egypte Contemporaine, 20 (1929), pp. 287-305.

Jullien, L. "Chrônique agricole de l'année 1929." L'Egypte Contemporaine, 21 (1930), pp. 193-213.

Jullien, L. "Chrônique agricole de l'année 1932." L'Egypte Contemporaine, 24 (1933), pp. 477-491.

Jullien, L. "Chrônique agricole de l'année 1933." L'Egypte Contemporaine, 25 (1934), pp. 519-529.

Jullien, L. "Chrônique agricole de l'année 1934." L'Egypte Contemporaine, 26 (1935), pp. 527-538.

Kalecki, Michal. "Intermediate Regimes in The Last Phase in the Transformation of Capitalism. N.Y.: Monthly Review Press, 1972.

Kamal, Michel. "Féodaux, Paysans Riches, et Fellahs." Démocratie Nouvelle. (April, 1968).

Kamil, Izz al-Din. Al-Zira'ah Al-Mikanikiyyah. (Mechanized Agriculture). Cairo: 1976.

Karmack, Andrew. The Tropics and Economic Development: A Provocative Inquiry into the Poverty of Nations. Baltimore: Johns Hopkins University, 1976.

Khedr, Mohammad. The Impact of Drainage on Agricultural Production in the A.R.E. Unpublished Ph.D. dissertation. Minufiyya University, 1979.

Koval, Andrew J., and Bahgat, Ahmed A. "Ten Horsepower Agriculture." Paper presented to the Symposium on Appropriate Mechanization of Small Farms in Africa, Cairo, 1980.

Laclau, Ernesto. "Feudalism and Capitalism in Latin America." New Left Review, 67.

Lakani, S. Bey. "Cotton: Estimation of the Crop and Measurement of Its Elasticity." L'Egypte Contemporaine, 17 (1926), pp. 265-279.

Lakani, S.I. "The Possibility of Levying an Income Tax in Egypt." L'Egypte Contemporaine, 29, (1938), pp. 181-200.

Lambert, A. "Divers modes de faire-valoir des terres en Egypte." L'Egypte Contemporaine, 29 (1938), pp. 181-200.

Lambert A. "Les Salaires dans l'entreprise agricole égyptienne." L'Egytpe Contemporaine, 33 (1943).

Lancret, Michel Ange. "Mémoire sur le système d'imposition territoriale et sur l'administration des provinces de l'Egypte dans les dernières années du guvernement des Mamlouks," Description de l'Egypte. Etat Moderne. Vol. I, Part 1. Paris: 1909.

Landes, David S. Bankers and Pashas. International Finance and Economic Imperialism in Egypt. Cambridge, Mass.: Harvard University Press, 1958.

Lappe, Francis Moore. Diet for a Small Planet. N.Y.: Ballantine, 1971.

Latham, Michael C., Robert B. McGandy, Mary B. McConn and Frederick J. Stare. Scope Manual on Nutrition. Kalamazoo, Michigan: Upjohn Co., 1970.

Lecarpentier, G. L'Egypte Moderne. Paris: 1914.

Lefebvre, Georges. The French Revolution. Two vols. John H. Stewart and James Friguglietti, trans. N.Y.: Columbia University Press, 1964.

Leiper R.T. "La production experimentale des Bilharzioses Egyptiennes." Bulletin de l'Institut Egyptien. Cairo: March, 1917.

Lewis, Bernard. The Emergence of Modern Turkey. London and N.Y.: Oxford University Press, 1961.

Levi, I.G. "La distribution du crédit en Egypte." L'Egypte Contemporaine, 9 (1918), pp. 113–132.

Levy, Rueben. The Social Structure of Islam. Cambridge: Cambridge University Press, 1962.

Lipton, Michael. "The Theory of the Optimizing Peasant." Journal of Development Studies, 4 (April, 1968), pp. 327–351.

Little, Tom. High Dam at Aswan: The Subjugation of the Nile. New York: The John Day Company, 1965.

Lokke, Carl L. "The French Agricultural Mission to Egypt." Agricultural History, 10 (July, 1936.)

Lozach and Hug. L' Habitat Rural en Egypte. Cairo: Imprimerie de l'Institut Francais d'archeologie orientale du Caire, 1930.

Luce, R. Duncan, and Howard Raiffa. Games and Decisions. N.Y.: John Wiley, 1957.

Mabro, Robert. The Egyptian Economy 1952–1970. London: Oxford, 1974.

Mabro, Robert. "Employment and Wages in Dual Agriculture." Oxford Economic Papers, 23 (Nov., 1971), pp. 401–417.

Mansfield, Peter. The British in Egypt. New York: Holt Rinehart & Winston, 1971.

Martin G. and I.G. Levy. "Le marché égyptien et l'utilité de la publication des mercuriales." L'Egypte Contemporaine, 1 (1910), pp. 441–489.

Mayfield, James B. Local Institutions and Egyptian Rural Development. Cornell University Rural Development Committee Monograph. 1978.

Marham, Jesse. The Fertilizer Industry: Study of an Imperfect Market. Nashville, Tenn.: Vanderbilt University Press, 1958.

Marx, Karl. Capital, Vols. I and III. N.Y.: International Publishers, 1967.

Marx, Karl. The Eighteenth Brumaire of Louis Bonaparte. N.Y.: International Publishers, 1963.

Karl Marx on Colonialism and Moderization. Avineri, Shlomo, ed. N.Y.: Doubleday, 1969.

Marei, S.A. "Overturning the Pyramid." Ceres, 2, 1969.

Masson, Paul. Histoire du commerce français dans le Levant au XVIIIe siècle. Paris: 1911.

Mathews, Joseph J. Egypt and the Formation of the Anglo-French Entente of 1904. Phila.: University of Pennsylvania Press, 1939.

May, Jacques. The Ecology of Malnutrition in the Far and Near East. N.Y.: Hafner Publishing Co., 1961.

Mayfield, James B. Rural Politics in Nasser's Egypt. Austin, Tex.: The University of Texas Press, 1971.

Mboria, Lefter. La Population de l'Egypte. Cairo: 1938.

McCarthy, Justin A. "Nineteenth Century Egyptian Population," in Elie Kedourie, ed., The Middle Eastern Economy: Studies in Economics and Economic History. London: Frank Cass and Co., Ltd., 1976.

McCoan, J.C. Egypt as It Is. London: Cassell, Petter, and Galpin, 1877.

Mead, Donald C. Growth and Structural Change in the Egyptian Economy. Homewood, Ill.: R.D. Irwin, 1967.

Mellor, John W. "Food Price Policy and Income Distribution in Low Income Countries." Economic Development and Cultural Change. 27,1 (October, 1978).

Merton, S. "The Distribution of Cotton Seed." Agricultural Journal of Egypt, VII (1918), pp. 35-48.

Milner, Viscount. England in Egypt. London, 1892.

Minost, E. "L'Egypte économique et financiere." L'Egypte Contemporaine, 24 (1933) pp. 45-67.

Minost, E. "Essai sur la agricole." L'Egypte Contemporaine, 21 (193), pp. 535-583.

Minost, E. "Essai sur la richesse foncière de l'Egypte." L'Egypte Contemporaine, 21 (1930), pp. 334-355.

Mohi-el-Din, Amr. "Under-Employment in Egyptian Agriculture." in Manpower and Employment in Arab Countries: Some Critical Issues. Geneva: I.L.O., 1975.

Monost, E. "L'action contre la crise." L'Egypte Contemporaine, 22 (1931) pp. 409-457.

Mohi-el-din, Yehia. Egyptian Agriculture: A Case of Arrested Development. Unpub. Ph.D. Thesis. University of Wisconsin-Madison, 1966.

Moore, Barrington, Jr. Social Origins of Dictatorships and Democracy: Lord and Peasant in the Making of the Modern World.

Boston: Beacon Press, 1966.

Morishima, Michio. Marx's Economics: A Dual Theory of Value and Growth. Cambridge: Cambridge University Press, 1973.

Mosseri, V.M. "La fertilité de l'Egypte." L'Egypte Contemporaine, 17 (1926), pp. 93–124.

Mosseri, V.M. "Revue sommaire des travaux récents sur le maintien et l'amelioration de la qualité des cotons egyptiens." L'Egypte Contemporaine, 17 (1926), pp. 393–433.

Murray, Keith A.H. "Feeding the Middle East in War Time." Royal Central Asian Journal, 32 (July–October, 1945).

"Nadwah 'an al-mikanikiyyah al'zira'iyyah al-misriyyah." (Debate on Farm Mechanization). L'Egypte Contemporaine. #311 (January, 1968).

Nagy, I.E. Die Landwirtshaft im heutigen Aegypten und ihre Entwicklungsmoeglichkeiten. Vienna: Scholle–Verlag, 1936.

Nagi, Mostafa. Labor Force and Employment in Egypt. N.Y.: Praeger, 1971

Nahas, J.F. Situation économique et sociale du fellah égyptien. Paris: 1901.

Nassif, E. "L'Egypte, est-elle surpeuplée?" L'Egypte Contemporaine, 33 (1942).

Nour, Mustafa. Les rapports entre l'évolution démographique et l'evolution économique en Egypte. Fribourg: 1959.

Nutting, Anthony. Nasser. New York: E.P. Dutton & Co., Inc., 1972.

O'brien, Patrick. "The Long Term Growth of Agricultural Production in Egypt: 1821–1962," in P.M. Holt, ed., Political and Social Change in Modern Egypt. London and N.Y.: Oxford University Press, 1968, pp. 162–195.

O'brien, Patrick. The Revolution in Egypt's Economic System. London and N.Y.: Oxford University Press, 1966.

Owen, E.R.J. "Agricultural Production in Historical Perspective," in P. Vatikiotis, ed., Egypt Since the Revolution. N.Y.: Praeger, 1968.

Owen, E.R.J. Cotton and the Egyptian Economy, 1820–1914. London and N.Y.: Oxford University Press, 1969.

Owen, E.R.J. "Egypt and Europe: From French Expedition to British Occupation" in Roger Owen and Bob Sutcliffe, eds. Studies in the Theory of Imperialism. London: 1972.

Owen, E.R.J. "The Development of Agricultural Production in Nineteenth Century Egypt—Capitalism of what Type?" Unpublished paper, presented at the Conference on Economic History of the Near East, Princeton University Press, Princeton, N.J, June 16–20, 1974.

Owen, E.R.J. "The Influence of Lord Cromer's Indian Experience on British Policy in Egypt, 1883-1907." St.Anthony's Papers, No. 17. Oxford: Oxford University Press, 1965.

Owen, Roger. "Introduction: Resources, Population, and Wealth," in Thomas Naff and Roger Owen, eds., Studies in Eighteenth Century Islamic History. Carbondale, Ill.: Southern Illinois University Press, 1978.

Owen, Wyn. "Land and Water Use in the Egyptian High Dam Era." Land Economics 40,3 (August, 1964).

Pacific Consultants. New Lands Productivity in Egypt: Technical and Economic Feasability. Washington, D.C.: A.I.D. /NE-C-1645, 1980.

Parry, J.H. "Transport and Trade Routes" in The Cambridge Economic History of Europe, IV. E.E. Rich and C.H. Wilson, eds. pp. 155-219. Cambridge, 1967.

Piot Bey, J.B. "Coup d'oeil sur l'économie actuelle du bétail en Egypte." L'Egypte Contemporaine, 2 (1911), pp. 199-208.

Piot Bey, J.B. "Projet d'assurance mutuelle obligatoire contre la mortalité du betail en Egypte." L'Egypte Contemporaine, 1 (1910), pp. 369-377.

Platt, D.C.M. Finance, Trade and Politics in British Foreign Policy, 1815-1914. London: Oxford University Press, 1968.

Poliak, A.N. Feudalism in Egypt, Syria and the Levant, 1250-1900. London: Royal Asiatic Society, 1939.

Platt, K.B. "Land Reform in the U.A.R." A.I.D. Spring Review of Land Reform, 8, Washington, D.C.: 1970.

Platt, R.R., and Muhammad Bahy Hefny. Egypt: A Compendium. N.Y.: American Geographical Society, 1958.

Polanyi, Karl. The Great Transformation: The Political and Economic Origins of Our Time. Boston: Beacon Press, 1957.

Poli, Polo. "La culture du riz in Egypte." L'Egypte Contemporaine, 22 (1931), pp. 824-836.

Polieer, Leon. "Notes à propos de la loi des cinq feddans." L'Egypte Contemporaine, 4 (1913), pp. 501-517.

Radwan, Samir. Agrarian Reform and Rural Poverty: Egypt, 1952-1975. Geneva: ILO, 1978.

Radwan, Samir. Capital Formation in Egyptian Industry and Agriculture 1882-1967. London: Ithaca Press, 1974.

Radwan, Samir. "The State and Agrarian Change: A Case Study of Egypt, 1952-77." in Dharam Ghai, Azizur Rahman Khan, Eddy Lee, Samir Radwan, eds., Agrarian Systems and Rural Development. New York: Holmes & Meier, 1979.

Al-Fafi'i, Abd al-Rahman. Thawrah 1919. (The Revolution of 1919). Cairo: 1955.

Ransom, Roger L., and Richard Sutch. "Debt Peonage in the Cotton South After the Civil War. Journal of Economic History, XXXII (Sept., 1972), pp. 641-669.

Rashad Bey, Ibrahim. "The Cooperative Movement." L'Egypte Contemporaine, 30 (1933), pp. 485-492.

Raymond, Andre. Artisans et Commerçants au Caire au XVIIIe Siecle. Two vols. Damascus: 1973.

Reda, Khalil Saleh. The Underpriviledged of Egypt: Twelve Million Farmers. Unpublished M.S. thesis, University of Wisconsin-Madison, 1942.

Riad, Hassan. L'Egypte Nassérienne Paris: Minuit, 1964.

Ricardo, David. Principles of Political Economy. Pierro Sraffa, ed. Cambridge: Cambridge University Press, 1970.

Richards, Helmut. "Land Reform and Agribusiness in Iran." Middle East Research and Information Project Report, #43 (December, 1975).

Rifaat, M.A. The Monetary System of Egypt. London: 1935.

el Rifai, Husein Ali. La question agraire en Egypte. Paris: 1919.

Rivlin, Helen A.B. The Agricultural Policy of Muhammad 'Ali in Egypt. Cambridge, Mass: Harvard University Press, 1961.

Rizk, Ragaa. Urban Encroachment on Agricultural Land in Egypt, With Special Reference to Sharkia Governorate. Unpublished M. Sc. Thesis, Dept. of Agricultural Economics, Zagazig University, 1980.

Robinson, Joan. Essays in the Theory of Economic Growth. New York: St Martin's, 1968.

Robinson, Ronald, and John Gallaher with Alice Denny. Africa and the Victorians: The Climax of Imperialism. London: St. Martin's Press, 1961.

Rodinson, Maxime. Islam and Capitalism. Brian Pearce, trans. Austin, Texas: University of Texas Press, 1974.

Rosenberg, Nathan. "The Direction of Technical Change: Inducement Mechanisms and Focusing Devices." Economic Development and Cultural Change, 18 (Oct., 1969), pp. 1-24.

Rosenberg, Nathan, ed. The Economics of Technical Change. Baltimore: Penguin, 1971.

Rouillard, Germaine. La vie rurale dans l'empire byzantin. Paris: 1953.

Roumasset, James. Rice and Risk: Decision Making Among Low Income Farmers. Amsterdam: North Holland, 1976.

Ross, Lt. Col. Justin C. "Introduction," in William Willcocks, Egyptian Irrigation. London: F.Spon, 1889.

Rousso, J.R. "Le Remède contre le 'Boll weevil,'" L'Egypte Contemporaine, 14 (1923), pp. 223-249.

Rousso, J.R. "La sélection biologique de nos graines de coton." L'Egypte Contemporaine, 16 (1925), pp. 56-79.

Rural Employment Problems in the United Arab Republic. Geneva: I.L.O., 1969.

Russel, Paul F. "Public Health Factors: Part I: Malaria and Bilharzia," in L. Dudley Stamp, ed., A History of Land Use in Arid Regions. Paris: U.N.E.S.C.O., 1961, pp. 363-372.

Russel Pasha, Sir Thomas. Egyptian Service, 1902-1946. London: 1949.

Saab, Gabriel. The Egyptian Agrarian Reform, 1952-1962. London and N.Y.: Oxford University Press, 1967.

Saab, Gabriel. Motorisation de l'agriculture et developpement agricole au Proche Orient. Paris: S.E.D.E.S., 1950.

Sabry, Mohammad. La révolution égyptienne. Pais: 1919.

Saffa, Samir. "Exploitation économique et agricole d'un domaine rural égyptien." L'Egypte Contemporaine, 40 (1949), pp. 277-450.

Sahota, G.S. Fertilizer in Economic Development. N.Y.: Praeger, 1968.

St. John Bayle. Village Life in Egypt. London: Chapman and Hall, 1852.

Saleh, Muhammad. La Petite propriété rurale en Egypte. Grenoble: Joseph Allier, 1922.

Saul, S.B. Studies in British Overseas Trade, 1870-1914. Liverpool: Liverpool University Press, 1960.

Senior, N.W. Conversations and Journals in Egypt and Malta 1855-56. Two vols. London: Low, Marston, Searle and Rivington, 1882.

el Shafie, Mahmoud. Population Pressure on the Land and the the Problem of Capital Accumulation in Egypt. Unpub. Ph.D. thesis, University of Wisconsin-Madison, 1951.

Schanz, M. Cotton in Egypt and the Anglo-Egyptian Sudan. Manchester: 1913.

Shatz, J. "L'endettement et la crise des cultivateurs." L'Egypte Contemporaine, 33 (1942).

Shatz, J. "Les principales cultures égyptiennes." L'Egypte Contemporaine, 23 (1932), pp. 611-735.

Schoelh, Alexander. Aegypt den Aegyptern! Die politische und gesellshaftliche Krise der Jahre 1878-1882 in Aegypten. Zurich: 1972.

Schoelch, Alexander. "Constitutional Development in Nineteenth Century Egypt—A Reconsideration." Middle Eastern Studies, 10 (Jan., 1974), pp. 3-14.

286

Schmidt, Arno. Cotton Growing in Egypt. Manchester: International Federation of Master Cotton Spinners' and Manufacturers' Associations, 1912.

Scofield, William. "Foreign Investment in Egypt: Opportunities and Obstacles." Foreign Agriculture. (July 13, 1976).

Selim, Hussein Kamel. Twenty Years of Agricultural Development in Egypt. Cairo: 1940.

Shultz, T.W. ed. Distortions of Agricultural Incentives. Bloomington, Indiana: Indiana University Press, 1978.

Shultz, Theodore W. Transforming Traditional Agriculture. New Haven: Yale University Press, 1964.

Shaftik, M. Pahsa. Cause de la diminution en Egypte du rendement de la culture du coton. Cairo, 1926.

Shaw, Stanford J. The Financial and Administrative Organization and Development of Ottoman Egypt, 1517-1798. Princeton: Princeton University Press, 1962.

Shaw, Stanford J. "Landholding and Land Tax Revenues in Ottoman Egypt," in P.M. Holt, ed. Political and Social Change in Modern Egypt. London and N.Y.: Oxford University Press, 1964.

Shaw, Stanford J. Ottoman Egypt in the Age of the French Revolution. Cambridge, Mass.: Harvard University Press, 1964.

Singer, Charles, E.J. Holmyard, and A.R. Hall, eds. A History of Technology. Vol. 1. London and N.Y.: Oxford University Press, 1954.

Soliman, Mikhael. Création et avenir des syndicats agricoles en Egypte. Paris: Ernest Sagot et Cie., 1923.

Srivastava, Uma K., Rbert Crown, and Earl O. Heady. "The Green Revolution and Farm Income Distribution." Economic and Political Weekly, 6 (Dec., 1971), pp. A163-A172.

Stern, R.M. "The Price Responsiveness of Egyptian Cotton Producers." Kyklos, 12 (1959), pp. 375-384.

Stone, Russell. "Egypt," in W. Arthur Lewis, ed. Tropical Development, 1880-1913: Studies in Economic Progress. London: George Allen & Unwin Ltd., 1970.

Strachey, John. The End of Empire. London: Gollanz, 1959.

Strakosch, Siegfried. Erwachende Agrarlaender: Nationallandwirtshaft in Aegypten und im Sudan unter englishchen Einflusse. Berlin: 1910.

Street, James Harry. The New Revolution in the Cotton Economy: Mechanization and Its Consequences. Chapel Hill: University of North Carolina Press, 1957.

Strickland, C.F. "Agricultural Cooperatives in Egypt." L'Egypte Contemporaine 11 (1925), pp. 48-55.

Stuart, Villiers. Egypt After the War. London: John Murray, 1883.

Tanamli, A. "L'évolution de l'économie rurale égyptienne dans les cinquante dernières années. L'Egypte Contemporaine, 51 (1960), pp. 45-73.

Thorner, Daniel, and Alice. Land and Labour in India. London: 1962.

Tignor, R.L. "British Agricultural and Hydraulic Policy in Egypt, 1882-1892." Agricultural History, 37 (1962), pp. 63-74.

Tignor, Robert L. "Equity in Egypt's Recent Past." Mimeo, Princeton University, 1978.

Tignor, R.L. Modernization and British Colonial Rule in Egypt, 1882-1914. Princeton: Princeton University Press, 1966.

Timmer, C. Peter. "Fertilizer and Food Policies in LDC's" Food Policy 1, (1976).

Timmer, C. Peter. "Food Prices and Food Policy Analysis in LDC's." Food Policy, 5, 3 (August, 1980).

El-Tobgy, H.A. Contemporary Egyptian Agriculture. Cairo: Ford Foundation, 1976.

Todd, John A. "The Agricultural Drainage of the Egyptian Delta." Appendix I of the Unofficial Report of the Visit of the Delegation of the International Federation of Master Cotton Spinners' and Manufacturers' Associations To Egypt, October-November, 1912.

Todd, John A. The Banks of the Nile. London: Adams and Char-Black, 1913.

Tottenham, P.M. The Irrigation Service, Its Organization and Administration. Cairo: 1927.

U.S. Congress, Forty-Sixth Congress, First Session, Executive Document #5, Appendix, Report of the Agent and Consul General of the U.S. in Cairo, 1879.

U.S. Department of Agriculture. Global Food Assessment, 1980. Foreign Agricultural Economic Report no. 159. Washington, D.C.: 1980.

U.S. Department of Agriculture, cooperating with the U.S. Agency for International Development and the Egyptian Ministry of Agriculture. Egypt: Major Constraints to Increasing Agricultural Productivity. Washington, D.C.: USDA Foreign Agricultural Economic Report #120, 1976.

United Kingdom, British Board of Trade. Report of the British Goodwill Trade Mission to Egypt, November-December, 1945. London: 1946.

Van Vloten, M.Th.F. "La Motoculture et les avantages des grandes exploitations agricoles." L'Egypte Contemporaine, 1 (1910) pp. 648-655.

Vatikiotis, P.J. The Modern History of Egypt. N.Y.: Praeger, 1969.

Voll, Sarah Potts. "Egyptian Land Reclamation Since the Revolution." Middle East Journal, 34,22 (Spring, 1980).

Volney, C.F.C. Travels Through Syria and Egypt in the Years 1783, 1784 and 1785. Two vols. Dublin: 1793.

Wallace D. Mackenzie. Egypt and the Egyptian Question. London: 1883.

Wallace, Robert. "Opening Address on Egyptian Agriculture." University of Edinburgh, Agriculture Department Pamphlet, Oct. 22, 1891.

Wallerstein, Immanuel. The Modern World System: Capialist Agriculture and the Origins of the European World-Economy in the Sixteenth Century. N.Y.: Academic Press, 1974.

Warriner, Doreen. Land and Poverty in the Middle East. London and N.Y.: Oxford University Press, 1948.

Waterbury, John. "'Aish: Egypt's Growing Food Crisis." in Egypt: Burdens of the Past, Options for the Future. Bloomington, Indiana: Indiana Press, 1978.

Waterbury, John. Egypt: Burdens of the Past, Options for the Future. Bloomington, Indiana: University of Indiana Press, 1978.

Waterbury, John. Hydropolitics of the Nile Valley. Syracuse, New York: Syracuse University Press, 1979.

Waterbury, John. "Urbanization and Income Distribution in Egypt." Draft, mimeo, Princeton University Income Distribution Project. Princeton, New Jersey, 1979.

Wilmington, Martin W. The Middle East Supply Centre. Albany, New York: S.U.N.Y. Press, 1971.

World Bank. Arab Republic of Egypt: Agricultural Development Project (Minufiyya-Sohag); Staff Appraisal Report. 1977.

World Bank. Arab Republic of Egypt: Agricultural Development Project (Minufiyya-Sohag). Revised report, 1979.

World Bank. Land Reform. Sectoral policy paper. Washington, D.C.: 1975.

Wayment, Hilary, ed. Egypt Now. Cairo: 1943.

Wells, Sidney H. "L'enseignement agricole, industriel et commercial en Egypte," L'Egypte Contemporaine, 2 (1911), pp. 244-269.

Wharton, Clifton R., Jr., ed. Subsistence Agriculture and Economic Development. Chicago: Aldine, 1969.

Whitcombe, Elizabeth. Agrarian Conditions in North India. Vol. I. The United Provinces Under British Rule, 1860-1900. Berkelely and Los Angeles: University of California Press, 1972.

Willcocks, Sir William, and J.I. Craig. Egyptian Irrigation. Second Ed. London: F. Spon, 1913.

Williamson, Jeffrey G. "Review of Hayami and Ruttan." Journal of Economic History, XXXIII (June, 1973), pp. 484–487.

Wittfogel, Karl A. "The Hydraulic Civilizations" in William L. Thomas, Jr., ed., Man's Role in Changing the Pace of the Earth, Vol. I. Chicago: University of Chicago Press, 1956, pp. 152–164.

Wittfogel, Karl A. Oriental Despotism. New Haven: Yale University Press, 1956.

Wolf, Eric R. Peasant Wars of the Twentieth Century. N.Y.: Harper and Row, 1969.

Worthington, E.B. Middle East Science. London: H.M. Stationery Office, 1946.

Wright, Gavin and H. Kunreuther. "Cotton, Corn and Risk in the Nineteenth Century." Journal of Economic History, Sept., 1975, pp. 526–551.

Yallouz, A. "Chrônique législative et parlementaire, 1933–34." L'Egypte Contemporaine, 25 (1934), pp. 105–165.

Zaghloul, Fathi. "A Cost-of-Living Index for Rural Labourers, 1913–1961." Egyptian Ministry of Planning, Institute of National Planning Memo no. 557. Cairo: 1965.

el-Zalaki, Muhammad M.M. An Analysis of the Organization of Egyptian Agriculture and Its Influence onNational Ecnomic and Social Institutions. Unpub. Ph.D. thesis, University of California, Berkeley, 1941.

Zannis, Joseph. Le crédit agricole en Egypte. Paris: 1937.

Ziadeh, Farhat. Lawyers, the Rule of Law, and Liberalism in Modern Egypt. Palo Alto: Stanford University Press, 1968.

# Index